STUDY GUIDE TO ACCOMPANY

B I O L O G Y

Fifth Edition

STUDY GUIDE TO ACCOMPANY

B I O L O G Y

Fifth Edition

JOHN W. KIMBALL

HAROLD F. SEARS
UNIVERSITY OF SOUTH CAROLINA — UNION

RONALD W. WILSON
MICHIGAN STATE UNIVERSITY

PREPARED FOR
P.S. ASSOCIATES

A D D I S O N - W E S L E Y P U B L I S H I N G C O M P A N Y

Reading, Massachusetts · Menlo Park, California · London · Amsterdam · Don Mills, Ontario · Sydney

ISBN 0-201-10246-3
  CDEFGHIJ-AL-8987

PREFACE

A study of biology involves two fundamental aspects: the mastery of a great deal of descriptive material, terms, research studies, general concepts, and interrelationships; and the application of this body of information to the way you look at the world. John Kimball's text, which has been used by thousands of college students, is an excellent vehicle for both familiarizing you with the biological world and leading you to the point where you might say, "How remarkable! My knowing something about the living world has changed the way I view everything around me." It is our hope that this Study Guide will assist you in that demanding but rewarding process.

The Study Guide has several specific purposes. One is to help you to identify the main ideas and concepts presented in each text chapter. Another is to give you practice exercises to complete, which will reinforce the main ideas in the chapter and help you to anticipate test items you will eventually see on examinations. A third purpose is to present review questions that demand a logical, reasoned response, and that assist you in your goal of being able to see how different concepts presented in a single chapter (or in many chapters) can be tied together in an explanation of a single biological process.

DESIGN OF THE GUIDE

Each chapter of the Study Guide is organized around two main parts. Part I is entitled "Reviewing the Chapter." It includes a summary of the chapter that may be used in several ways: as a preview of what the chapter covers (to be read before you begin work on the text chapter), as a summary of what you have just read and what you will be expected to remember (to be read just after you complete the text chapter), and as a review (to be scanned in preparation for a quiz or examination). "Reviewing the Chapter" also contains a list of key terms; these represent the vocabulary that must be mastered in order to think, speak, and write knowledgeably about the biological concepts presented in a chapter. The definitions of these terms appear both in the text and in the glossary.

Part II, "Mastering the Chapter," covers four different approaches to mastering the contents of the text chapter.

The "Learning Objectives" list the several most important pieces of information—structures to describe, processes to compare, concepts to explain—within the chapter. This list will help you to focus on the main ideas of the chapter. By the time you have finished reading the text chapter and have completed the various review exercises in the Study Guide, you should be able to make clear and accurate written statements that show you have mastered each of the Learning Objectives.

"Reviewing Terms and Concepts" is a device for helping you to review the text chapter. These exercises follow the organization of the chapter and require the ability to recall specific terms and ideas covered in it. Answers to these fill-ins are given at the end of the chapter.

"Testing Terms and Concepts" is a list of 15 multiple-choice questions that also cover the entire text chapter. Some of these questions ask you to recall facts, terms, or definitions, as does "Reviewing Terms and Concepts." Other questions ask you to apply your knowledge to situations not specifically discussed in the text. It is important in biology, as in any field of study, that you not only learn basic facts and relationships, but that you understand them well enough to use them in new contexts.

Use this section not only as a review of the text chapter, and as a guide to how you should actively think about the material, but as an index of how well you have mastered the chapter. If you are able to answer all of the items correctly, then you should go directly to the last section of the chapter, "Understanding and Applying Terms and Concepts." If you cannot answer all questions, go back to the text to clear up the point(s) that gave you trouble.

The last section of each Study Guide chapter is entitled "Understanding and Applying Terms and Concepts." More than in previous exercises, you will be asked to demonstrate a broad understanding of the major ideas presented in the text. You will be expected to recall facts and ideas, associate these with related concepts, and use these ideas to interpret new situations. Again, correct answers are given at the end of the chapter.

USING THE STUDY GUIDE

Your biology course and your use of Biology will be more rewarding if you use the Study Guide systematically in conjunction with the text. We suggest that you begin with the Study Guide rather than the text.

Start by reading the "Chapter Highlights" at the beginning of each Study Guide chapter. Read it through slowly and carefully. This will be your "map" of the chapter; it will help you to organize the chapter in your mind before you begin reading the text, and will help you relate the chapter to previous chapters. Next read the text chapter, and think about each key term as you encounter it. When you are ready to test your comprehension of what you have studied, do the fill-in exercises, "Reviewing Terms and Concepts." On completing these exercises, turn back to the "Learning Objectives" and, treating these like essay questions, attempt to provide correct responses. Review the appropriate pages of the text for any objective you cannot meet. Now you are ready for "Testing Terms and Concepts" and "Understanding and Applying Terms and Concepts."

Sound easy? It isn't. It demands a commitment from you to put in at least an hour of extra work per chapter, and, of course, it requires you to follow carefully the directions outlined above. But if you put in this additional work, you should find not only that you will have an easier time on quizzes and examinations, but that your enjoyment of the course and the text will be significantly enhanced.

H.F.S.

R.W.W.

TABLE OF CONTENTS

Title	Page
Preface	i
Chapter 1: What is Life?	1
Chapter 2: Scientific Methods	9
Chapter 3: The Chemical Basis of Life: Principles	17
Chapter 4: The Molecules of Life	27
Chapter 5: The Cellular Basis of Life	37
Chapter 6: The Metabolism of Cells	48
Chapter 7: Energy Pathways in the Cell	58
Chapter 8: Photosynthesis	68
Chapter 9: Cell Division	78
Chapter 10: Genetics: The Work of Mendel	88
Chapter 11: Genes on Chromosomes	99
Chapter 12: The Chemical Nature of Genes	109
Chapter 13: Gene Expression	118
Chapter 14: The Organization of Genetic Information	127
Chapter 15: The Regulation of Gene Expression	135
Chapter 16: Reproduction in Plants	142
Chapter 17: Reproduction in Animals	152
Chapter 18: Early Development	163
Chapter 19: Later Development	171
Chapter 20: Heterotrophic Nutrition	180
Chapter 21: Gas Exchange in Plants and Animals	190
Chapter 22: The Transport of Materials in the Vascular Plants	199
Chapter 23: Animal Circulatory Systems	209
Chapter 24: The Immune Response	222

Chapter 25: Excretion and Homeostasis 233

Chapter 26: Responsiveness and Coordination in Plants 244

Chapter 27: Animal Endocrinology 253

Chapter 28: The Elements of Nervous Coordination 269

Chapter 29: The Nervous System 282

Chapter 30: Muscles and Other Effectors 291

Chapter 31: The Elements of Behavior 301

Chapter 32: Evolution: The Evidence 311

Chapter 33: Evolution: The Mechanisms 319

Chapter 34: The Origin of Life 330

Chapter 35: The Classification of Life 337

Chapter 36: The Prokaryotes (Kingdom Monera) 344

Chapter 37: The Protists and Fungi 356

Chapter 38: The Plant Kingdom 370

Chapter 39: The Invertebrates 378

Chapter 40: The Vertebrates 391

Chapter 41: Energy Flow through the Biosphere 403

Chapter 42: The Cycles of Matter in the Biosphere 413

Chapter 43: The Growth of Populations 424

Chapter 44: Interactions Between Species 432

Chapter 45: Human Ecology I: Pestilence 443

Chapter 46: Human Ecology II: Competing for Food 452

WHAT IS LIFE?

I. REVIEWING THE CHAPTER

A. CHAPTER HIGHLIGHTS

1.1 Why Study Biology?

Humans have a deep curiosity about themselves and about the rest of the living world. The study of biology can prepare one for a variety of satisfying careers, and some biological knowledge is necessary to understand many issues that affect human welfare.

1.2 The Characteristics of Life

Living matter is far more complex than nonliving matter. If we examine any part of a higher organism's body with a microscope, we find that it is made of cells, which are organized into tissues, which in turn form organs such as the stomach and kidney. Several organs— for example, stomach, liver, and intestine—work together as a system.

One of the most characteristic features of the chemical composition of living organisms on our planet is the presence of molecules containing carbon atoms. This feature is so characteristic that the chemistry of carbon compounds is called organic chemistry. It is very difficult for us to conceive of living matter without organic molecules.

A living organism continuously exchanges matter with its surroundings, in a process called metabolism. The organism takes in matter, reassembles it into new components, liberates and uses what energy may be in that matter, and releases unwanted materials back into the environment. The underlying assumption in the three biological experiments carried out by the Viking landers on Mars is that a fundamental property of life is metabolism and, therefore, if there is life on Mars, even the simplest and most primitive kind, it should reveal its presence by its capacity to carry on metabolism.

Reproduction is the "self-controlled" duplication of the structures characteristic of the organism. It leads to the growth of a single organism and the production of offspring, new individuals. Reproduction occurs in two major ways: Asexual reproduction preserves the status quo since it does not involve the union of genetic material from two individuals. Sexual reproduction produces new combinations of traits since two parents contribute to the formation of the new individual.

All living organisms respond to light, heat, gravity, and other environmental stimuli. They do so through the coordinated activities of their organs controlled by nerves (animals) and hormones (plants and animals).

More characteristic of life than anything else is the phenomenon of evolutionary change. Species gradually change over the course of generations, because new traits appear during sexual reproduction and become established and old ones disappear. Species also gradually split and form two or more new species.

B. KEY TERMS

cell	anabolism
tissue	catabolism
organ	stimulus
system	nerve
growth	hormone
asexual reproduction	effector
sexual reproduction	metabolism
subcellular	evolution
behavior	

II. MASTERING THE CHAPTER

A. LEARNING OBJECTIVES

When you have mastered the material in this chapter, you should be able to:

1. Define all key terms.

2. List three reasons for studying biology.

3. Discuss five properties that distinguish living from nonliving matter.

4. Distinguish between cells, tissues, organs, and systems.

5. State why it is difficult for earth scientists to conceive of life without organic molecules.

6. Distinguish between sexual and asexual reproduction and cite one advantage of each.

7. Distinguish between anabolism and catabolism.

8. Describe each of the three biology experiments designed for the Viking expedition to Mars.

B. REVIEWING TERMS AND CONCEPTS

1. To make effective decisions requires not only an understanding of human values in our lives but also a knowledge of the physical and (a) _____ principles that underlie our lives. Knowledge of human values comes from a study of those subject areas collectively called the (b) _____. Knowledge of the scientific principles must come from a study of the (c) _____, including biology, which is the science of (d) _____.

2. A microscopic examination of a higher animal reveals that its fundamental biological unit is the (a) _____. Many such structures are organized as (b) _____, which in turn form (c) _____. These structures may work together as a (d) _____.

3. All living organisms known to us on earth contain atoms of (a) _____, a feature so characteristic of life that the chemistry of (b) _____ compounds is called (c) _____ chemistry.

4. Living organisms continually exchange (a) _____ and (b) _____ with their environment. They continually acquire (c) _____ from the environment, assemble it into new components characteristic of the species, and give back different materials in a process called (d) _____. The synthesis of new matter out of raw materials gained from the environment is termed (e) _____. The breakdown of large organic molecules into smaller subunits is termed (f) _____.

5. (a) _____ reproduction preserves the status quo because it does not involve the union of material from two parents. (b) _____ reproduction produces new combinations of traits two parents contribute to the formation of the new individual.

6. In one of the Viking biological experiments, Martian soil was collected and exposed to carbon dioxide and carbon monoxide. The synthesis of complex organic molecules from such simple compounds would have been evidence that living organisms were carrying out the fundamental process of (a) _____. In another experiment, radioactive nutrients were added to some Martian soil and later the sample was tested for breakdown products, such as carbon dioxide. The appearance of breakdown products would have been evidence of (b) _____.

7. All living organisms respond to changes in light, heat, gravity, and other environmental (a) _____. They do so through the coordinated activities of their organs controlled by (b) _____ and (c) _____.

8. Evolution is a gradual change in the characteristics of a (a) _____ over many generations. Evolutionary change occurs because new traits arise during (b) _____ reproduction and are established in the species. (c) _____ evolution enables organisms to live more successfully in a given environment than their ancestors were able to.

3

C. TESTING TERMS AND CONCEPTS

1. Biology is the scientific study of
    a) all animals.
    b) all living things.
    c) all things that contain carbon.
    d) all things of concern to man.

2. The fundamental biological unit is the
    a) organ.
    b) organism.
    c) cell.
    d) tissue.

3. Tissues of various kinds are organized into
    a) organs.
    b) systems.
    c) cells.
    d) organisms.

4. Aggregates of cells form
    a) organs.
    b) systems.
    c) tissues.
    d) organisms.

5. The chemistry of carbon compounds is termed
    a) organic chemistry.
    b) inorganic chemistry.
    c) biochemistry.
    d) metabolic chemistry.

6. The process by which a living organism synthesizes matter out of
    environmental raw materials and breaks down matter as a source of
    energy is termed
    a) catabolism.
    b) metabolism.
    c) growth.
    d) responsiveness.

7. In the Viking experiment in which a soil sample was incubated with
    a dilute soup of radioactive organic molecules, and in which the
    atmosphere above the sample was regularly monitored for the
    appearance of radioactive gases such as carbon dioxide, the
    appearance of radioactive gases would have been evidence for
    a) reproduction.
    b) evolution.
    c) catabolism.
    d) anabolism.

8. Our understanding of the universe suggests that life on Mars, or
    on any other planet, probably
    a) exists in a wide variety of sizes, as on Earth, and is composed
       of one or more cells.

4

b) duplicates itself by some sort of sexual reproduction.
c) locomotes over or under the surface of the soil and responds to stimuli from its environment.
d) carries out both synthetic or constructive chemical reactions and degradative or destructive reactions.

9. When two parents contribute to the formation of a new individual, with the result that new combinations of traits are produced, the reproductive process is said to be
a) evolutionary.
b) genetic.
c) sexual.
d) asexual.

10. If a starfish is cut in half, each half will develop into a complete individual. This illustrates the process of
a) budding.
b) organic reproduction.
c) sexual reproduction.
d) asexual reproduction.

11. From the point of view of an organism, changes in light, heat, gravity, and mechanical contact are best termed environmental
a) stimuli.
b) effectors.
c) conditions.
d) characteristics.

12. Chemical regulators that play a part in coordinating the actions of organisms are known as
a) hormones.
b) nerves.
c) stimuli.
d) glands.

13. Evolution is a process that occurs within
a) organs.
b) systems.
c) organisms.
d) species.

14. Of the organisms described below, which would be capable of the most rapid evolution?
a) a plant that reproduces sexually only once a century
b) a plant that lives hundreds of years and rarely reproduces at all, sexually or asexually
c) an animal that has an unusually inaccurate method of reproduction, so that mistakes in copying the genes occur frequently
d) an animal that has a kind of sexual reproduction that is unusually accurate, so that genetic information is passed on with almost no mistakes

15. When evolution enables a species to survive more successfully in a given environment than did the ancestral species, the evolution

that has occurred is said to be
a) successful.
b) balanced.
c) divergent.
d) adaptive.

D. UNDERSTANDING AND APPLYING TERMS AND CONCEPTS

1. If biological knowledge is required to resolve various political
    issues, why should ordinary citizens, as well as politicians, have
    this knowledge? (a) _____
    _____
    List four current issues that do require some knowledge of biology.
    b) _____
    c) _____
    d) _____
    e) _____

2. In the spaces following these statements write the appropriate
    letter: (L) for living, (N) for nonliving, (B) for both living
    and nonliving, depending on which is being described.
    a) Material making up a substance in a state of order _____
    b) The presence of tissues _____
    c) Has the capacity to produce more units of its kind _____
    d) Is composed of atoms _____
    e) Is composed of molecules _____
    f) Is capable of carrying on metabolism _____
    g) Contains carbon compounds _____
    h) The material making up the substance contains essentially the
       same atoms it contained a year ago. _____
    i) Responds in a coordinated way to environmental change _____

3. Number the following structures from the smallest and simplest
    (1) to the largest and most complex (8).
    _____ a) cell
    _____ b) subcellular structure
    _____ c) system
    _____ d) atom
    _____ e) organ
    _____ f) molecule
    _____ g) organism
    _____ h) tissue

4. Living things are said to be very complex; this is one of our
    five "characteristics of life." The planet Jupiter is also complex
    in its structure and in the reactions and events that take place
    within it. Is the planet Jupiter alive? (a) _____
    Explain. (b)_____
    _____
    _____

5. In what way is the growth of an icicle like the growth of a
    living organism? (a) _____
    _____

6

Why is this <u>not</u> considered to be living growth? (b) _____

_____

_____

6. Dogs evolved from a wolflike ancestor, and some breeds are very
    wolflike today. How then can you explain the existence of such
    diverse types of dogs as the Great Dane, dachshund, and poodle?

_____

_____

ANSWERS TO CHAPTER EXERCISES

Reviewing Terms and Concepts

1. a) biological b) humanities c) sciences d) life

2. a) cell b) tissues c) organs d) system

3. a) carbon b) carbon c) organic

4. a) matter b) energy c) matter d) metabolims e) anabolism
    f) catabolism

5. a) Asexual b) Sexual

6. a) anabolism b) catabolism

7. a) stimuli b) nerves c) hormones

8. a) species b) sexual c) Adaptive

Testing Terms and Concepts

1. b	6. b	11. a
2. c	7. c	12. a
3. a	8. d	13. d
4. c	9. c	14. c
5. a	10. d	15. d

Understanding and Applying Terms and Concepts

1. a) Voters must understand the issues in order to choose the best
    representative. b) development of nuclear power c) pollution
    laws d) tobacco price supports e) licensing controversial
    drugs or birth control techniques

2. a) B b) L c) B d) B e) B f) L g) B h) N i) L

3. a) 4 b) 3 c) 7 d) 1 e) 6 f) 2 g) 8 h) 5

4. a) no b) A cell's organic molecules are hundreds of times larger
    and more complex than Jupiter's inorganic molecules. These
    organic molecules are arranged in a greater variety and number of
    ways, into the various subcellular components, and they interact

chemically in many more ways. One cell is more complex than this entire planet.

5. a) The icicle does increase in size, and more material is taken in from the environment than is given back. b) This growth is not "self-controlled,"; but arises from the external factors of gravity and freezing temperature.

6. Many dog breeds are examples of family lines that have been greatly affected by humans. Humans have determined which characteristics, such as large size or elongated shape, are desirable, which individual dogs will live and breed, and therefore which genes will be passed on to future generations and become common.

---

*involving energy expenditure and a genetic blueprint.

SCIENTIFIC METHODS

I. REVIEWING THE CHAPTER

A. CHAPTER HIGHLIGHTS

2.1 Introduction

Good scientists, like any number of other people, must work hard and employ their basic common sense. More than other people however, scientists must be skeptical of their own work and of that of others, always looking for the flaw in the argument, and they must be willing to tolerate much uncertainty in their professional lives. It usually takes a long time to be reasonably confident of one's conclusions, and one can never be absolutely sure.

Science is basically a four-step process. The scientist makes observations about some part of the natural world. He or she formulates a tentative explanation of why things are as he or she has observed. The scientist tests possible explanations with experiments in new situations and modifies or refines the explanation. Finally, the scientist reports her discovery to the rest of the scientific community.

2.2 Scientific Observations

Observations are the facts of science. For instance, many humans react to poison ivy with a form of allergy called contact sensitivity, and many do not. Other chemicals elicit this reaction in humans and in other animals, and a "model" of contact sensitivity has been developed, involving the mouse and its reaction to DNFB.

Painting the skin with DNFB sensitizes the mouse and a second application elicits a reaction. Further study has shown that intravenous injection of DNFB four days prior to surface exposure prevents the development of contact sensitivity.

These studies included careful instrumentation so observations would be accurate and quantified, use of several animals and statistical analysis to be aware of individual variation, and "control" experiments that show the reaction is a consequence of the specific experimental treatment and not chance, contamination, or other unconsidered aspects of the animals' rearing.

2.3 Scientific Explanations

A hypothesis is a tentative explanation for a set of observed phenomena. For instance, maybe an intravenous injection causes something to be produced that actively supresses the ability of the animal to become allergic to the chemical.

2.4 Testing Hypotheses

A test is another set of observations directed at the specific hypothesis under consideration. For instance, if tolerance is an active response following injection, then the tolerant animal must have something in it that a control animal would not have. This "something" (in the blood or lymphatic system perhaps) should confer tolerance on an uninjected animal. This experiment must involve the same care (careful measurement, several observations, statistical treatment of data, and controls) as did the original observational work.

2.5 Reporting Scientific Work

Only when new facts and hypotheses are taken up by the entire community of interested scientists do they become part of science. This is accomplished by word of mouth and then by publication in a refereed journal.

Biological papers consist of an introduction, a materials and methods section complete enough to allow another scientist to repeat the work, the specific results, a discussion of those results, acknowledgments of those who helped with the study, references to related published literature, and a summary or abstract.

2.6 Reproducibility of Scientific Work

Scientific work must be reproducible by the same worker and by others. If a study cannot be repeated, it is assumed that the initial study and its conclusions were wrong.

2.7 Building on the Work of Others

Although a few scientists have made major discoveries alone and rather suddenly, most scientific work simply adds observations or ideas to an existing body of knowledge, which then grows gradually and slowly. This is another illustration of the fact that science is a group activity.

2.8 Basic versus Applied Science

Applied science must adhere to the same rules and standards as basic science does; however, the motivation is usually different. Applied science seeks to solve a specific, practical problem, while basic research studies the natural world out of a purer curiosity. Each fuels the other, and both are necessary to scientific advancement.

B. KEY TERMS

contact sensitivity	discussion
standard deviation	acknowledgments

10

hypothesis   references

introduction   summary

materials and methods   abstract

results

II. MASTERING THE CHAPTER

A. LEARNING OBJECTIVES

When you have mastered the material in this chapter, you should be able to:

1. Define all key terms.

2. List two characteristics more typical of scientific activity than of other human efforts and explain why they are important.

3. List the four major steps of a scientific study.

4. Define and explain the importance of "control" experiments, animal "models," and "null" hypotheses.

5. Calculate the mean and standard deviation of a group of measurements.

6. Describe the organization of a typical scientific paper.

7. Discuss the relationship of basic to applied science.

B. REVIEWING TERMS AND CONCEPTS

1. Science is basically a four-step process. The scientist makes (a) _____ about some part of the natural world. He or she formulates a (b) _____. The scientist then (c) _____ his (d) _____ with experiments in new situations. Finally, the scientist (e) _____ his discovery to the rest of the scientific community.

2. (a) _____ are the facts of science. Gathering them should involve careful (b) _____ so observations will be accurate and quantified, use of (c) _____, _____ and (d) _____ _____ to be aware of individual variation, and (e) _____ experiments that show the reaction is a consequence of the specific experimental treatment and not of unconsidered aspects of the animals' rearing.

3. A (a) _____ is a tentative explanation for a set of observed phenomena. A (b) _____ is another set of observations directed at the specific (c) _____ under consideration.

11

4. Only when new facts and hypotheses are taken up by the scientific community do they become part of science. This is accomplished by word of mouth and then by publication in a (a) _____ _____. Biological papers consist of an (b) _____ , a _____ and a _____ section, the specific (c) _____ , a (d) _____ , (e) _____ of those who helped with the study, (f) _____ to related published literature, and a (g) _____ or (h) _____ .

5. Scientific work must be (a) _____ by the same worker and by others. If a study cannot be repeated, it is assumed that the initial study and its conclusions were (b) _____ .

6. Although a few scientists have made major discoveries alone and rather suddenly, most scientific work simply adds (a) _____ or (b) _____ to an existing body of knowledge, which then grows gradually and slowly. This is another illustration of the fact that science is a (c) _____ _____ .

7. Applied science must adhere to the same rules and standards as basic science does; however, the (a) _____ is usually different. Applied science seeks to solve (b) _____ _____ , while basic research studies the natural world out of a purer curiosity. Each fuels the other, and both are necessary to scientific advancement.

C. TESTING TERMS AND CONCEPTS

1. The scientific method
    a) is a specific procedure or series of steps followed by successful scientists.
    b) yields proofs of scientific principles or theories.
    c) demands that one be unusually critical of one's own work and of that of others.
    d) is a type of thinking not used by most nonscientists.

2. Which of the following is not an important part of scientific activity?
    a) gathering the opinions of other scientists working on related problems
    b) predicting what one ought to see in circumstances not yet examined
    c) collecting large numbers of observations about a particular phenomenon
    d) publishing one's work in a journal or symposium

3. Contact sensitivity
    a) has not been alleviated by simple trial and error techniques.
    b) is a reaction to a relatively few complex compounds manufactured by plants and animals as chemical defenses.
    c) involves destruction of cells by a variety of biological poisons.
    d) is a biological process that requires the active participation of the victim's own body systems.

4. Reasons for using an animal "model" in medical research do <u>not</u> include which of the following?
   a) Models are cheaper.
   b) Models are more convenient.
   c) Models often yield more valuable data.
   d) Much research on humans would be immoral.

5. Which of these questions <u>cannot</u> be asked using a scientific method?
   a) Should the United States expand its use of nuclear power?
   b) Can the United States expand its use of nuclear power?
   c) What will result if the United States does not develop its nuclear power?
   d) What will the consequences of a nuclear meltdown be?

6. Experiments are performed more than once for all <u>but</u> <u>one</u> of the following reasons. Which one?
   a) to permit statistical analyses and conclusions about populations rather than about individuals
   b) to permit statistical analysis and discussion of variation among individuals
   c) to permit instrumentation and therefore greater accuracy of measurement
   d) to recognize and eliminate rare, chance (and therefore meaningless) outcomes

7. If one is testing the value of a new drug in curing a particular disease, the group of subjects exposed to the disease and expected to show full symptoms is the
   a) positive control group.
   b) negative control group.
   c) active experimental group.
   d) passive experimental group.

8. A number of measurements are made of some bodily function, and the mean is calculated. For a scientist to conclude that a treatment is affecting that bodily function, it must result in a measurement that is
   a) within one standard deviation of the mean.
   b) at least one standard deviation lower than the mean.
   c) at least one standard deviation higher than the mean.
   d) at least two standard deviations from the mean.

9. The statement that half the flowers in a particular garden are yellow and the other half are red is an example of a(n)
   a) hypothesis.
   b) observation.
   c) explanation.
   d) generalization.

10. In an experiment, the control group is treated the same as the experimental group, except that the control group
    a) receives the treatment being evaluated.
    b) does not receive the treatment being evaluated.
    c) receives no treatment at all.
    d) none of these—a control is for purposes of duplication.

11. An hypothesis is a(n)
    a) possible explanation.
    b) summary of a group of observations.
    c) experiment with controls.
    d) generally accepted conclusion.

12. Scientific papers rarely contain
    a) a table of contents.
    b) an introduction.
    c) acknowledgments of help.
    d) a summary.

13. The observations one makes in using a scientific method are
    usually published in what section of a scientific paper?
    a) methods
    b) results
    c) materials
    d) discussion

14. Basic research
    a) involves the development of new instrumentation.
    b) focuses on the physics and chemistry of a problem.
    c) focuses on practical problems or short-term goals.
    d) focuses on fundamental problems about how nature works.

15. Imagine that you have observed a variety of flies, gnats, mosquitos
    and bees, and you find all have six legs. You hypothesize that
    insects only develop six legs, never more nor less. What should
    the next stage in your scientific study be?
    a) You should expand your study to other related animals, like
       spiders and centipedes.
    b) You should urge others to examine your insect material.
    c) You should formulate a prediction about the number of legs on
       other insects, like butterflies or grasshoppers.
    d) You should collect and study a variety of other insects.

D. UNDERSTANDING AND APPLYING TERMS AND CONCEPTS

1. The scientific method is not useful to scientists only. Show how
    the method could be used in the field of literature or art by
    describing a hypothetical study. What might an artist do in
    taking each of the following steps?
    a) observation _____
    b) hypothesis _____
    c) test _____
    _____
    _____

2. It is said that, "Science done in secret is not science at all."
    Why is it so important that scientific results and conclusions be
    published, or at least somehow communicated to others; so important
    that, if it is not done, it is simply not science? _____
    _____
    _____

14

3. The ears of 12 mice were painted with a chemical and the increase in size measured. One series of experiments yielded increases of 0, 2, 4, 6, 8, and 10 mm, and another yielded 0, 1, 2, 8, 9, and 10 mm. Calculate the mean for one of these sets of data. a) _____. Which set would be considered better data, allowing one to be more confident that the mean was biologically meaningful? b)_____. Why? c)_____
    _____.

4. One winter a scientist drinks three glasses of orange juice per day each time he begins sniffling with a cold, and four colds clear up within three days of their onset. The scientist concludes that vitamin C helps cure colds. From a scientific point of view, is this conclusion warranted? Why or why not? _____
    _____
    _____

ANSWERS TO CHAPTER EXERCISES

Reviewing Terms and Concepts

1. a) observations   b) tentative explanation   c) tests
    d) explanations   e) reports

2. a) Observations   b) instrumentation   c) several animals
    d) statistical analysis   e) control

3. a) hypothesis   b) test   c) hypothesis

4. a) refereed journal   b) introduction, materials, methods
    c) results   d) discussion   e) acknowledgments   f) references
    g) summary   h) abstract

5. a) reproducible   b) wrong

6. a) observations   b) ideas   c) group activity

7. a) motivation   b) practical problems

Testing Terms and Concepts

1. c	6. c	11. a
2. a	7. a	12. a
3. d	8. d	13. b
4. c	9. b	14. d
5. a	10. b	15. c

Understanding and Applying Terms and Concepts

1. a) Tabulate the word frequencies in known works of Shakespeare. b) Suggest that training and experience has resulted in biased and unique performances by this artist. c) Predict that known work by Bacon will have word frequencies significantly different from that discovered in Shakespeare and that an unsigned, suspected

work by Shakespeare will have word frequencies not significantly different from that in his known works. (Of course, many different answers are possible here.)

2. This relates to the importance of repeatability in science. If research cannot be repeated by others, it is not accepted, and work that isn't publically described cannot be repeated.

3. a) 5 mm b) the first set c) The standard deviation of the first set is less. The greater variation in the second set suggests that variables other than the experimental variable are acting.

4. Most importantly, no controls were used. Something else in the orange juice could have been the active agent, or, his colds might have lasted three days no matter what he did. Also, only one subject was used, the scientist himself, and the study was not published or replicated.

THE CHEMICAL BASIS OF LIFE:
PRINCIPLES

I. REVIEWING THE CHAPTER

A. CHAPTER HIGHLIGHTS

3.1 Forms of Matter

The chemical organization of the many kinds of living things is remarkably similar. And the chemical organization of living and non-living matter alike is based on the same materials and principles.

Living matter ceaselessly interacts with the nonliving environment, which includes the lithosphere (rocks and soil), hydrosphere (water), and atmosphere. The earth's habitable zone is called the biosphere. These four spheres are composed of mixtures of identifiable ingredients. To understand their chemistry, the scientist must have means of separating the components of these mixtures, isolating and identifying them.

3.2 Elements

The elements of a compound, such as table salt, which consists of the elements sodium and chlorine, are strongly associated and hence more difficult to separate than are the components of a mixture. While the components of a mixture may occur in random proportions, the elemental components of compounds always occur in fixed proportions; e.g. water consists of two parts hydrogen to one part oxygen (H_2O).

Of the 90 most common elements, only 25 or so are contained in living matter. Living matter has the capacity of taking in from the environment those elements it needs, as it needs them, and of concentrating them.

3.3 Atoms

Atoms, the smallest parts of an element that can enter into combination with other elements, contain three basic building blocks—electrons (negatively charged), protons (positively charged), and neutrons (neutral). Virtually all of the mass of any atom is contained in its nucleus of protons and neutrons. Because the atoms of a given element have a characteristic number of protons and neutrons, each element has a fixed atomic weight. Atoms of the same element that differ in their atomic weight are called isotopes.

Some elements combine readily and others with difficulty or not at all. It is the number and arrangement of the electrons in an atom's outermost shell that establishes the chemical behavior of an element.

3.4 Chemical Bonds

The greater an affinity an atom has to gain additional electrons, the greater its "electronegativity." Atoms that form compounds are held together by chemical bonds. Strongly electronegative atoms can pull one or more electrons away from weakly electronegative atoms. The atom gaining the electron(s) becomes a negative ion; the atom losing the electron(s) becomes a positive ion. The mutual attraction of two oppositely charged ions holds the ions together as a compound by an ionic bond; sodium combined with chlorine (table salt) is an example. Two atoms with similar electronegativities brought together share pairs of electrons and are said to form covalent bonds; carbon and hydrogen bonded as the compound methane is an example.

3.5 The Hydrogen Bond/3.6 Water Lovers and Water Haters/3.7 Acids and Bases

Because the opposite ends of water molecules (and certain other molecules) have opposite electrical charges (polar molecules), weak bonds, called hydrogen bonds, can exist between adjacent water molecules. Hydrogen bonding accounts for many of water's remarkable properties.

Acids and bases are important biological compounds since they play key roles in the synthesis and degradation of compounds. An acid can be thought of as any substance that donates protons in a reaction and a base as any substance that accepts protons.

3.8 Molecular Weight and the Mole

It is sometimes important for a scientist to prepare a solution of a known number of molecules. He or she first determines the molecular weight of the substance to go into solution and then weighs out an equivalent number of grams of the substance, called 1 mole, and finally adds water to make one liter of solution. This is called a 1-molar solution. There are always 6×10^{23} molecules of any substance in one mole.

3.9 pH

The acid or base strength of a solution can be expressed on the pH scale, a logarithmic scale which ranges from a value of 1 (most acid) through 7 (neutral) to 14 (most basic). A substance of pH 2, for instance, is 10 times less acidic than a substance of pH 1.

3.10 Chemical Changes/3.11 Bond Energy/3.12 Oxidation-Reduction Reactions

Chemical reactions occur when there are rearrangements of their atomic components, e.g., water can be decomposed into atoms of hydrogen and oxygen. During all chemical reactions, reactants are converted into products. When substances combine, they may be oxidized or reduced. An atom of a substance that loses an electron is said to have been oxidized; an atom that gains an electron is reduced. The most common kind of biological oxidation is caused by the removal of hydrogen atoms from a substance. Life continues to exist on earth because it is able to exploit oxidation-reduction (redox) reactions.

Since living organisms require energy, it is important to understand the term as well as its implication to life. Energy is defined as the capacity to do work. Chemical energy is stored in the form of chemical bonds. The amount of energy obtained from a chemical will be the amount of energy released when the bond is broken. In chemistry, the term "free energy" refers to the bond energy available to do work.

We frequently measure energy by a unit called a kilocalorie (kcal). It is the amount of heat to raise one liter or water one degree Celcius. If energy is released during a chemical reaction, the free energy value is negative—if it is required, the value is positive.

B. KEY TERMS

morphology	electronegativity
physiology	covalent bond
lithosphere	polar covalent bond
hydrosphere	hydrogen bond
atmosphere	hydrophilic
biosphere	apolar
mixture	polar molecule
compound	acid
element	base
atom	carboxyl
molecule	group
dialysis	amino group
electron	mole
proton	pH
neutron	reactant
atomic number	product
atomic weight	bond energy

isotope ΔG

dalton   oxidation

ion   reduction

valence   redox

ionic bond

II. MASTERING THE CHAPTER

A. LEARNING OBJECTIVES

When you have mastered the material in this chapter, you should be able to:

1. Define all key terms.

2. Cite three methods for separating the components of a mixture.

3. Cite the principle involved in each of the methods cited in objective two.

4. Distinguish between a compound and an element.

5. Distinguish between a mixture and a compound.

6. List the three basic components of atoms and describe their electrical charge and relative masses.

7. Explain the basis of assigning atomic weights to atoms of different elements.

8. Distinguish between atomic weight and atomic number.

9. Describe the difference between ionic bonding and covalent bonding, and cite one example of each.

10. State how a polar covalent bond differs from an ordinary covalent bond, and cite the chemical significance of polar covalent bonding.

11. Distinguish between an acid and a base.

12. What is meant by a "mole," a 1-molar solution?

13. What does it mean when we say water has a pH of 7.0?

14. What is bond energy? How is it related to "free energy"?

15. What is the relationship of electronegativity to "redox" reactions?

B. REVIEWING TERMS AND CONCEPTS

1. The four principal "spheres" of the environment include the (a) _____, _____, _____, and _____. The four spheres are composed of mixtures of various ingredients. About (b) _____ different chemical elements occur naturally, of which about (c) _____ are found in living organisms.

2. Table salt is a compound composed of the two (a) _____ sodium and chlorine. Compounds are (b) _____ (more/less) difficult to break down than are mixtures. While the components of a mixture nearly always occur in (c) _____ (random/fixed) proportions, the element-components of compounds always occur in (d) _____ (random/fixed) proportions.

3. The fundamental unit of an element is called a(n) (a) _____. Atoms contain three basic building blocks, (b) _____, _____, and _____. Of the three, (c) _____ do not carry an electrical charge. Virtually all of the mass of an atom is contained in its (d) _____ and is contributed by its (e) _____ and _____. Atoms tend to have the same number of (f) _____ as _____ and, therefore, are electrically _____. Charged atoms are called (h) _____.

4. Because the atoms of a given element have a characteristic number of protons and neutrons, each element has a fixed (a) _____ _____. Atoms of the same element that _differ_ in atomic weight are called (b) _____.

5. It is the number and arrangement of electrons in an atom's (a) _____ shell that establish the chemical or reactive behavior of an element. The more massive an atom, the (b) _____ (lesser/greater) the number of electron shells it has. Those elements with a full outer shell (c) _____ (do/do not) react readily with other atoms.

6. Atoms forming compounds are held together by chemical (a) _____. The mutual attraction of oppositely charged ions holds them together as a compound by an (b) _____ bond. Two atoms with similar electronegativities brought together (c) _____ pairs of electrons and are said to form (d) _____ bonds.

7. The greater the affinity an atom has to gain (a) _____, the greater its electronegativity. Atoms that form compounds are held together by chemical (b) _____. The mutual attraction of electrically but oppositely charged atoms holds the atoms together by (c) _____ bonding. Two atoms with similar electro-negativities are held together by (d) _____ bonding. Weak bonds called (e) _____ bonds hold polar molecules together.

8. An acid can be thought of as any substance that (a) _____

21

_____ protons in a reaction, a base as any substance that (b) _____ protons.

9. When a scientist prepares a solution of a known number of molecules, he or she first determines the (a) _____ weight of the substance to go into solution and then weighs out an equal number of grams of the substance, which is defined as one (b) _____, and finally adds water to make one (c) _____ of solution. This is called a (d) _____ solution.

10. The acidity or alkalinity of a substance can be expressed on the pH scale, which ranges from a value of 1 (most (a) _____) through 7 (which is neither (b) _____ nor _____) to 14 (most (c) _____). A substance of pH 2 is (d) _____ times less (e) _____ than a substance of pH 1.

11. When water is decomposed into oxygen and hydrogen, it undergoes (a) _____ change, because its (b) _____ have been rearranged. During all chemical reactions, reactants are converted into (c) _____, during which (d) _____ energy is involved. The greater the (e) _____ of two atoms bonding, the (f) _____ the bond energy involved.

12. (a) _____ _____ is the energy that can be harnessed to do work. If energy is required for a chemical reaction to occur, the value of "ΔG" is (b) _____.

13. When substances combine, they may be oxidized or (a) _____. An atom of a substance that loses an electron is said to have been (b) _____; an atom that gains an electron is said to have been (c) _____. The most common kind of biological oxidation is caused by the removal of (d) _____ atoms from a substance.

C. TESTING TERMS AND CONCEPTS

1. The subunit of an atom that has no charge is the
    a) electron.
    b) proton.
    c) neutron.
    d) isotope.

2. A dalton is
    a) 1/12 the mass of an atom of $_{14}^{12}C$.
    b) 1/12 the mass of an atom of ^{12}C.
    c) the mass of an electron.
    d) the same as molecular weight.

3. The chemical properties of an atom are determined by the number and arrangement of its
    a) protons.
    b) neutrons.
    c) electrons.
    d) isotopes.

4. Electrically charged atoms are held together by
   a) covalent bonds.
   b) ionic bonds.
   c) polar covalent bonds.
   d) hydrogen bonds.

5. Water molecules are held together by
   a) ionic bonding.
   b) covalent bonding.
   c) hydrogen bonding.
   d) hydrophobic bonding.

6. Whenever a substance donates protons to some other substance, we describe the proton donor as
   a) an acid.
   b) a base.
   c) an oxidizing agent.
   d) hydrophilic.

7. A solution with a pH of 2 is how many times less or more acid than a solution of pH 1?
   a) 1 times less acid
   b) 2 times less acid
   c) 20 times more acid
   d) 10 times less acid

8. Any oxidizing agent entering a chemical reaction is
   a) oxidized.
   b) reduced.
   c) neutralized.
   d) acidified.

9. Polar molecules not only attract each other, but they also attract
   a) isotopes.
   b) hydrophobic compounds.
   c) ions.
   d) nonpolar molecules.

10. The shared pairs of electrons in the compound methane constitute
    a) ionic bonding.
    b) covalent bonding.
    c) hydrogen bonding.
    d) hydrophilic bonding.

11. The distribution of partial charges around the water molecule is
    a) symmetrical.
    b) the same as ionic bonding.
    c) asymmetrical.
    d) the same as redox potential.

12. Certain elements have an atomic number expressed as a whole number plus a fraction because its atoms
    a) have uneven numbers of protons and electrons.
    b) occur as isotopes.

c) have uneven numbers of electron shells.
d) our inability to weigh them accurately.

13. Living organisms differ from nonliving things because they
a) contain elements different than those in the nonliving world.
b) are capable of changing one element into another.
c) never contain isotopes.
d) are capable of concentrating certain elements.

14. Comparing the isotopes of a particular element, it can be said that
a) isotopes have different kinds of protons.
b) chemically, isotopes behave the same.
c) isotopes have different electrical charges.
d) chemically, isotopes always have greater oxidizing potentials.

15. If elements combine and heat is given off
a) the free energy change is positive.
b) the free energy change is negative.
c) both elements were oxidized.
d) energy was "used up" in the reaction.

D. UNDERSTANDING AND APPLYING TERMS AND CONCEPTS

1. You have 100 grams of a hypothetical compound A (mole. wt.= 100) and 200 grams of B (mole. wt.= 200). Which sample contains the greatest number of molecules? _____

2. What is the principal factor that determines the magnitude of bond energy for a particular kind of bond? _____

3. In biological research a frequently used isotope of carbon is ^{14}C. How does it differ from ^{12}C? (a) _____
How is it similar? (b) _____
Why is it useful to biologists? (c) _____

4. Trace the energy you are now using to study back to its origin.
a) _____
b) _____

5. Explain the relationships between calories, food, and exercise.
a) _____
b) _____
c) _____

6. $C_2H_5OH + O_2 \quad CH_3COOH + H_2O$
   (M)   (N)   (O)   (P)

a) M and N are called _____.
b) In its conversion to O, M has been _____.

7. What determines how atoms of a given element will react? _____
_____

8. The <u>valence</u> of an atom is determined by the number of electrons that it must _____.

9. A certain compound has a pH of 7.0. It is _____ times more acidic than one with a pH of 9.

10. For a particular type of chemical bond, e.g., C=O, the energy values may vary between compounds. Why? _____
_____.

ANSWERS TO CHAPTER EXERCISES

Reviewing Terms and Concepts

1. a) lithosphere, hydrosphere, atomosphere, biosphere
    b) 90
    c) 25

2. a) elements b) more c) random d) fixed

3. a) atom b) electrons, protons, neutrons c) neutrons d) nucleus
    e) neutrons, protons f) electrons, protons g) neutral h) ions

4. a) atomic weight b) isotopes

5. a) outermost b) greater c) do not

6. a) bonds b) ionic c) share d) covalent

7. a) electrons b) bonds c) ionic d) covalent e) hydrogen

8. a) donates b) accepts

9. a) molecular b) mole c) liter d) 1-molar

10. a) acid b) acid base c) basic d) 10 e) acidic

11. a) chemical b) atoms c) products d) bond e) electronegativity
    f) greater

12. a) Free energy b) positive

13. a) reduced b) oxidized c) reduced d) hydrogen

Testing Terms and Concepts

1. c	6. a	11. c
2. a	7. d	12. b
3. c	8. b	13. d
4. b	9. c	14. b
5. c	10. c	15. b

Understanding and Applying Terms and Concepts

1. Both will contain the same number.

2. electronegativity

3. a) ^{14}C has two more neutrons
   b) same number of protons and electrons
   c) ^{14}C is radioactive and can be "traced"

4. a) energy in foods from chemical bonds
   b) bonds in food form when plants capture solar energy

5. a) Calories measure energy content of food.
   b) Exercise requires energy.
   c) If we take in more calories than we use, the excess is stored,
      i.e., additional weight.

6. a) reactants
   b) oxidized

7. the number and arrangement of electrons

8. gain, lose, or share

9. 100

10. because of other atoms that may be attached to compounds in
    question

4

THE MOLECULES OF LIFE

I. REVIEWING THE CHAPTER

A. CHAPTER HIGHLIGHTS

4.1 Mydrocarbons

        Carbons, more than any other element, is characteristic of life,
for it can bond to four other atoms simultaneously, these may be atoms
of other elements or additional carbon atoms. It is this ability to
link to other atoms in so many ways that makes carbon such a fundamental
element of living things. An examination of some typical hydrocarbons,
the simplest organic molecules, will demonstrate the versatility of
the carbon atom.

4.2 Lipids

        Lipids are more complex organic compounds. Among them are fats,
composed of glycerol and fatty acids. Fats are a rich store of potential
energy because of the many weak covalent bonds. Fats do not mix with
water because they are hydrophobic. When a polar group containing a
phosphate group substitutes for a fatty acid in a fat molecule, a
phospholipid is produced that is amphiphilic, or hydrophobic at one
end and hydrophilic at the other. This property, as will be shown
later, is very important in the structuring of living materials as well
as exemplifying how the carbon atom can be bond in a reversible manner
to other atoms and molecules. Bile salts, another group of lipids,
belong to a lipid group called steroids—sex hormones being examples
of the latter.

4.3 Carbohydrates

        Carbohydrates are organic compounds containing carbon, hydrogen,
and oxygen, the general formula being (CH_2O). Starch and cellulose
are two common carbohydrates. Both are polymers constructed of many
repeating units (called monomers) of the sugar glucose. Such chainlike
carbohydrates are referred to as polysaccharides. Glucose is a six-
carbon sugar called a hexose, as are the sugars galactose and fructose.
Sugars, such as sucrose and maltose, that are constructed from two
simple sugar units are called disaccharides. Carbohydrates are
important sources of energy, particularly glucose, which is highly
soluble in cellular fluids. Starches are not soluble and thus are
good sources of stored energy. Cellulose, because of the way glucose
units are polymerized, is a rigid polysaccharide and is the major
structural component of plant cell walls.

4.4 Proteins

Proteins make up about 50% of the dry weight of earth's living matter. Proteins are sources of energy, they are structural units, and they serve as catalysts called enzymes. Proteins are polymers composed of amino acid units. An amino acid in turn is composed of a hydrogen atom, an amino group, a carboxyl group, and one of 20 different "R" groups. It is the nature of the "R" group that establishes the identity of a particular amino acid. These "R" groups also establish many of the physical, chemical, and biological properties of the proteins. The amino group is basic and so attracts protons, while the carboxyl group is a proton donor and thus an acid. The amino acid monomers are united into chains when the carboxyl group of one amino acid unites via a so-called peptide bond to the amino group of the next. During the formation of this bond, a molecule of water is eliminated. When proteins are digested, the peptide bonds are enzymatically broken and a molecule of water used in a reaction generally called hydrolysis.

The sequence of amino acids along with the position of any disulfide bridges determine the primary structure of proteins. The secondary structure of a protein is the 3-dimensional arrangement of the amino acid chain—the helical shape of alpha=keratin is an example. The total conformation of a globular protein is called its tertiary structure. The presence in a protein of distinct subunits—for example, the four chains of hemoglobin—is called the protein's quaternary structure.

While the function of a protein depends on the protein's three-dimensional structure, that structure, in turn, depends on the precise sequence of amino acids, its primary structure. An in turn, the precise sequence of amino acids is established by the genetic code.

4.5 Nucleic Acids

The genetic code is embodied in and expressed by macromolecules called nucleic acids, of which there are two major kinds—DNA and RNA. Both are unbranched linear polymers. Their monomers are called nucleotides, which consist of a pentose sugar attached to a phosphate group and a nitrogen-containing base. The bases which make up DNA are thymine, adenine, guanine, and cytosine. In RNA, uracil is used in place of thymine. Again, as with the amino acids in proteins, the sequence of the nucleotides is critical in the biological role of the nucleic acids. As monomers, the nucleotides are involved in energy transfer and they participate in the activities of certain enzymes and hormones.

B. KEY TERMS

organic	starch
lipid	cellulose
fatty acid	isomer

28

glycerol	protein
phospholipid	amino acid
amphiphilic	enzyme
steroid	peptide bond
carbohydrate	polypeptide
macromolecule	hydrolysis
glucose	prosthetic group
monomer	denatured
polymer	nucleic acids
polysaccharide	nucleotide/nucleoside
disaccharide	

II. MASTERING THE CHAPTER

A. LEARNING OBJECTIVE

When you have mastered the material in this chapter, you should be able to:

1. Define all key terms.

2. State the two properties that enable carbon to serve as the "backbone" for the formation of an almost limitless variety of molecules.

3. State the chemical property of fats that makes them rich stores of potential energy.

4. Cite the chemical basis for the amphiphilic property of certain lipids.

5. Cite three different types of lipids and a biological role for each. What common property allows these three rather diverse compounds to all be classified as lipids?

6. List three differences between mono and/or disaccharides and polysaccharides. What are the biological implications of the differences?

7. Explain why cellulose, although it is like starch in being a polysaccharide with glucose as the monomer, differs so much in its physical properties from starch.

29

8. List four functions of protein and cite as many examples as you can.

9. Discuss how the four units that are attached to the alpha carbon of amino acids are related to the structure and function of proteins.

10. Describe what determines the primary, secondary, tertiary, and quaternary structures of proteins.

11. Distinguish between simple proteins and conjugated proteins.

12. Cite the two major kinds of nucleic acids and identify the sub-units of which they are composed.

13. Differentiate between a nucleotide and a nucleoside.

14. List three roles played by nucleic acids.

B. REVIEWING TERMS AND CONCEPTS

1. The simplest organic molecules are (a) _____ composed only of carbon and hydrogen.

2. Among the lipids are fats, composed of glycerol and (a) _____. Fats are a rich store of potential energy because of their _____ bonds resulting from small electronegativity difference between atoms. When a polar group containing a phosphate group substitutes for a (c) _____ _____ in a fat molecule, a (d) _____ is produced.

3. Carbohydrates contain atoms of (a) _____, (b) _____, and (c) _____. Starch and cellulose are polymers constructed of many repeating units called (d) _____, which are units of the sugar (e) _____. Sugars such as sucrose and maltose, constructed from two simple sugar units, are called (f) _____. (g) _____ is a polysaccharide and perhaps the single most abundant organic molecule in nature. (a) _____ is a polysaccharide and has much the same role in animals that (i) _____ does in plants.

4. Proteins are polymers composed of (a) _____ units, which in turn are composed of a hydrogen atom, an (b) _____ group, a (c) _____ group and an (d) _____ group. What distinguishes one amino acid from another is the nature of its (e) _____ group. In a given protein the amino group is basic and so attracts (f) _____, while the carboxyl group is a (g) _____ donor and thus an (h) _____.

5. Amino acids are assembled into polypeptides by means of (a) _____ bonds. The (b) _____ group of one amino acid interacts with the (c) _____ group of the next with the elimination of a molecule of (d) _____ between them. The sequence of amino acids in a protein, coupled with the knowledge of the location of any

30

disulfide bridges in the protein, is called the (e) _____
structure of the protein. The precise, three-dimensional arrangement
of amino acids in the alpha-keratin polypeptide is called the (f)
_____ helix and is an example of the (g) _____ structure
of a protein. The total conformation of a (h) _____ protein is
called its tertiary structure. The presence in a protein of distinct
subunits is called the protein's (i) _____ structure.

6. While the function of a protein depends on the protein's (a)
_____ structure, that structure depends on the precise
sequence of (b) _____ _____, which in turn is established
by the (c) _____ code. Proteins have several biological functions;
they are sources of (d) _____, they are involved in cell (e) _____,
and some are (f) _____ which catalyze most biochemical reactions.

7. The hereditary information is manifested via (a) _____
_____ which are polymers of (b) _____ and are of two types, (c)
_____ and _____. The sugar portion of DNA is (d) _____.
The nitrogen containing units of nucleic acids are called (e) _____.
DNA and RNA both contain (f) _____ bases. The bases found in DNA
are (g) _____, _____, and _____. RNA contains (h) _____
of the four bases used in DNA but differs in that the fourth is (i) _____
_____ instead of (j) _____. Nucleotides are also important in
reactions involving (k) _____ transfer as well as participating in
certain (l) _____ and/or (m) _____ mediated reactions.

C. TESTING TERMS AND CONCEPTS

1. Each of the 20 amino acids differ from the others because of the
   a) prosthetic group.
   b) "R" group.
   c) disulfide bridge.
   d) carboxyl group.

2. The precise order of amino acids in a protein is established by
   a) a nucleic acid
   b) its tertiary structure.
   c) the location of disulfide bridges.
   d) another protein.

3. In which kind of organic compound does carbon bind only to hydrogen
   or other carbon atoms?
   a) carbohydrate
   b) lipid
   c) nucleic acid
   d) hydrocarbon

4. A trip to the grocery store will reveal a wide variety of fat
   containing products, e.g., butter, cereal, hamburger. The fats in
   these products are different because of their
   a) glycerols.
   b) fatty acids.
   c) phosphates.
   d) sodium stearates.

5. A protein that can mix with a lipid at one end and water at the other is said to be
   a) hydrophilic.
   b) hydrophobic.
   c) amphiphilic.
   d) isomeric.

6. An unknown compound has been isolated from a corn plant. Which of the following is it likely not to be?
   a) starch
   b) lipid
   c) nucleic acid
   d) glycogen

7. A student is given four samples and told that they are glucose, lactose, insulin, and RNA. They are in test tubes marked 1, 2, 3, and 4 but she does not know which compound is in which tube. She is instructed to identify the contents of each tube. Her first test is to hydrolyze a portion of each tube. Her results reveal that all but the contents of tube 3 give a positive result. She should therefore conclude that tube 3 contains _____.
   a) glucose
   b) lactose
   c) insulin
   d) RNA

8. In her next test, the student finds that only tube 2 gives a positive result when tested for the presence of sulfur. She should report that tube B contains _____.
   a) glucose
   b) lactose
   c) insulin
   d) RNA

9. The last test performed by the student shows that the compound in tube A contains phosphate. Her conclusion is that tube 1 contains

   _____.
   a) glucose
   b) lactose
   c) insulin
   d) RNA

10. Which of the above mentioned test tubes would contain a compound classified as a monomer?
    a) tube 1
    b) tube 2
    c) tube 3
    d) tube 4

11. The relationships A=T and C=G are found in
    a) nucleotides.
    b) nucleosides.
    c) RNA.
    d) DNA.

12. Which of the following is <u>least</u> likely to be used as an energy storage molecule?
    a) starch
    b) cholesterol
    c) glycogen
    d) tristearin

13. A difference between DNA and RNA is that
    a) the carbohydrate portion of the two is different.
    b) the former contains nucleotides and the latter, nucleosides.
    c) the position of the phosphate bonds is different.
    d) DNA contains guanine and RNA does not.

14. The presence in a protein of distinct subunits, like the four chains of hemoglobin, gives rise to the so-called
    a) tertiary structure.
    b) prosthetic structure.
    c) quaternary structure.
    d) amphiphilic structure.

15. Destroying all but the primary structure of a protein destroys the protein's functional properties and the protein is said to be
    a) hydrolyzed.
    b) degraded.
    c) denatured.
    d) conjugated.

16. The prosthetic group is a feature of
    a) all proteins.
    b) only "simple" proteins.
    c) conjugated proteins.
    d) only plant proteins.

D. UNDERSTANDING AND APPLYING TERMS AND CONCEPTS

1. There are millions of different proteins made from the same twenty amino acid "building blocks". Using only three amino acids, show how you could make at least 10 proteins. (Use four amino acids per protein.)
    Example: $AA_1-AA_2-AA_3-AA$.

   _____

2. By now you may be wondering if protein "$AA_1-AA_2-AA_3-AA_1$" is the same as protein "$AA_1-AA_3-AA_2-AA_1$". Note the second is just turning the first, end for end. The answer is no, they are <u>not</u> the same protein. Explain. _____

   _____

3. This chapter has introduced you to the major classes of biologically important compounds. You have also been given examples of each. For each of the following pairs, cite at least two <u>similarities</u> and two <u>differences</u> between the two.

glucose-tristearin_____

starch-hemoglobin_____

Cholesterol-adenine_____

4. For purposes of comparison, which of the following pairs is
    analogous to optical isomers?

    a) your right hand and right foot

    b) your right hand and left foot

    c) your right hand and left hand

    d) your right and your friend's right hand

5. Hair is largely composed of protein. Based upon information in
    this chapter, how would you explain the physical/chemical basis
    of "perming" hair? (If this term is not familiar to you, ask
    around.) _____
    _____
    _____

6. Here is a short portion of a DNA molecule: 3' CTAGGT 5'. Which
    of the following is the same as the one given and why?

    a) 5' TGGATC 3'

    b) 5' CTAGGT 3'

7. Some proteins are very water soluble and others are virtually
    insoluble. Give a plausible reason for this observation. _____
    _____
    _____.

8. Suppose that you placed the following molecules in water. Using
    a diagram, show what the likely result would be. Shading indicates
    the hydrophobic end.

ANSWERS TO CHAPTER EXERCISES

Reviewing Terms and Concepts

1. a) hydrocarbons

2. a) fatty acids b) weak c) fatty acid d) phospholipid

3. a) carbon b) hydrogen c) oxygen d) monomers e) glucose

f) disaccharides g) Cellulose h) Glycogen i) starch

4. a) amino b) acid c) amino, carboxyl d) "R" e) "R" f) protons
    g) proton h) acid

5. a) peptide b) carboxyl c) amino d) water e) primary f) alpha
    g) secondary h) globular i) quaternary

6. a) three-dimensional b) amino acids c) genetic d) energy
    e) structure f) enzymes

7. a) nucleic acids b) nucleotides c) DNA d) RNA e) deoxyribose
    f) bases g) four h) thymine, adenine, guanine, cytosine
    i) three j) uracil k) thymine l) energy m) enzyme n) hormone

Testing Terms and Concepts

1.	b	6.	d	11.	d
2.	a	7.	a	12.	b
3.	d	8.	c	13.	a
4.	b	9.	d	14.	c
5.	c	10.	c	15.	c
				16.	c

Understanding and Applying Terms and Concepts

1. AA_1-AA_2-AA_1-AA_2, AA_3-AA_1-AA_2-AA_2, etc. It is highly likely that
    none of the ten that you constructed is the same as these two.
    The reason, of course, is that there are many different ways of
    combining the three amino acids into a four-unit protein. In fact,
    this is the point, using only twenty different amino acids it is
    possible to make an astronomically large number of different pro-
    teins.

2. Proteins have carboxyl and amino "ends." Thus, by analogy, reading
    a protein is like reading a word. "Protein" does not mean the
    same as "nietorp."

3. Glucose-Tristearin
    Similarities: 1) Both are monomers. 2) Both are made of only
    carbon, hydrogen and oxygen.

    Differences: 1) Glucose is soluble in water, but tristearin is
    not. 2) Tristearin can be hydrolyzed into several subunits
    (glycerol, and stearic acid), glucose cannot.

    Starch-Hemoglobin
    Similarities: 1) Both are polymers. 2) Both can be hydrolyzed
    into monomers.

    Differences: 1) All the monomers of starch are alike, while
    hemoglobin has twenty different monomers in its structure. 2)
    The major role of starch is energy storage, and the major role of
    hemoglobin is oxygen/carbon dioxide transport.

Cholesterol-Adenine

Similarities: 1) Both demonstrate "ring-shaped" molecular structure. 2) Both have unsaturated or so-called double bonds.

Differences: 1) Cholesterol is converted to many other compounds. Adenine participates in the genetic machinery as well as being involved with the functioning of several enzyme and hormone mediated reactions. 2) In addition to carbon, oxygen, and hydrogen, adenine also has nitrogen atoms in its chemical make-up.

4. c

5. The heat and/or chemicals used in perming break many of the nonpeptide bonds and allow the hair to take on a new shape. After the treatment, new bonds are formed and thus hold the hair in the new shape.

6. a) Nucleic acids are "read" from the 5' end to the 3' end.

7. The solubility, as well as many other properties of proteins, is affected by the nature of the various "R" groups of the component amino acids.

8.

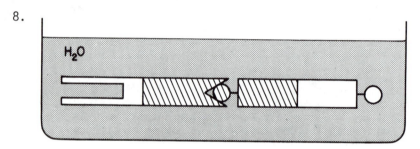

5

THE CELLULAR BASIS OF LIFE

I. REVIEWING THE CHAPTER

A. CHAPTER HIGHLIGHTS

5.1 The Cell As The Unit of Structure of Living Things

Single cells of certain kinds are living, fully independent
organisms, such as the amoeba (but are too small to be seen without
the aid of a microscope). Most organisms, however, are composed of
many cells that are incapable of independent existence. Such cells
are specialized to carry out one or a few tasks for the organism of
which it is part.

5.2 The Cell Membrane

All cells are enclosed within a cell membrane, which serves as an
interface between the cell's internal and external environments. These
membranes contain about 50% lipid and 50% protein. The phospholipid
portion is arranged in a bilayer that permits certain molecules to
enter and leave the cell while blocking certain others. Such membranes are said to be differentially permeable. While extrinsic
proteins occur both on the outer and inner surfaces of cell membranes,
intrinsic proteins reside within the lipid bilayer. Membranes also
occur within the cell, compartmentalizing various intracellular
structures. These membranes are also triple-layered and differ from
the cell membrane chiefly in the kinds of lipids and proteins forming
them. In general, the cell can be said to consist of two major
portions—the nucleus, or control center, and the surrounding cytoplasm.

5.3 The Nucleus

The nucleus is bounded by a pair of porous membranes, within which
hereditary units called chromosomes are suspended. The major proteins
associated with chromosomes are the histones. The nucleus also
contains DNA and RNA. The nucleus is essential to the proper
functioning of a cell. Remove it and the cell dies.

5.4 The Cytoplasm

Everything within the cell, except the nucleus, is called the
cytoplasm. The electron microscope reveals elaborate patterns of
membranes and membrane-bounded compartments within the cytoplasm.
These clearly defined structures are called organelles. By means of
a centrifuge the organelles in a preparation of cytoplasm can be
isolated for study.

37

5.5 Mitochondria

One kind of organelle is the mitochondrion. Mitochondria are
spherical or rod-shaped bodies numbering up to 1000 or more in some
cells. These membrane-bounded structures have a second, inner, folded
membrane. Their chief function, through the action of enzymes, is the
oxidation of food substances, so the mitochondria convert the potential
energy of different food materials into a form of energy the cell can
use. Mitochondria have been aptly called the powerhouse of the cell.

5.6 Chloroplasts

Chloroplasts are another class of organelle, these occurring only
in plants and some algae. They are usually disk-shaped and, like
mitochondria, contain rich supplies of phospholipids and proteins; but
they also contain the green pigment chlorophyll. It is the chlorophyll
that traps the energy of sunlight and enables green plants to synthesize
their own food through photosynthesis.

5.7 Ribosomes

Ribosomes are among the smallest organelles. These nearly
spherical bodies are the site at which protein synthesis takes place.
Some of the proteins synthesized in ribosomes are released directly
to the fluid of the cytoplasm, the protein hemoglobin in red blood
cells being an example. Other proteins are packaged in some type of
membrane-bounded organelle that is secreted from the cell.

5.8 The Endoplasmic Reticulum/5.9 The Golgi Apparatus/5.10 Lysosomes
5.11 Peroxisomes

The endoplasmic reticulum and Golgi apparatus both are elaborate
systems of membranes. The former, containing many ribosomes on their
outer surface, are the sites of protein synthesis. Some of the proteins synthesized are then transferred to the Golgi apparatus, where
additional carbohydrate may be added. The protein-filled sacks of
the Golgi apparatus then may migrate to the cell surface and be
discharged to the outside. Other protein-filled sacs may be retained
within the cell as lysosomes, which contain enzymes and digest a
variety of waste and unwanted products within the cell. Peroxisomes
resemble lysosomes in size and also are filled with enzymes, one of
which catalyzes the breakdown of hydrogen peroxide.

5.12 Vacuoles

Vacuoles are fluid-filled sacs that store food material or wastes
in the cytoplasm. In plants, vacuoles may also contain pigments.

5.13 The Functions of Intracellular Membranes

Intracellular membranes serve at least two major functions: (a)
to establish a numbe- of compartments separating their contents from
the rest of the cell, e.g. chromosomes in the nucleus, and digestive
enzymes in lysosomes; and (b) to provide for the orderly spatial

organization of enzymes and pigments, e.g. in mitochondria and
chloroplasts. Membranes are also capable of regulating the transport
of ions and molecules in only one direction.

5.14 Microfilaments/5.15 Intermediate Filaments/5.16 Microtubules/ 5.17 Centrioles/5.18 Cilia and Flagella

Microfilaments are long, thin fibers composed of the protein actin
and associated with cell movement, e.g. the pinching off of new cells
during cell division. Intermediate filaments are smaller then microtubules and larger than microfilaments. Despite their differences in
protein composition, the intermediate filaments are important "framework" structures in the cell. Centrioles are made of microtubules
and are involved in cell division. Microtubules are straight, hollow
cylinders of the protein tubulin and they are thought to provide some
rigidity to certain parts of cells. They also form the spindle, which
distributes chromosomes to each daughter cell during cell division.

Many cells have whiplike extensions called cilia and flagella.
In microorganisms, these structures are used for locomotion. In
animal cells cilia function to move materials past the cell.

5.19 Cell Coatings

Usually, cells have some exterior coating. In many animal cells
it serves to cement adjacent cells together. In many algae and in all
higher plants, the cell coating is made of cellulose, which forms a
rigid, boxlike cell wall.

5.20 Prokaryotes and Eukaryotes

Bacteria and blue-green algae have cells lacking the membrane-
bounded nucleus and membrane-bounded organelles. Such cells are called
prokaryotes, in contrast to the eukaryotes which have a membrane-
bounded nucleus and organelles.

5.21 Differentiation

Through mitosis, certain single-celled organisms divide and
produce two new independent individuals, each fully capable of carrying
out all life functions. There also are cell colonies made up of
clusters of independent cells that can detach, divide, and start up
new colonies. In other colonies, some of the cells cannot live an
independent life. Such cells have become specialists that contribute
certain functions to the complex organism and receive, in turn, certain
benefits that they cannot achieve for themselves.

5.22 Animal Tissues

In the human organism, cell differentiation has reached a stage
where there probably are some 100 distinguishable kinds of differentiated
cells. One or more kinds of cells may be organized into tissues, of
which there are several kinds. Epithelial tissues, for instance,
occur at the boundary between cell masses and a cavity. Connective

tissues include supporting connective tissue, such as cartilage and bone. Binding connective tissue includes tendons and ligaments. Fibrous connective tissue serves as packing and binding material for most of our organs. In addition, there is muscle tissue, such as skeletal, smooth, and cardiac, associated with movement. And there is nervous tissue, composed chiefly of neurons, associated with communication.

5.23 Plant Tissues

Plant tissues include the meristem, which is responsible for cell division; protective tissue occurring on the surface of roots, stems, and leaves; parenchyma tissue, with food storage as a chief function; collenchyma tissue, which provides mechanical support for the plant; sclerenchyma tissue, also associated with mechanical support; xylem tissue, which transports water and minerals from the roots to the leaves; and phloem tissue, which transports food and hormones throughout the plant.

5.24 Cell-To-Cell Junctions

Three types of junctions can occur at points where cells contact one another: tight junctions, which act as barriers to substances that might move through intercellular spaces; adhering junctions, which are strong attachment points between adjacent cells; and gap junctions, which serve as pathways between certain cells and allow certain ions and low molecular weight compounds to pass from cell to cell.

B. KEY TERMS

cell membrane	intercellular membrane
glycoproteins	microfilaments
phospholipid bilayer	intermediate filaments
differential permeability	cilia
intrinsic proteins	microtubules
extrinsic proteins	basal bodies
nucleus	flagella
chromatin	spindle
chromosomes	mitosis
nucleolus	prokaryote
cytoplasm	eukaryote
cell coatings	differentiation

organelles	epithelial tissue
mitochondria	connective tissue
matrix	nervous tissue
cristae	muscle tissue
chloroplasts	protective tissue
ribosomes	meristem
endoplasmic reticulum	collenchyma
RER	sclerenchyma
SER	xylem
Golgi apparatus	phloem
lysosomes	tight junction
peroxisomes	adhering junction
vacuoles	gap junction

II. MASTERING THE CHAPTER

A. LEARNING OBJECTIVES

When you have mastered the material in this chapter, you should be able to:

1. Define all key terms.

2. Distinguish between an amoeba (for instance) and a nerve cell (for instance) in the context of cell function and survival.

3. State how the structure of the phospholipid bilayer of a cell membrane utilizes the amphipathic property of phospholipid molecules.

4. Cite four structural units of the cell nucleus and state one function of each.

5. State the principal function of mitochondria.

6. State the principal function of chloroplasts, and why life as we know it on earth could not exist without these structures.

7. State the principal function of ribosomes.

8. Distinguish between RER and SER.

9. Cite three functions of the Golgi apparatus in plant and animal cells.

10. State two principal functions of lysosomes.

11. Cite two major functions of intracellular membranes.

12. State the general function of microfilaments.

13. Cite two functions of microtubules.

14. Cite similarities between the structure of cilia, flagella, basal bodies, and centrioles and the structure of microtubules.

15. Cite two functions of cell coatings.

16. Distinguish between prokaryotes and eukaryotes, giving an example of each.

17. Distinguish between independent unicellular organisms, colonial organisms, and multicellular organisms, citing an example of each.

18. Cite four kinds of tissue in multicellular animals and state at least one function of each.

19. Cite six kinds of tissue in multicellular plants and state at least one function of each.

20. Cite the three kinds of cell-to-cell junctions, and state the essential feature of each.

21. What is the significance of Boveri's experiment?

B. REVIEWING TERMS AND CONCEPTS

1. The amoeba is a unicellular organism that is fully independent; but most organisms are (a) _____-cellular and incapable of an independent existence. Such cells are said to be (b) _____ to carry out one or a few tasks.

2. All cells are enclosed within a cell (a) _____, which is composed of about 50% (b) _____ and 50¢ _____. The phospholipid portion is arranged in a (c) _____-layer, which permits certain molecules only to enter and leave the cell. Such membranes are said to be differentially (d) _____.

3. Intracellular membranes differ from the cell membrane chiefly in the kinds of (a) _____ and _____ forming them. In general, the cell can be said to consist of two major portions—the (b) _____, or control center, and the surrounding (c) _____.

4. The membrane-bounded nucleus contains hereditary units called (a) _____. The nucleus is essential to the well-being of the cell. Everything within the cell, except the nucleus, is called

42

the (b) _____. This material contains a variety of structures collectively called (c) _____, which can be isolated for study by means of a piece of equipment called the (d) _____.

5. One kind of organelle is a membrane-bounded structure with an elaborate inner membrane organized into folds and its chief function is that of converting the (a) _____ energy of food materials into energy the cell can use. These structures are called (b) _____.

6. Another organelle, one occurring only in (a) _____ plants and some algae and containing the pigment chlorophyll, is the (b) _____. This organelle enables these organisms to synthesize their own food through the process of (c) _____.

7. Protein synthesis occurs in those organelles called (a) _____. These structures are found both scattered through the cytoplasm and attached to the (b) _____ reticulum, which may release some of the proteins synthesized for transfer to the (c) _____ apparatus. Protein-filled sacs then may migrate to the cell (d) _____ and be discharged to the outside. Other protein-filled sacs may be retained within the cell as structures called (e) _____. These contain enzymes that (f) _____ a variety of waste and unwanted products within the cell. Resembling lysosomes in size and function are other organelles called (g) _____.

8. (a) _____ are irregular-shaped sacs that may contain food or waste materials. In plants, they may also contain (b) _____.

9. Intracellular membranes establish a number of compartments that isolate certain potentially "dangerous" structures, such as (a) _____, containing digestive juices called (b) _____.

10. Microfilaments are long thin fibers composed of the protein (a) _____ and associated with cell (b) _____ and (c) _____. Microtubules are straight, hollow, cylinders made of the protein (d) _____, and provide certain parts of the cell with (e) _____. They also form the (f) _____, which distributes (g) _____ to each daughter cell during cell division.

11. Many cells have short, whiplike extensions called (a) _____, while others have longer extensions called (b) _____.

12. Plant cells have a cell coating composed of (a) _____, which makes the boxlike cell walls of plants rigid.

13. Bacteria and blue-green algae are included in that group of cells classified as (a) _____ because they lack a membrane-bounded nucleus and membrane-bounded organelles. Cells that do have those structures are classified as (b) _____.

14. In the human organism, cell differentiation is highly advanced. Specialized cells may be grouped together to form tissues, such as

(a) _____ tissue, which occurs at the boundary between cell masses and a cavity. Cartilage and bone are examples of (b) _____ tissue. Binding connective tissue includes (c) _____ and liga-ments. Three major kinds of muscle tissue are (d) _____, _____, and _____.

15. In plants, new plant growth occurs in the (a) _____ tissue region. Water and minerals are transported up the plant through the (b) _____ tissue, while sugar and hormones are distributed throughout the plant through the (c) _____ tissue.

16. Those cell-to-cell junctions presenting a barrier to the passage of molecules and ions through the spaces between cells are termed (a) _____ junctions. Those providing strong mechanical attachments are termed (b) _____ junctions. (c) _____ junctions allow for the passage of ions and small molecular weight compounds between cells.

C. TESTING TERMS AND CONCEPTS

1. All cells have
    a) cell membranes.
    b) chloroplasts.
    c) lysosomes.
    d) cell walls.

2. Lysosomes are associated with
    a) protein synthesis.
    b) intracellular digestion.
    c) polysaccharide synthesis.
    d) cell division.

3. In plants, new cells are produced by the
    a) collenchyma.
    b) meristem.
    c) parenchyma.
    d) xylem.

4. Which of the following is not a term used for the material left after all the organelles have been removed from the cytoplasm?
    a) supernatant
    b) cytosol
    c) hyaloplasm
    d) sediment

5. The so-called energy "currency," ATP, is produced in the
    a) mitochondria.
    b) chloroplasts.
    c) ribosomes.
    d) microtubules.

6. Those proteins packed within a membrane are synthesized by the
    a) SER.
    b) mitochondria.

44

c) RER.
d) vacuole.

7. In plants, cellulose is synthesized within the
   a) Golgi apparatus.
   b) nucleus.
   c) mitochondria.
   d) chloroplast.

8. Peroxisomes are associated with the
   a) degradation of hydrogen peroxide.
   b) synthesis of ribosomes.
   c) conversion of carbohydrates into lipids.
   d) all oxidation reactions.

9. Cilia and flagella are fundamentally the same, derived from
   a) lysosomes.
   b) basal bodies.
   c) Golgi apparatus.
   d) vacuoles.

10. The establishment of numerous compartments within the cell is a
    function of
    a) lysosomes.
    b) Golgi apparatus.
    c) intracellular membranes.
    d) adhering junctions.

11. The cell wall of plant cells gains it stiffness by being composed
    of
    a) cellulose.
    b) collagen.
    c) microtubules.
    d) tubulin.

12. In many cells, microtubules are associated with
    a) cytoplasmic streaming.
    b) membrane formation.
    c) chromosome formation.
    d) ribosome synthesis.

13. Which tissue would not be found in a dog?
    a) nervous
    b) epithelium
    c) collenchyma
    d) collective

14. In a cell actively synthesizing protein, which of the following
    would be present in the largest number?
    a) nuclei
    b) mitochondria
    c) peroxisomes
    d) ribosomes

15. During the healing process, damaged tissue is observed to die and decay. Which organelle would you expect to be most active in this process?
    a) peroxisomes
    b) mitochondria
    c) lysosomes
    d) basal bodies

D. UNDERSTANDING AND APPLYING TERMS AND CONCEPTS

1. Cell membranes are not like screen, i.e., allowing things to pass simply on the basis of size. They are therefore said to be _____ _____.

2. What prevents the lysosomes from digesting normal cells? _____ _____

3. Glucose is "burned" in the_____.

4. The portion of the protein that is next to lipid layers of membrane is likely to be_____.

5. The _____ is/are located inside the nucleus and are involved with ribosome production.

6. The cell has a framework much like a building. In cells the "girders" are _____, _____ _____, and _____ _____.

7. In an animal, which tissue is involved with coordinating the activities of the rest? _____

8. In which type of plant tissue would you be most likely to find mitosis occuring at a given time? _____

9. A leaf cell would differ most from a fish skin cell in that the former would have _____ and _____ _____.

10. Sucrose (a relatively low mole. wt. substance) might pass most readily between cells via_____.

ANSWERS TO CHAPTER EXERCISES

Reviewing Terms and Concepts

1. a) multi   b) differentiated

2. a) membrane   b) lipid, protein   c) bi   d) permeable

3. a) lipids, proteins   b) nucleus   c) cytoplasm

4. a) chromosomes   b) cytoplasm   c) organelles   d) centrifuge

46

5. a) potential b) mitochondria

6. a) green b) chloroplast c) photosynthesis

7. a) ribosomes b) endoplasmic c) Golgi d) membrane e) lysosomes
    f) digest (break down) g) peroxisomes

8. a) Vacuoles b) pigments

9. a) lysosomes b) enzymes

10. a) actin b) motion c) locomotion d) tubulin e) rigidity
    f) spindle g) chromosomes

11. a) cilia b) flagella

12. a) cellulose

13. a) prokaryotes b) eukaryotes

14. a) epithelial b) supportive c) tendons d) skeletal, smooth,
    cardiac

15. a) meristematic b) xylem c) phloem

16. a) tight b) adhering c) Gap

Testing Terms and Concepts

1. a	6. d	11. a
2. b	7. a	12. a
3. b	8. a	13. c
4. d	9. b	14. d
5. a	10. c	15. c

Understanding and Applying Terms and Concepts

1. differentially permeable

2. The digestive enzymes are enclosed in a membrane.

3. mitochondria

4. hydrophobic

5. nucleolus

6. microfilaments, microtubules, intermediate filaments

7. nervous tissue

8. meristem

9. cell walls, chloroplasts

10. gap junctions

6

THE METABOLISM OF CELLS

I. REVIEWING THE CHAPTER

A. CHAPTER HIGHLIGHTS

6.1 The Cellular Environment

The cellular environment always is a liquid, whether the cell is
an amoeba in a pond or a living cell of an oak tree. Our body cells
are bathed in an interstitial fluid derived from the blood. In general,
the fluid bathing all living cells is called extracellular fluid, or
ECF.

6.2 The Composition of the ECF

The ECF is composed mostly of water, which contains dissolved
gases such as oxygen and carbon dioxide; a number of inorganic ions,
including sodium (Na^+), potassium (K^+), chloride (Cl^-), and calcium
(Ca^{++}); also included are certain foods, vitamins, and hormones. The
pH of the extracellular fluid and its termperature are also important
to the cell's welfare. One major advantage multicellular organisms
have over unicellular organisms is an ability to regulate the ECF in
a more or less steady state, called homeostasis.

6.3 Diffusion

The continuous random motion of molecules free to move about in
a given volume of space causes those molecules to be more or less
evenly distributed within that space, through a process called diffusion. The random motion of the molecules of two substances separated
by a porous membrane eventually will cause both substances to diffuse
through the membrane and reach equal concentrations on both sides of
the membrane.

6.4 Osmosis

Osmosis is a special case of diffusion and occurs through membranes said to be differentially permeable, since they permit the
passage of certain molecules or ions but not that of certain others.
Water tends to pass through a cell membrane from the region of higher
concentration of water to the region of lower concentration. The
greater the difference in concentration, the greater the tendency for
osmosis to occur and thus the greater the osmotic pressure.

Water enters the cells of pond plants by osmosis until the water
pressure within the cell's rigid cell wall equals that of the osmotic

pressure, at which time osmosis ceases. But freshwater animals lack rigid cell walls, so they must have a means of actively pumping out water to prevent excessive amounts from building up and rupturing their cell membranes. Such an environment is said to be hypotonic.

The water concentration in the cytoplasm of marine plants and marine invertebrates is about the same as that in the sea. So these organisms exist in a state of osmotic equilibrium with their watery environment. The external water solution is said to be isotonic.

A land plant placed in sea water quickly wilts, or loses water. This is because a given volume of sea water contains a smaller number of water molecules than a given volume of the cytoplasm of the plant. The sea water is said to be hypertonic to the cytoplasm. As the plant's cells continue to lose water through osmosis, the cells soon shrink, a condition called plasmolysis. Ocean fish avoid this problem by drinking sea water and then desalting it, using metabolic energy. Diffusion of substances that require special enzymes, called permeases, is called facilitated diffusion.

6.5 Active Transport

Certain cells contain concentrations of a given molecule or ion in concentrations far higher than those in the external environment. To move materials into a cell against a concentration gradient requires an expenditure of metabolic energy, a phenomenon called active transport. It occurs in plants and animals alike. The mechanisms of active transport are not well understood, but it is clear that the cell membrane, through the action of enzymes, plays an important part in running the metabolic "pump." In addition to being able to recognize specific ions and molecules the pumps require energy in order to perform their task.

6.6 Endocytosis/6.7 Exocytosis

Cells also transport material from the ECF into the cytoplasm by engulfing the material and so "swallowing" it, a process called endocytosis. Amoebas ingest food in this way, and white blood cells ingest foreign matter in our blood in the same manner. Those proteins and other macromolecules too large and too hydrophilic for simple diffusion enter the cell through endocytosis. Exocytosis is the reverse of endocytosis and accounts for the expulsion of protein and other macromolecules from the cell.

6.8 Cell Chemistry

Most chemical reactions require an input of energy to start the reaction going, called energy of activation. Such energy is needed to break the bonds of the reactants. In combustion of wood or glucose, for example, energy of activation (heat from a match) is needed only to start the reaction, not to maintain it.

6.9 Enzymes

In living organisms, enzymes reduce the energy of activation needed. An enzyme binds temporarily to one or more of the reactants and by so doing lowers the amount of activation energy needed to start the reaction. The bonds holding an enzyme to its substrate are weak. The substrate molecule must fit into a complementary surface on the enzyme molecule in somewhat the same way a key fits in a lock. Because enzymes are very specific for the reactions they catalyze, a given enzyme usually can trigger only one or a few reactions.

Enzyme action may be inhibited by the addition of molecules of a substance the shape of which closely approximates that of the substrate, a phenomenon called competitive inhibition. A reaction may also be inhibited if a certain nonprotein substance called a cofactor is absent. Cofactors may be metal ions—Na^+ or Fe^{++}—or small organic molecules called coenzymes, vitamin B_1 being an example. The activity of enzymes may also be affected strongly by changes in temperature and pH, both of which affect an enzyme's tertiary structure and hence its ability to fit its substrate.

6.10 Regulation of Enzymes

It is essential that cells have means of regulating the activity of their enzymes. To do so, several methods are employed by living organisms. Enzymes are prevented from digesting their cells by certain inhibitory factors. For example, enzymes such as proteinases can be inhibited by being maintained in an inactive form within the cell until exposed to conditions of low pH outside the cell, at which time they become activated. Enzymes remaining within the cell can be inhibited by regulatory molecules that attach to the enzyme molecule, and by so doing alter the enzyme's shape enough to inhibit the enzyme from attaching to its substrate. This phenomenon is called an allosteric effect. Enzyme regulation processes are of extreme importance in homeostasis.

B. KEY TERMS

metabolism	active transport
ECF	turgor pressure
hormones	endocytosis
homeostasis	phagocytosis
concentration gradient	exocytosis
diffusion	cytosol
osmotic pressure	co-factor
osmosis	competitive inhibition
plasmolysis	prosthetic group
facilitated diffusion	enzyme specificity

permeases   energy of activation

hypertonic   allosteric effect

hypotonic   feedback inhibition

isotonic   precursor activation

II. MASTERING THE CHAPTER

A. LEARNING OBJECTIVES

When you have mastered the material in this chapter, you should be able to:

1. Define all key terms.

2. List the four major components of the extracellular fluid, and give four examples of required inorganic ions.

3. Distinguish between simple diffusion and osmosis as a special case of diffusion, and cite an example of each.

4. List four factors affecting the rate at which materials diffuse through a cell membrane.

5. Distinguish between osmotic conditions in a freshwater and sea water environment for a freshwater plant cell, an amoeba, and a bony fish.

6. Explain how the amoeba manages to exist in a hypotonic environment.

7. Distinguish between osmosis and active transport, citing an example of each.

8. Distinguish between endocytosis and exocytosis and give two examples of each.

9. State the importance of energy of activation in the context of chemical reactions.

10. List three ways an enzyme may be bonded to a substrate.

11. Explain what would happen if enzymes were not regulated. Use the proteinase pepsin as an example.

12. Describe the difference between feedback inhibition and precursor activation.

13. Describe how the "allosteric effect" operates.

B. REVIEWING TERMS AND CONCEPTS

1. The cell environment always is a (a) _____, whether the cell is part of an oak tree or part of you. The fluid bathing living cells is called the (b) _____ and is derived in vertebrates from the (c) _____. This fluid is composed mostly of (d) _____, but also contains dissolved gases such as (e) _____, inorganic (f) _____, foods, vitamins, and (g) _____ _____. The pH and (h) _____ are also important conditions of the ECF. Multicellular organisms have the ability to regulate the ECF in a more or less steady state, called (i) _____.

2. The continuous random motion of molecules free to move about in a given volume of space causes those molecules to be more or less evenly distributed within that space, through a process called (a) _____. (b) _____ is a special case of diffusion, i.e. the diffusion of water, and occurs through membranes said to be (c) _____ permeable, since they permit the passage of certain molecules or ions (in this case, water) but not that of certain others. Water tends to pass through a cell membrane from the region of (d) _____ concentration to the region of (e) _____ concentration of water molecules. The greater the difference in (f) _____, the greater the tendency for (g) _____ to occur, and thus the greater the (h) _____ pressure.

3. Water enters the cells of freshwater plants by (a) _____ until the water pressure inside the cell's rigid cell wall equals that of the (b) _____, at which time (c) _____ ceases. But freshwater animals lack rigid cell walls, so they must have a means of (d) _____ out water. Such an environment is said to be (e) _____-tonic.

4. The water concentration in the sea is about the same as that in the cytoplasm of marine plants and invertebrates. So these organisms exist in a state of osmotic (a) _____ with their watery environment. The environment is said to be (b) _____-tonic.

5. A land plant placed in sea water wilts quickly by losing water. This is because a given volume of sea water contains a (a) _____ number of water molecules than a given volume of plant cytoplasm. The sea water is said to be (b) _____-tonic to the cytoplasm. Ocean fish avoid this problem by drinking sea water and then (c) _____ it, using (d) _____ energy.

6. An expenditure of (a) _____ energy is required for a cell to contain a higher concentration of an ion or molecule than does its external environment. This process is called (b) _____ transport. The mechanisms of active transport are not well understood, but it is clear that the cell (c) _____, through the action of (d) _____, plays an important part in running the metabolic "pump." Diffusion that depends upon special membrane transport mechanisms, such as permeases, is said to be (e) _____ diffusion.

7. Cells can transport material from the ECF into the cytoplasm by engulfing the material in a process called (a) _____.

(b) _____ blood cells ingest foreign matter in this way. (c) _____ and other macromolecules too large and too hydro- (d) _____ for simple (e) _____ enter the cell through endocytosis. (f) _____ is the reverse of this process and accounts for the expulsion of protein and other macromolecules from the cell.

8. Most chemical reactions require an input of (a) _____ to start the reaction going, called the (b) _____ of (c) _____, which is needed to break the (d) _____ of the reactants.

9. In living organisms (a) _____ permit reactions to occur at low temperatures. An enzyme binds temporarily to one or more of the (b) _____ and by so doing (c) _____ (raises/lowers) the amount of activation energy needed to start the reaction. A (d) _____ molecule must fit into a complementary surface on the enzyme molecule in somewhat the same way as a key fits into a lock. Enzymes tend to be very (e) _____ (specific/general) for the reactions they catalyze.

10. Enzyme action may be (a) _____ by the addition of molecules of a substance the shape of which closely approximates that of the substrate, a phenomenon called (b) _____ _____ __. Enzymes may also be inactive unless a certain nonprotein substance, called a (c) _____, is present. These substances may be a (d) _____ ion or small organic molecules called (e) _____.

11. The activity of enzymes may also be affected strongly by changes in (a) _____ and _____, both of which affect an enzyme's tertiary structure and hence its ability to fit its substrate.

12. Some digestive enzymes can be prevented from digesting their cells by being maintained in an inactive form until exposed to conditions of proper (a) _____ when outside the cell. Within the cell, certain molecules, called (b) _____ molecules, may attach to an enzyme and alter its (c) _____. This can inhibit the enzyme from attaching to its substrate or enforce its binding. In either case this phenomenon is called an (d) _____ effect.

C. TESTING TERMS AND CONCEPTS

1. Sea water in relation to the cytoplasm of a land organism is said to be
   a) isotonic.
   b) hypotonic.
   c) hypertonic.
   d) allosteric.

2. Movement of material through a cell membrane against a concentration gradient is achieved through
   a) active transport.
   b) diffusion.
   c) osmosis.
   d) turgor pressure.

3. Materials taken into a cell by endocytosis are made available to the cytosol through the action of
   a) plasmolysis.
   b) osmosis.
   c) lysosomes.
   d) mitochondria.

4. The energy required to trigger a chemical reaction is termed energy of
   a) reaction.
   b) activation.
   c) conversion.
   d) production.

5. Enzymes work by
   a) lowering the energy of activation.
   b) doubling the energy of activation.
   c) raising the energy of activation.
   d) eliminating the energy of activation.

6. Enzyme action may be regulated by all except
   a) feedback inhibition.
   b) precursor activation.
   c) competitive inhibition.
   d) competitive activation.

7. Of the following substances, the one that cannot pass through a cell membrane by diffusion is
   a) hydrophobic molecules.
   b) proteins.
   c) hydrophilic molecules.
   d) ions.

8. The speed with which osmosis takes place in marine plants is a measure of the difference in the
   a) weight of molecules and ions involved.
   b) strength of charge or ions involved.
   c) concentration of molecules or ions involved.
   d) density of the molecules involved.

9. The forces that usually hold an enzyme to its substrate are
   a) hydrophilic regions of the two substances.
   b) covalent bonds.
   c) hydrogen and ionic bonds.
   d) hydrophobic.

10. Some chemical reactions will not occur without the presence of a small organic cofactor called a(n)
    a) activator.
    b) coenzyme.
    c) stimulator.
    d) initiator.

11. Which of the following is not likely to be found in the ECF?

a) vitamins
b) ions
c) hormones
d) DNA

12. Facilitated diffusion is characterized by all of the following except
a) does not occur in plants.
b) requires energy.
c) uses permeases.
d) concentrates a substance against a gradient.

13. A process involving the secretion of large amounts of material at once from a cell is called
a) phagocytosis.
b) plasmolysis.
c) endocytosis.
d) exocytosis.

14. The higher the water concentration inside a plant cell, the
a) lower the turgor pressure.
b) greater the tendency for water to diffuse into the cell.
c) higher the turgor pressure.
d) more likely it is to wilt.

15. In order to function as a "pump," an enzyme must have all the following features except
a) a site to bind to the substance being transported.
b) the ability to move across the membrane.
c) a method of utilizing energy.
d) a method of reproducing itself.

D. UNDERSTANDING AND APPLYING TERMS AND CONCEPTS

1. An enzyme involved in moving ions or molecules through a membrane is called a(n) _____.

2. Too much fertilizer is said to "burn" plants and results in wilting and often death. It is likely that cells in these plants have undergone _____.

3. Why is it harmful to drink sea water? _____
_____
_____

4. How does an enzyme facilitate the "burning" of glucose? _____
_____

5. What is the term for a coenzyme that is tightly bound to an enzyme? _____
_____

Decide whether the following statements are true (T) or false (F).

_____ 6. Every living cell in your body is bathed in liquid.

_____ 7. Some materials in the ECF are toxic if the concentrations are higher enough.

_____ 8. Movement of ions/molecules in and out of cells is always a strictly physical event.

_____ 9. By analogy, when you blow up a balloon, you are increasing its turgor pressure.

_____ 10. Endocytosis occurs at the tissue level.

_____ 11. In synthesizing glucose, the $\triangle$G would have a negative value.

_____ 12. In order to digest starch, it must come in contact with a hydrolase.

_____ 13. Enzymes function, at least in part, by distorting the chemical bonds of the substrate.

_____ 14. When several enzymes work in a sequential fashion, they often are arranged "side-by-side."

_____ 15. Control of enzyme activity is only controlled at the level of the actual enzyme.

ANSWERS TO CHAPTER EXERCISES

Reviewing Terms and Concepts

1. a) liquid b) ECF c) blood d) water e) oxygen (or CO_2) f) ions

2. a) diffusion b) Osmosis c) differentially d) higher e) lower
   f) concentration g) osmosis h) osmotic

3. a) osmosis b) osmotic pressure c) osmosis d) pumping e) hypo

4. a) equilibrium b) iso

5. a) smaller b) hyper c) desalting d) metabolic

6. a) metabolic b) active c) membrane d) enzymes e) facilitated

7. a) endocytosis b) White c) Proteins d) philic e) diffusion
   f) Exocytosis

8. a) energy b) energy c) activation d) bonds

9. a) enzymes b) reactants c) lowers d) substrate e) specific

10. a) inhibited (or stopped) b) competitive inhibition c) cofactor
    d) metal e) coenzymes

11. a) temperature, pH

12. a) pH   b) regulatory   c) shape   d) allosteric

Testing Terms and Concepts

1.	c	6.	d	11.	d
2.	a	7.	b	12.	a
3.	c	8.	c	13.	d
4.	b	9.	c	14.	c
5.	a	10.	b	15.	d

Understanding and Applying Terms and Concepts

1. permease

2. plasmolysis

3. the high salt concentration "draws" water from your cells

4. lowers the energy of activation

5. prosthetic group

6. T

7. T

8. F

9. T

10. F

11. F

12. T

13. T

14. T

15. F

ENERGY PATHWAYS IN THE CELL

I. REVIEWING THE CHAPTER

A. CHAPTER HIGHLIGHTS

7.1 Anabolism and Catabolism

Nutrition that involves dependence upon preformed organic molecules is called heterotrophic nutrition. The organisms using this kind of food are called heterotrophs, and the first step in their digestive processes is to break down the solid materials into small soluble molecules. For the organism to function, energy still locked up in the products of digestion—sugars, amino acids, fatty acids, and glycerol—must be liberated.

During catabolism, energy is released and the energy-rich end products of digestion are further broken down into the energy-poor products CO_2, H_2O, and NH_3. During anabolism, the reverse occurs—the simple end products of digestion being reassembled into macromolecular components, polysaccharides, lipids, proteins, and nucleic acids. Anabolic reactions require the input of energy with a resulting gain in free energy.

7.2 Glycolysis/7.3 ATP and NAD/7.4 Glycolysis: Priming The Pump/ 7.5 Glycolysis: The First Oxidation

When glucose is broken down in the absence of oxygen, the process is called glycolysis. Yeast cells, for example, catabolize glucose to ethanol and carbon dioxide in a process called alcoholic fermentation. Overworked muscle cells derive part of their energy requirements through a process called lactic acid fermentation. Eleven enzymes located in the cytosol are involved in glycolysis.

Two coenzymes are also important in glycolysis; ATP, which stores energy in a high-energy phosphate bond, and NAD, which is important in many oxidation reactions because of its ability to be reduced. Glycolysis not only yields energy, but also sets the stage for additional energy-yielding reactions. In particular, glycolysis pro- duces pyruvic acid.

7.6 Lactic Acid Fermentation from Pyruvic Acid

In overworked muscle cells, oxygen is inadequate for cellular respiration, and lactic acid is produced. Most of the energy in the original fuel (glucose) is still trapped in the waste product lactic acid. Although it is not a highly efficient process, the fermentation of lactic acid does yield free energy in the form of ATP, which is made available to cells short of oxygen for cellular respiration.

7.7 Alcoholic Fermentation

Alcoholic fermentation also is a wasteful process, most of the energy originally in the glucose being stored in the end product ethanol. The reduction of pyruvic acid to lactic acid or ethanol is an anabolic, energy-consuming process. The inefficiency of these two fermentations is overcome in the presence of ample oxygen. When, early in evolution, organisms came to use oxygen as the final acceptor of electrons removed from carbohydrates and to completely oxidize pyruvic acid, cellular respiration resulted.

7.8 Cellular Respiration/7.9 The Citric Acid Cycle/7.10 The Respiratory Chain

The crucial events of cellular respiration are (a) the complete oxidation of pyruvic acid by the stepwise removal of all the hydrogen atoms, leaving three molecules of CO_2 ; and (b) the passing on of the electrons removed as part of these hydrogen atoms to molecular oxygen. Once inside the mitochondria the pyruvic acid is cleaved to yield carbon dioxide, and a 2-carbon fragment that is attached to another coenzyme, Coenzyme A. The acety-Coenzyme A now enters the citric acid cycle, where the final breakdown to CO_2 occurs. The CO_2 is released with the concomittant transfer of the electrons to oxygen through the respiratory chain. The respiratory chain consists of several coenzymes that vary stepwise in their redox potentials.

7.11 Coupling Electron Transport to the Synthesis of ATP

The chemiosmotic theory proposes a mechanism for the coupling of electron flow to the synthesis of ATP. The theory relates both chemistry and electron microscopy to the importance of the spatial arrangements of biochemical reactions.

7.12 The Respiratory Balance Sheet: Materials/7.13 The Respiratory Balance Sheet: Energy

When one molecule of glucose is fully oxidized during cellular respiration, and when the ATPs made available through glycolysis are included, a total of 36 molecules of ATP are made available by the oxidation of that one molecule. The combustion of one mole of glucose releases 686 kcal of energy, almost all of it as heat. But the cellular respiration of glucose traps some of this energy by the synthesis of ATP. Each molecule of ATP stores, as a conservative estimate, 7.3 kcal. Therefore, 36 ATPs store some 263 kcal of free energy, a yield of 38% of the free energy of the glucose, and a performance far superior to the meager 7% yield from the fermentation of glucose.

7.14 The Storage Battery of Life

The mitochondria act like discharging storage batteries. If we bring two substances of differing redox potential together so that the

opportunity of electron flow between them exists, we have created a
battery. In the mitochondria just such a situation exists between
carbon (reduced to the extent occurring in carbohydrates like glucose)
and oxygen, the most electronegative substance in the system. Using
the storage battery analogy, it is possible to think of and equate
free energy changes within a cell to redox potentials.

7.15 What about Other Fuels?

The biochemical pathway of cellular respiration is a versatile
one and facilitates the breakdown of foods other than glucose. Fats,
fatty acids, and amino acids may all take part. So the respiration
of excess fats and proteins in the diet is brought about, and with no
special mechanisms of cellular respiration being needed.

7.16 Control of Cellular Respiration

Glycolysis and cellular respiration must be precisely controlled,
considering the cell's changing needs for ATP, oxygen, and the changing
fuel sources. Among the control systems operating are feedback
inhibition and precursor activation, both of which depend upon the
presence of allosteric enzymes.

7.17 How These Discoveries Were Made

The complex biochemical steps of glycolysis and cellular
respiration often are studied by following the fate of a certain compound in a reaction as it is being produced or used up. The procedures
frequently employ the use of radioactive or "tagged" molecules.

7.18 The Uses of Energy

Glycolysis and cellular respiration transform the free energy of
food into the free energy stored in the high-energy phosphate bonds of
ATP. The ATP, in turn, serves as the immediate source of energy for
all the energy-activities of the organism, such as the performance of
mechanical work (e.g. locomotion), active transport, heat production
(particularly important in the case of warm-blooded organisms), and
anabolism, or the synthesis of macromolecules.

7.19 Anabolism

All anabolic transformations are catalyzed by enzymes. The products of anabolism serve a number of essential functions. Glycogen
and fats and even proteins serve as reserve fuels for later catabolism.
The synthesis of new nucleic acids enables the organism to store and
express additional copies of its library of genetic information. Protein, protein-carbohydrates, and protein-lipid molecules are essential
structural components of the organism, both intracellular and extracellular. When synthesis of these materials proceeds at a faster rate
than their degradation, the organism grows.

7.20 Summary

This chapter shows how energy from catabolic processes is obtained from certain chemicals so that the energy might be transferred to anabolic processes.

B. KEY TERMS

heterotrophic	alcoholic fermentation
digestion	Coenzyme A (Kreb's cycle)
anabolism	citric acid cycle
catabolism	respiration
glycolysis	respiratory chain
ATP	cytochrome enzymes
NAD	cristae
FAD	chemiosmotic theory
"high energy" bonds	allosteric enzyme
lactic fermentation	active transport

II. MASTERING THE CHAPTER

A. LEARNING OBJECTIVES

When you have mastered the material in this chapter, you should be able to:

1. Define all key terms.

2. State the two "fates" of the small organic molecules that are the products of digestion once they enter the cytoplasm of cells, and give one example of each.

3. Distinguish between the aerobic and anaerobic breakdown of glucose, citing the end products when yeast cells catabolize glucose and when muscle cells catabolize glucose.

4. Cite the chief function of ATP and of NAD in metabolism.

5. Cite three major ways in which cells handle the pyruvic acid produced by glycolysis.

6. Cite the two crucial events of cellular respiration.

7. Cite three links in the respiration chain during which a reactant is oxidized, and say in what organelle the reaction occurs.

8. State the reactants and products of cellular respiration.

9. State why the chemical reaction for cellular respiration occurs with - G.

10. Cite the three places in the respiratory chain where a drop in potential yields enough energy to synthesize a molecule of ATP.

11. Cite two instances in which intermediate compounds formed during cellular respiration link glucose metabolism with the metabolism of other food molecules.

12. Cite two examples of feedback inhibition in the control of cellular respiration.

13. Cite four energy requiring activities in living organisms, and give an example of each.

14. Cite three major anabolic activities.

15. Explain why it is a good strategy to have anabolic and catabolic pathways separated.

B. REVIEWING TERMS AND CONCEPTS

1. (a) _____ organisms require preformed organic molecules for their nutritional requirements. Solid food is broken down by a process called (b) _____. In this process little energy is released, because most of the (c) _____ energy is still in molecules of (d) _____, _____, (e) _____ acids, and _____ acids—the products of this process.

2. During (a) _____, energy is released and the energy-rich end products of digestion are further broken down into the energy- (b) _____ products CO_2, H_2O, and NH_3. During (c) _____, the reverse occurs—the simple end products of digestion being reassembled into macromolecular components, polysaccharides, (d) _____, _____, and (e) _____ acids. Anabolic reactions require the (f) _____ of energy with a resulting (g) _____ in free energy.

3. When glucose is broken down without the presence of oxygen, to provide energy for the cell, the process is called (a) _____. In this way, yeast cells break glucose down into (b) _____ and (c) _____ _____, a process called (d) _____ fermentation. Overworked muscle cells also use glycolysis to meet their energy needs, the end product being (e) _____ _____ to metabolize glucose, and the final products are (g) _____ _____ and (h) _____.

4. ATP, adenosine triphosphate, is a (a) _____ with high-energy (b) _____ bonds. When broken by (c) _____, these bonds release large amounts of (d) _____ energy.

62

5. (a) _____ fermentation also is a wasteful process since most of the energy originally stored in the glucose remains in the end-product (b) _____. The reduction of pyruvic acid to (c) _____ acid or (d) _____ is an anabolic, energy-(e) _____ process.

6. The crucial events of cellular (a) _____ are (1) the complete (b) _____ of pyruvic acid by the stepwise removal of all the (c) _____ atoms, leaving three molecules of CO_2; and (2) the passing on of the (d) _____ removed as part of these hydrogen atoms to molecular (e) _____.

7. When one molecule of (a) _____ is fully oxidized during cellular respiration, and when the ATPs made available through glycolysis are included, a total of 36 molecules of (b) _____ are made available by the oxidation of that one molecule.

8. The combustion of one mole of (a) _____ releases 686 kcal of energy, almost all of it as (b) _____. But the cellular respiration of glucose traps some of this energy by the synthesis of (c) _____. Each molecule of ATP stores, as a conservative estimate, (d) _____. Therefore, 36 ATPs store some 263 kcal of (e) _____ energy, a yield of 38% of the free energy of the glucose, and a performance far (f) _____ to the meager 7% yield from the (g) _____ of glucose.

9. (a) _____ and cellular respiration transform the (b) _____ energy of food into the (c) _____ energy stored in the high-energy (d) _____ bonds of ATP. The ATP, in turn, serves as the immediate source of energy for all the energy-requiring activities of the organism.

10. In addition to glucose, (a) _____, _____ and other (b) _____ are also important energy sources. Fortunately, there are common features to many of the catabolic pathways; one is the use of (c) _____ as an entry point for catabolites of several different molecules into the citric acid cycle.

11. As with most cellular reactions, catabolism is regulated. For example, ATP slows the activity of phosphofructokinase in a process called (a) _____ inhibition. Therefore, the enzyme is said to be (b) _____ because its activity is influenced by a molecule other than the substrate.

12. Some major uses of energy obtained during catabolism are: (a) _____ _____, _____ _____, and _____. Anabolic reactions are for the (b) _____ of macromolecules such as (c) _____ _____, and _____.

C. TESTING TERMS AND CONCEPTS

1. At each oxidation in cellular respiration, the product in each case contains, in relation to the reactant,

a) more free energy.
b) less free energy.
c) the same amount.
d) a varying amount.

2. Energy released by oxidation during cellular respiration is associated with the synthesis of
a) ATP
b) glucose
c) glycogen
d) FAD

3. In addition to glucose, cells may tap as a source of energy all of the following except
a) fats.
b) proteins.
c) amino acids.
d) ions.

4. An increasing concentration of ATP inhibits the activity of phosphofructokinase, thus slowing the production of
a) ATP.
b) ADP.
c) NAD.
d) FAD.

5. Muscle cells "desparate" for oxygen are provided an energy source of ATP by
a) lactic acid fermentation.
b) alcoholic fermentation.
c) aerobic respiration.
d) PEP oxidation.

6. When cells catabolize glucose aerobically, the end products are
a) alcohol and CO_2.
b) NAD and H_2O.
c) H_2O and CO_2.
d) ATP and CO_2.

7. Anabolism is the process involving the
a) degradation of organic molecules.
b) synthesis of macromolecules.
c) conversion of CO_2, H_2O, and NH_3.
d) transport of electrons.

8. When fats are used as fuel, the glycerol portion of the molecule enters the glycolytic pathway when converted to
a) PGAL.
b) malic acid.
c) GDP.
d) PEP.

9. Anabolism, like catabolism, requires
a) more reactants than products.
b) ribosomes.

c) enzymes.
d) mitochondria.

10. The coenzyme is important as an electron acceptor during glycolysis is
   a) NAD.
   b) FAD.
   c) Coenzyme A.
   d) cytochrome B.

11. Glycolysis and the citric acid cycle enzymes are found in these respective parts of the cell:
   a) mitochondria/cytosol.
   b) cytosol/mitochondria.
   c) cytosol/endoplasmic reticulum.
   d) Golgi apparatus/mitochondria.

12. The passage of electrons down the respiratory chain and the synthesis of ATP have been related by the
   a) allosteric theory.
   b) precursor activation theory.
   c) cristae theory.
   d) chemiosmotic theory.

13. The redox potential of NAD compared to oxygen is
   a) higher.
   b) lower.
   c) the same.
   d) different from case to case.

14. The difference between the free energy needed to break the bonds of the reactants and the free energy liberated in the formation of the product bonds is a
   a) net free energy yield.
   b) total free energy yield.
   c) net free energy deficit.
   d) total free energy deficit.

15. The reactions in the citric acid cycle illustrate an important biological phenomenon, namely that living cells process things
   a) randomly.
   b) orderly.
   c) sequentially.
   d) uniquely.

D. UNDERSTANDING TERMS AND CONCEPTS

The following is a series of hypothetical reactions.

$$A \xrightarrow{1} B \xrightarrow{2} C \xrightarrow{3} D$$

1. If A is radioactive and enzyme 3 is inhibited, where would you find most of the radioactivity? _____

2. If B is radioactive and enzyme 1 is inhibited, where would you expect to find most of the label? _____

3. Biological oxidation via the "electron transport train" involves
   a substance taking electrons from another substance with (a)
   _____ electronegativity and passing them on to a
   substance with (b) _____ electronegativity.

4. Which organism would get the most free energy from glucose, one
   that is anaerobic or one that is aerobic? _____

Decide whether the following statements are true (T) or false (F).

____ 5. NAD contains "high energy" phosphate bonds.

____ 6. Plants make their own sugars via photosynthesis and are thus
        said to be heterotrophic.

____ 7. An example of hydrolysis is the breaking down of protein
        into amino acids.

____ 8. Without enzymes catabolism would be impossible.

____ 9. Lactic acid builds up in over-worked muscle because of a
        lack of oxygen.

____ 10. Cristae are the folded membranes inside mitochondria.

____ 11. The theory of chemiosmosis states that enzymes of the
         respiratory chain must interact directly with the enzymes
         involved in ATP synthesis.

____ 12. The physical arrangement of the enzymes that participate in
         respiration is more important than that of those involved
         in glycolysis.

____ 13. There is no way to equate the electronegativity of a molecule
         with the energy it contains.

____ 14. Living things utilize ATP in order to prevent their own
         decay.

____ 15. Living things conform to the statement "You can not get
         something for nothing."

ANSWERS TO CHAPTER EXERCISES

Reviewing Terms and Concepts

1. a) Heterotrophic b) digestion c) bond d) sugars, glycerol
   e) fatty, amino

2. a) catabolism b) poor c) anabolism d) lipids, proteins
   e) nucleic f) input g) gain

3. a) glycolysis h) ethanol c) carbon dioxide d) alcoholic
    e) lactic acid f) oxygen g) carbon dioxide h) water

4. a) nucleotide (coenzyme) b) phosphate c) hydrolysis d) free

5. a) Alcoholic h) ethanol c) lactic d) ethanol e) consuming

6. a) respiration h) oxidation c) hydrogen d) electrons e) oxygen

7. a) glucose b) ATP

8. a) glucose b) heat c) ATP d) 7.3 kcal e) free f) superior
    g) fermentation

9. a) Glycolysis b) free c) free d) phosphate

10. a) fats, proteins h) carbohydrates c) Coenzyme A

11. a) feedback b) allosteric

12. a) mechanical work, active transport, heat production, anabolism
    b) synthesis c) polysaccharides, nucleic acids, proteins

Testing Terms and Concepts

1.	h	6.	c	11.	b
2.	a	7.	b	12.	d
3.	d	8.	a	13.	a
4.	a	9.	c	14.	a
5.	a	10.	a	15.	c

Understanding and Applying Terms and Concepts

1.	c	12.	T
2.	d	13.	F
3.	a) lower b) greater	14.	T
4.	aerobic	15.	T
5.	F		
6.	F		
7.	T		
8.	T		
9.	T		
10.	T		
11.	F		

8

PHOTOSYNTHESIS

I. REVIEWING THE CHAPTER

A. CHAPTER HIGHLIGHTS

8.1 Early Experiments

Living cells need a continuous supply of energy. The ultimate
source of such energy is the organic molecules produced by green plants
and algae in the process called photosynthesis. These plants convert
carbon dioxide and water, with the energy of sunlight and in the
presence of the green pigment chlorophyll, into the carbohydrate glucose,
releasing the gas oxygen as a by-product. Green plants and algae are
classed as autotrophs, while those other organisms depending on green
plants and algae as a source of energy-rich organic molecules are
termed heterotrophs.

Early experimenters could explore the phenomena of photosynthesis
only by indirect means. For example, simple methods were employed to
show that weight gain in plants was not due solely to uptake from the
soil. It was also shown that in the dark plants behave like animals
in that they remove oxygen from the air, but that in light they produce oxygen.

8.2 The Pigments

There are two types of chlorophyll, termed chlorophyll a and
chlorophyll b. It has been possible to determine which wavelengths
of visible light each type of chlorophyll most readily absorbs. Both
absorb visible light most strongly in the red and blue regions of the
spectrum. They reflect most strongly in the green wavelengths, and
that is why green plants appear green to our eyes. Other pigments,
which absorb strongly in the blue region and reflect reds and oranges,
are the carotenoids.

8.3 Chloroplasts

Those organelles containing the photosynthetic pigment chlorophyll
are called chloroplasts, and their structure shows similarities to
mitochondria. Chloroplasts produce more chloroplasts, and develop
from proplastids. The presence of DNA in chloroplasts helps account
for their curious autonomy.

8.4 The Leaf

In higher plants, chloroplasts are generally confined to the cells of young stems, immature fruits, and leaves. The leaves are the real photosynthetic factory of green plants. They consist of an upper epidermis, covered with a waxy secretion called cutin, beneath which is the palisade layer of cells, where most photosynthetic activity occurs. Beneath this layer is the spongy layer, where food is stored and where the exchange of gases between the leaf and the environment occurs. Openings in the leaf surface (particularly the lower epidermis), called stomata, regulate the amount of water vapor, carbon dioxide, and oxygen entering and leaving the leaf. This is accomplished by the change in shape of guard cells, which are sensitive to the amount of carbon dioxide contained within the leaf.

8.5   Factors Limiting the Rate of Photosynthesis

The rate of photosynthesis in a green leaf increases with increased light intensity, but not indefinitely. Temperature change also affects the rate of photosynthesis, as does the concentration of carbon dioxide. For example, an increased rate of photosynthesis with increased temperature does not occur if the supply of CO_2 is limited.

8.6   The Dark Reaction

There are two major sets of reactions during photosynthesis, the light reactions and the dark reactions. The dark reactions include the uptake of CO_2 and its reduction by hydrogen atoms. The oxygen produced in the light reactions is derived from water, not from the carbon dioxide. So one role played by light in photosynthesis is the separation of the hydrogen and oxygen atoms in water molecules.

8.7   The Light Reactions/8.8   Photosystems I and II

The light reactions include two distinct processes—one energized by long wavelengths (photosystem I) and the other by shorter wavelengths (photosystem II). In photosystem I, energy harvested by one set of pigments is transferred to a form of chlorophyll a called P_{700}. In photosystem II the absorption of light leads to the oxidation of the form of chlorophyll a known as P_{680}, which is a more powerful oxidizing agent than P_{700}. As P_{680} acquires electrons from water molecules, the molecules are split and oxygen is liberated. It is the reducing activities of the light reactions of photosynthesis that leads to the synthesis of the ATP that drives the dark reactions.

8.9   C4 Plants

The so-called C4 plants are able to trap carbon dioxide by a C4 pathway. This system, while not major, is important because it operates at very low CO_2 concentrations. It also allows the plants to recycle some of the CO_2 that they respire.

8.10   Summary

The energy used by all living things ultimately comes from the sun. Solar energy is trapped in chemical bonds during the process called photosynthesis. Respiration and photosynthesis are complementary processes in several respects, particularly in that electron transport is at the heart of both.

B. KEY TERMS

autotroph   palisade layer

photosynthesis   spongy layer

chlorophyll a   lower epidermis

chlorophyll b   guard cells

carotenoids   stomata

action spectrum   cutin

absorption spectrum   fluorescense

chloroplasts   P680

proplastids   P700

lamellae   reaction center pigments

thylakoids   photosystems I and II

grana   photophosphorylation

stroma   cyclic photophosphorylation

NADP

upper epidermis

II. MASTERING THE CHAPTER

A. LEARNING OBJECTIVES

When you have mastered the material in this chapter, you should be able to:

1. Define all key terms.

2. Distinguish between photosynthesis and respiration.

3. Distinguish between an action spectrum and absorption spectrum.

4. Cite the two ways in which chloroplasts are produced in a green plant.

5. Cite six structures of a green leaf and state the principal function of each.

6. State how increased light intensity and an increase in temperature affect the light and dark reactions during photosynthesis, and what occurs when the CO_2 supply is limited.

7. State the general overall reaction occurring during the dark reaction phase of photosynthesis.

8. State the principal function of light in photosynthesis.

9. State the source of fluorescence in light-energized chlorophyll.

10. Cite the principal reactions in photosystem I and in photosystem II, and state how the former is dependent on the latter.

11. Explain how the structure of the chloroplast facilitates photo-synthesis.

B. REVIEWING TERMS AND CONCEPTS

1. Living cells need a supply of energy, the ultimate source of such energy being (a) _____ molecules produced by (b) _____ plants in a process called (c) _____. These plants convert carbon dioxide and (d) _____, with the energy of (e)_____ and in the presence of the green pigment, (f) _____, into the carbohydrate (g) _____, releasing the gas (h) _____ as a by-product.

2. Green plants are classed as (a) _____-trophs, while those other organisms depending on green plants as a source of energy are termed (b) _____.

3. There are two types of chlorophyll, termed chlorophyll (a) _____ and chlorophyll _____. Both absorb visible light most strongly in the (b) _____ and _____ regions of the spectrum and reflect most strongly in the (c) _____ portion . Other pigments, which absorb strongly in the blue region and reflect reds and oranges, are the (d) _____.

4. Those structures containing the photosynthetic pigment chlorophyll are called (a) _____, and their structure shows similarities to (b) _____. Chloroplasts produce more chloro-plasts, and develop from (c) _____. The presence of (d) _____ in chloroplasts helps account for their curious autonomy.

5. In higher plants, chloroplasts are generally confined to the cells of young (a) _____, immature (b) _____, and (c) _____. The (d) _____ are the real photosynthetic factory of green plants. They consist of an (e)_____ epidermis with a waxy secretion called (f) _____, beneath which is the (g) _____ layer of cells, where most (h) _____ activity occurs. Beneath this layer is the (i) _____ layer, where (j) _____ is stored and where the exchange of (k) _____ between the leaf and the environment occurs.

6. Windowlike openings in the lower epidermis, called (a) _____, regulate the amount of (b) _____ vapor, (c) _____ dioxide, and (d) _____ entering and leaving the leaf. This is accomplished by the change in shape of (e) _____ cells, which are sensitive to the amount of (f) _____ _____ contained within the leaf.

7. The rate of photosynthesis in a green leaf increases with increased (a) _____ intensity, but not indefinitely. (b) _____ change also affects the rate of photosynthesis, as does the concentration of (c) _____ _____.

8. There are two major sets of reactions during photosynthesis, the (a) _____ reactions and the _____ reactions.

9. The dark reactions of photosynthesis occur within the (a) _____ of the chloroplasts and involve the (b) _____ of carbon dioxide. The reactions require large amounts of (c) _____ NADP (NADPH) and ATP. During the light reactions (d) _____ is split, (e) _____ reduced, and (f) _____ liberated. (g) _____ pigments absorb solar energy and pass it on to the P_{680} and P700 chlorophylls. Thus, these latter two are said to be (h) _____ center pigments because they are the only two to be oxidized.

10. The light reaction is divided into two processes, (a) _____, which is energized by light of (b) _____ wavelengths, and photosystem II, which is energized by (c) _____ wavelengths. The (d) _____ potential of photosystem II is great enough to remove (e) _____ from water. The light is also important for (f) _____ or the formation of (g) _____ from ADP and inorganic phosphate. The enzymes of photosystems I and II are associated with the (h) _____ membranes of the chloroplasts.

11. Some of the (a) _____ incorporated by plants into sugars comes from the 4- (b) _____ pathway. This system is advantageous when the carbon dioxide concentration is (c) _____; as is the case when the (d) _____ are closed during hot, dry days.

C. TESTING TERMS AND CONCEPTS

1. Chloroplasts and the principal photosynthetic activity occur in the
   a) spongy layer.
   b) lower epidermis.
   c) palisade layer.
   d) upper epidermis.

2. In photosystem I, chloroplasts use light to produce
   a) glucose.
   b) water.
   c) NADPH.
   d) CO_2.

3. Light is one factor involved in photosynthesis. Two others are
   a) CO_2 and O_2.
   b) O_2 and CO.
   c) temperature and CO_2.
   d) temperature and O_2.

4. Light energy striking a chlorophyll molecule in a solution induces fluorescence by
   a) raising electrons from a low to a higher energy level.
   b) causing electrons to migrate around the molecule.
   c) lowering electrons from a low to a lower energy level.
   d) causing electrons to migrate from chlorophyll.

5. The oxygen released during photosynthesis derives from
   a) water.
   b) CO_2.
   c) glucose.
   d) ATP.

6. During the operation of photosystem I
   a) P_{700} is oxidized and NADP+ is reduced.
   b) NADPH is reduced, as is P_{700}.
   c) P_{680} is oxidized, as is NADPH.
   d) P_{680} is reduced along with ATP.

7. Those pigments that help fill in the absorption gaps so that a larger part of the sun's spectrum can be used for photosynthesis are the
   a) carotenoids.
   b) chloroplasts.
   c) chlorophyll a.
   d) FAD and FADH.

8. The general accomplishment of the dark reaction is the
   a) synthesis of ATP.
   b) synthesis of PGAL.
   c) reduction of CO_2 by hydrogen.
   d) splitting of water.

9. When light is absorbed by photosystem II
   a) P_{700} is reduced and CO_2 is liberated.
   b) P_{680} is reduced and O_2 is liberated.
   c) P_{680} is converted to P_{700}.
   d) P_{700} is converted to P_{680}.

10. The phytol chain is characterized by all of the following except
    a) it is hydrophobic.
    b) it contains many unsaturated double bonds.
    c) it is important in the "dark" reaction.
    d) it would be found in the chloroplasts.

11. Which of the following is not a membrane structure?
    a) lamellae
    b) thylakoids

c) grana
d) stroma

12. If a plant was placed in the light with an atmosphere containing CO_2 that was labeled with radioactive carbon, you would soon find labeled
a) ATP.
b) O_2.
c) FAD.
d) glucose.

13. Which reaction(s) would you expect to find in a plant during the night?
a) glucose + oxygen   carbon dioxide + water + energy
b) carbon dioxide + water + energy   glucose + oxygen
c) both a and b
d) neither a nor b

14. Which reaction(s) would you expect to find in the same plant during the day?
a) glucose + oxygen⟶carbon dioxide + water + energy
b) carbon dioxide + water + energy⟶glucose + oxygen
c) both a and b
d) neither a nor b

15. Cyclic photophosphorylation is so named because
a) no outside source of electrons is required.
b) the photons of light are recycled.
c) the ATP is used over and over.
d) the light causes the ATP to assume a ring (cyclic) structure.

D. UNDERSTANDING TERMS AND CONCEPTS

1. Put the following dark reactions in their proper order by writing 1, 2, 3, and so on in the blank spaces.

_____ A reduction by NADPH leads to the formation of PGAL.
_____ A six-carbon sugar molecule is formed.
_____ Two PGAL molecules travel up the pathway of glycolysis and ultimately produce one molecule of glucose.
_____ The leaf takes in CO_2.
_____ Two molecules of 3-phosphoglyceric acid are formed.
_____ Ribulose diphosphate combines with CO_2.
_____ Four molecules of DPGA are formed.

Now put the following light reactions in their proper order. Treat the two photosystems independently.

2. Photosystem I

_____ NADPH is formed.
_____ Substance X is reduced.
_____ NADP+ is reduced.
_____ The energy harvested by antenna pigments is transferred to a molecule of P_{700}.

74

_____ Ferredox in is reduced.

3. Photosystem II

_____ Electrons from water molecules reduce P_{680}.
_____ The P_{700} in Photosystem I is reduced.
_____ The absorption of light oxidizes P_{680}.

4. Pretend that you can see a molecule of CO_2 in the atmosphere outside a leaf. Trace its path until it is incorporated into glucose.

_____
_____
_____

5. In a laboratory, how could we actually visualize this "molecular trip"? What is the technique called? _____

_____

6. Why are most "grow lights" reddish rather than greenish? _____

_____

7. You are riding down the street with a friend who has not read your text and asks you what the bumper sticker "Have you thanked a green plant today?", means. What would you say? _____

_____
_____
_____

8. Explain to your friend how cotton jeans are the end product of a "living chemical factory." Include plant biochemistry and plant anatomy in your answer. _____

_____
_____
_____

9. Tell your friend about the paper in this book in the context of question 8. _____

_____
_____
_____

ANSWERS TO CHAPTER EXERCISES

Reviewing Terms and Concepts

1. a) organic b) green c) photosynthesis d) water e) sunlight
   f) chlorophyll g) glucose h) oxygen

2. a) auto b) heterotrophs

3. a) a, b b) red, blue c) green d) carotenoids

4. a) chloroplasts b) mitochondria c) proplastids d) DNA

5. a) stems b) fruits c) leaves d) leaves e) upper f) cutin
    g) palisade h) photosynthetic i) spongy j) food k) gases

6. a) stomata b) water c) carbon d) oxygen e) guard f) carbon
    dioxide

7. a) light b) Temperature c) carbon dioxide

8. a) light, dark

9. a) stroma b) fixation c) reduced d) water e) NAD f) oxygen
    g) Antenna h) action

10. a) photosystem I b) longer c) shorter d) redox e) electrons
    f) photophosphorylation g) ATP h) thylakoid

11. a) carbon dioxide b) carbon c) low d) stomata

Testing Terms and Concepts

1. c	6. a	11. d
2. c	7. a	12. d
3. c	8. c	13. a
4. a	9. b	14. c
5. a	10. c	15. a

Understanding and Applying Terms and Concepts

1. a) 6 b) 3 c) 7 d) 1 e) 4 f) 2 g) 5

2. a) 5 b) 2 c) 4 d) 1 e) 3

3. a) 2 b) 3 c) 1

4. a) passes guard cells into a stoma
    b) passes through the spongy layer to the palisade layer
    c) enters palisade cell
    d) through the cytoplasm to a chloroplast

5. By using radioactive ^{14}C, we could follow the $^{14}CO_2$ by auto-
    radiography

6. Photosynthesis is more efficient in the red region than the green
    region—recall the action/absorption spectra.

7. The energy we require for work, maintenance, etc. comes to us from
    the sun through plants, where solar energy is trapped in chemical
    bonds during photosynthesis. This is true even in the case of
    carnivores (meat eaters) since the animals that are ingested have
    consumed plants or plant products or other animals that did. Plants
    are also crucial in producing the O_2 that we require to burn the
    food we eat.

8. Carbon dioxide and water are incorporated into glucose through

photosynthesis. The glucose has several fates, one of which is being incorporated into cellulose. Cellulose is the major chemical in plant cell walls. Cotton is the fibrous cell wall material from a cotton plant.

9. Paper, as you probably know, is a product of wood. Wood is largely xylem tissue of the tree. Once again, cellulose is cell wall material.

CELL DIVISION

I. REVIEWING THE CHAPTER

A. CHAPTER HIGHLIGHTS

9.1 The Genetic Continuity of Cells

Two quite distinct methods of producing offspring are found among living things. One, called sexual reproduction, is the production of new individuals by two parents, each contributing hereditary information to the new individual. The other, called asexual reproduction, is the production of a new individual by only one parent. The most common method of asexual reproduction among unicellular organisms is fission, during which the single cell divides into two organisms of about equal size. Yeast cells reproduce asexually by budding, during which the new individuals produced are of unequal size. Plants may reproduce by vegetative reproduction, in which one of the plant's parts gives rise to an entire new plant. Fungi and certain plants reproduce asexually by producing spores, lightweight structures that are carried far and wide on air currents. Some plants and animals reproduce asexually by fragmentation, a process in which the organism breaks up into several parts, each part regenerating an entire new organism.

9.2 Mitosis

Mitosis occurs in four consecutive phases. During prophase, the nucleoli and nuclear membrane begin to disappear, while the chromosomes begin to appear in doublet form. At the next stage, metaphase, a structure called the spindle of microtubules appears and the chromosomes become oriented in relation to it. Anaphase begins when the duplicated chromosomes separate from each other, the two halves of each doublet moving to opposite poles on the spindle. During telophase the nucleoli reappear and a nuclear membrane begins to form around the chromosomes. In plants, a cell plate forms at the equator and a cell wall is secreted on each side of the cell plate, thus completing cell division. Mitosis is the method of growth in plants and animals alike.

9.3 The Nature of Sexual Reproduction

In sexual reproduction a sex cell, called a gamete, of each parent fuses with its opposite number and so combines hereditary information from each parent in the offspring, called the zygote. The essence of sexual reproduction is newness, variation, as a result of combining the hereditary material—DNA—from both parents.

9.4 Sexual Reproduction in Bacteria

Although bacteria are unicellular and posses no specialized organs, sexual reproduction of some bacteria has been demonstrated. The evidence for this is seen in the production of offspring that can exist under conditions impossible for either parent to survive and thus indicating an exchange of genetic material has occurred. This case illustrates the essence of the transfer of genetic information from one individual to another.

9.5 Meiosis

Each species of organism has a characteristic number of chromosomes, the numbers being even numbers. Cells containing a full complement of homologous pairs of chromosomes are said to contain the diploid or 2n number of chromosomes. During mitosis, each new cell produced contains the same number (2n) of chromosomes as the parent cell. During meiosis, two consecutive cell divisions take place, but the chromosomes are duplicated only during the first division. This means that the four daughter cells (gametes) produced after the second division each contain the haploid (n) number of chromosomes.

Because chromosomes are assorted randomly during meiosis, and because crossing over and exchange of segments may occur between homologous pairs, wide variation in the gametes produced by a given parent occurs. It is this variation that makes the individuals produce by sexual reproduction differ genetically from their parents.

9.6 Summary

Organisms that reproduce asexually are genetically identical to the parent. The process in which the genetic material is divided in asexual reproduction is called mitosis. The same process takes place when more cells are added to tissue, as in growth or replacement.

Sexual reproduction involves two parents. If gametes are involved, they are produced by meiosis, that is, the process in which the chromosome number is reduced, and the randomization and recombination of genetic material takes place.

B. KEY TERMS

asexual reproduction

clone	fission
chromosome	sexual reproduction
mitosis	gametes
homologous pair	isogametes
chromatin	heterogametes
doublet	sperm

79

interphase	egg
prophase	diploid/haploid
metaphase	chiasmata
anaphase	synapsis
telophase	meiosis
centromere	crossing over
spindle	interkinesis
cell plate	hermaphroditic
asters	self-fertilization
cytokinesis	cross-fertilization
	random assortment

II. MASTERING THE CHAPTER

A. LEARNING OBJECTIVES

When you have mastered the material in this chapter, you should be able to:

1. Define all key terms.

2. Distinguish between sexual and asexual reproduction.

3. Distinguish between budding and sporulation.

4. Distinguish between fragmentation and grafting as means of asexual reproduction, and cite one example of each.

5. Distinguish between mitosis and meiosis.

6. List the four chief phases in mitosis, and describe what happens during each.

7. State the essence of sexual reproduction and explain how that essence is achieved.

8. State how meiosis prevents the accumulation of an ever-increasing number of chromosomes in individuals.

9. State when and how crossing over occurs.

10. Cite the two processes occurring during meiosis that make it unlikely that any two gametes from a given parent will be identical.

B. REVIEWING TERMS AND CONCEPTS

1. Two quite distinct methods of producing offspring are found among living things. One, called (a) _____ reproduction, is the production of new individuals by two parents, each contributing (b) _____ information to the new individual. The other, called (c) _____ reproduction, is the production of a new individual by only one parent.

2. The most common method of asexual reproduction among unicellular organisms is (a) _____, during which the single cell divides into two organisms of about (b) _____ size. Yeast cells reproduce asexually by (c) _____, during which the new individuals produced are of (d) _____ size. Plants may reproduce by (e) _____ reproduction, in which one of the plant's parts gives rise to an entire new plant. Fungi and certain plants reproduce asexually by producing (f) _____, lightweight structures that are carried far and wide on air currents. Some plants and animals reproduce asexually by (g) _____, a process in which the organism breaks up into several parts, each part regenerating an entire new organism.

3. Mitosis occurs in four consecutive phases. During prophase, the (a) _____ and (b) _____ membrane begin to disappear, while the (c) _____ begin to appear in doublet form. At the next stage, (d) _____, a structure called the (e) _____ appears and the (f) _____ become oriented in relation to it. (g) _____ begins when the duplicated chromosomes separate from each other, the two halves of each doublet moving to opposite poles on the (h) _____. During (i) _____ the (j) _____ reappear and a nuclear membrane begins to form around the chromosomes.

4. In sexual reproduction a sex cell, called a (a) _____, of each parent fuses with its opposite number and so combines (b) _____ information from each parent in the resulting cell, called the (c) _____.

5. Each species of organism has a characteristic number of (a) _____, the numbers being even numbers. Cells containing a full complement of (b) _____ pairs of chromosomes are said to contain the (c) _____ or 2n number of chromosomes. During (d) _____, each new cell produced contains the same number (2n) of chromosomes as the parent cell. During (e) _____, two consecutive cell divisions take place, but the chromosomes are (f) _____ only during the first division. The resulting cells have the (g) _____ or n number of chromosomes.

6. Because chromosomes are assorted randomly during meiosis, and because (a) _____ pairs may (b) _____ over each other and so exchange segments, (c) _____ in the gametes produced by a given parent occurs. It is this variation that makes the individuals produced by (d) _____ reproduction differ (e) _____ from their parents.

C. TESTING TERMS AND CONCEPTS

1. During mitosis, the chromosomes begin to appear at
    a) prophase.
    b) anaphase.
    c) metaphase.
    d) telophase.

2. When two gametes unite, the resulting zygote is
    a) 2n.
    b) n.
    c) haploid.
    d) doublet.

3. When a cell divides by meiosis, the number of cells produced is
    a) two.
    b) four.
    c) eight.
    d) sixteen.

4. A fertilized egg is a
    a) heterogamete.
    b) isogamete.
    c) zygote.
    d) chiasmata.

5. The chief advantage of sexual reproduction in terms of genetic
    material is
    a) constancy.
    b) structural improvement of the offspring.
    c) new combinations.
    d) assurance of a next generation.

6. The spindle consists of an array of
    a) microfilaments.
    b) microtubules.
    c) chromatin.
    d) chromosomes.

7. The attachment points between two nonsister chromatids are called
    a) centromeres.
    b) asters.
    c) synapses.
    d) chiasmata.

8. Sexual reproduction in bacteria differs from its counterpart in a
    higher plant because
    a) no new varieties are produced as a result.
    b) meiosis does not occur in bacteria.
    c) DNA is not involved.
    d) no energy is involved in the case of bacteria.

9. Homologous chromosomes "pair-up" during the metaphase of
    a) mitosis.

b) meiosis I.
c) meiosis II.
d) mitosis and meiosis I.

10. If chromosomes contain DNA, all <u>except</u> which of the following could be expected?'
a) An egg and sperm would contain equal amounts of DNA.
b) A gamete would have twice as much DNA as a body cell.
c) A gamete would have half as much DNA as a body cell.
d) All body cells would have the same amount of DNA.

11. After mitosis, the resulting cell with respect to the parent cell has
a) the same kind and number of chromosomes.
b) the same kind but different number of chromosomes.
c) the same number but different kinds of chromosomes.
d) an indeterminate kind and number of chromosomes because of random assortment.

12. The diploid number of chromosomes in a human is 46. Which of the following is <u>not</u> correct?
a) N equals 23.
b) There are 23 chromatids in an egg.
c) There are 92 chromatids present during metaphase of meiosis I.
d) A sperm contains 46 chromosomes.

13. Which of the following processes involves meiosis?
a) spore formation
b) budding
c) sperm formation
d) vegetative reproduction

14. If you treated cells undergoing meiosis with "tagged" thymidine when would you expect the most labeling of chromosomes to occur?
a) interphase
b) interkinesis
c) prophase I
d) teleophase II

15. Most of the genetic variation associated with meiosis occurs during
a) prophase I.
b) metaphase I.
c) metaphase II.
d) interkinesis.

D. UNDERSTANDING AND APPLYING TERMS AND CONCEPTS

Here is a nucleus of an hypothetical cell:

1. What stage of division is represented? _____

2. What is the diploid number for this organism? _____

3. What is the N number for this organism? _____

4. Draw the nucleus of the cell at:

   a) metaphase of mitosis

   b) metaphase of meiosis I

   c) metaphase of meiosis II

   d) telophase of mitosis

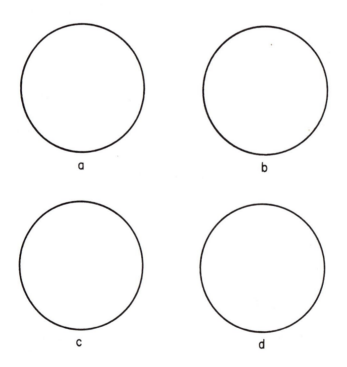

84

5. You have been told that random assortment is important in the development of genetic variation, here is your opportunity to show it for yourself. Assume that the shaded chromosomes drawn above are paternal and unshaded are maternal. How many possible arrangements of the chromosomes are there at metaphase of meiosis I. Draw four of them.

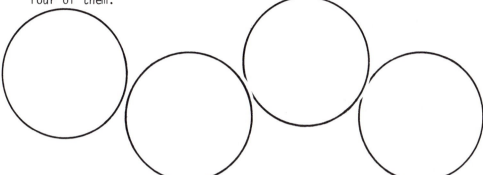

6. How many different kinds of gametes will result from the above arrangements? _____

7. Assigning letters to the chromosomes as follows, list all the different gametes that will result from meiosis in the above.

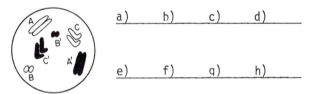

a) _____ b) _____ c) _____ d) _____

e) _____ f) _____ g) _____ h) _____

8. After meiosis, will the following cell give rise to normal gametes. Why?

9. After meiosis, will the following cells give rise to the same kinds of gametes? Why?

ANSWERS TO CHAPTER EXERCISES

Reviewing Terms and Concepts

1. a) sexual   b) hereditary   c) asexual

2. a) fission   b) equal   c) budding   d) unequal   e) vegetative   f) spores
    g) fragmentation

3. a) nucleoli   b) nuclear   c) chromosomes   d) metaphase   e) spindle
    f) chromosomes   g) Anaphase   h) spindle   i) teleophase   j) nucleoi

4. a) gamete   b) hereditary   c) zygote

5. a) chromosomes   b) homologous   c) diploid   d) mitosis   e) meiosis
    f) duplicated   g) haploid

6. a) homologous   b) cross   c) variation   d) sexual   e) genetically

Testing Terms and Concepts

1.	a	6.	b	11.	a
2.	a	7.	d	12.	d
3.	b	8.	b	13.	c
4.	c	9.	b	14.	a
5.	c	10.	b	15.	b

Understanding and Applying Terms and Concepts

1. prophase (mitosis or meiosis I)

2. six

3. three

4.

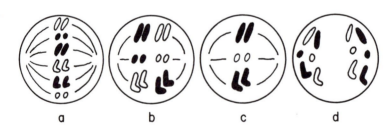

| a | b | c | d |

5. 4

6. eight

7. a) ABC b) ABC' c) AB'C d) AB'C' e) A'B'C' f) A'B'C' g) A'BC'
    h) A'BC

8. No, the resulting gametes would not contain one member of each
    homologous pair even though they would each contain 3 chromosomes.

9. Yes, a cell is three dimensional and we have been working with them
    in only two dimensions. Therefore, what we have displayed linearly
    would be circular inside the cell. Hence, it is a matter of
    perspective how we align them.

10

GENETICS:
THE WORK OF MENDEL

I. REVIEWING THE CHAPTER

A. CHAPTER HIGHLIGHTS

10.1 The Theory of the Inheritance of Acquired Characteristics

When living things reproduce asexually, their offspring develop
into exact copies of their parents so long as they are raised under
similar environmental conditions. When living things reproduce sexually,
their offspring develop combinations of traits different from one another
and different from those of either parent.

Lamarck's theory of the inheritance of acquired characteristics
was an early attempt to account for the observed adaptations of plants
and animals to the environment. For example, the long neck of the
giraffe supposedly developed because giraffes had to stretch their
necks to browse on leaves and each generation passed on to its offspring a slightly longer neck. No evidence in support of this theory
has yet been discovered. According to Weismann, multicellular organisms
are made up of two major classes of cells: The germplasm contains the
gamete-producing cells, while the somaplasm contains all the other body
cells.

10.2 Mendel's Theory: The Background/10.3 Mendel's Experiments

In his genetics experiments, Mendel worked with garden pea plants
that had definite and inheritable characteristics, such as flower
color, seed coat texture, tallness as opposed to dwarfness, and so on.
In one of his first experiments, Mendel crossed a round-seed variety
with a wrinkled-seed variety and found that in all cases the offspring
produced round seeds, so roundness was said to be dominant over
wrinkledness. Because the second generation was produced by dissimilar
parents, Mendel called it the hybrid generation, also designated the
F_1 generation.

When Mendel allowed members of the F_1 generation to fertilize
themselves, the offspring were a mixture of round and wrinkled in a
ratio of 3 round to 1 wrinkled. The reappearance of wrinkled seeds in
the F_2 generation could mean only that at least some of the F_1 plants
were also carrying a "factor" (gene) for the wrinkled-seed condition.

10.4 Mendel's Hypotheses/10.5 How Hypotheses Are Judged/10.6 The
      Testcross: A Test of Mendel's Hypothesis

To account for his results, Mendel made several assumptions. One was that when the gametes are prepared, (genetic) factors separate and are distributed as units to each gamete. This is called Mendel's law of segregation.

Mendel found that round-seed parent (P1) individuals were homozygous for round-seededness, or had two genes for the condition (RR). Likewise, wrinkled-seed P1 individuals were homozygous (rr). Three different combinations are thus possible (RR, Rr, rr) in the F2 generation, and Mendel found that the ratio was 1:2:1. But because of the dominance of R over r, there is no way to distinguish visibly between homozygous seeds containing RR and those heterozygous individuals containing Rr. Both would have round seed coats, termed the same phenotype, though they would not have the same genotype.

10.7 Dihybrids—The Law of Independent Assortment/10.8 Mendel's Theory: The Sequel

Mendel also cross-bred pea plants differing in two characteristics— round seed coats and yellow cotyledons with those having wrinkled seed coats and green cotyledons. All F_1 dihybrid offspring were round-seeded and had yellow cotyledons. Mendel next allowed F_1 individuals to fertilize themselves and found that F_2 offspring appeared with not two, but four, different phenotypes—some round-green, some wrinkled-yellow, plus those resembling the two P_1 types—in a ratio of 9:3:3:1, respectively. This indicated that the smooth-yellow traits and the wrinkled-green traits were not linked, but that P_1 genes for each trait were free to sort independently. This fact came to be known as Mendel's law of independent assortment. Little attention was paid to Mendel's scientific contributions during his lifetime. However, a third of a century after his work was published three other scientists independently and nearly simultaneously "rediscovered the gene."

10.9 Continuous Variation: The Multiple-Factor Hypothesis

Mendel worked with clear-cut qualitative characteristics, such as round versus wrinkled, in his studies. Many phenotypic features are not as easily distinguished and often vary continuously along a gradient, for example, height or weight. When phenotypes intermediate to the parents arise, the phenomenon is referred to as incomplete dominance. Often the trait is influenced by more than just one pair of genes. The multiple-factor hypothesis helps account for quantitative changes throughout a particular range.

10.10 Summary

Genes are responsible for the traits displayed by all organisms. The theory has grown from Mendel's research and states that in diploid organisms an individual gets one gene for every trait from each parent. The theory is powerful because it allows testable predictions to be made.

B. KEY TERMS

inheritance of acquired characteristics

germplasm

somaplasm

stamens

pistils

hybrid

F_1, F_2 generations

dominant

recessive

gene

law of segregation

allele

heterozygous

homozygous

genotype

phenotype

hypothesis

theory

inductive

deductive

law

law of independent assortment

incomplete dominance
multiple-factor hypothesis

theory of polygenic inheritance

II. MASTERING THE CHAPTER

A. LEARNING OBJECTIVES

When you have mastered the material in this chapter, you should be able to:

1. Define all key terms.

2. State the essence of the theory of the inheritance of acquired characteristics and why it is not universally accepted by geneticists.

3. List the series of five assumptions Mendel made in his hypothesis to account for his results in cross-breeding garden pea plants.

4. Distinguish between a phenotype and a genotype, and give an example of each.

5. Distinguish between a heterozygous and homozygous genotype.

6. Distinguish between a hypothesis and a theory.

7. Do a monohybrid cross, that is, determine the kinds of gametes produced by the parents, and put the gametes together in the proper combinations following fertilization.

8. Cite the circumstances under which a test cross is normally performed.

9. Do a dihybrid cross, as in number 7 above.

10. Describe what occurs in a situation in which there is a lack of dominance.

11. Describe what occurs when a trait is being influenced by more than one pair of genes.

B. REVIEWING TERMS AND CONCEPTS

1. When living things reproduce (a) _____, their offspring develop into exact copies of their parents so long as they are raised under similar (b) _____ conditions. When living things reproduce (c) _____, their offspring develop traits different from one another and different from those of either (d) _____.

2. According to Weismann, multicellular organisms are made up of two major classes of cells: The (a) _____ contains the gamete-producing cells, while the (b) _____ contains all the other body cells.

3. In one of his first experiments, Mendel crossed a round-seed variety with a wrinkled-seed variety of garden pea and found that in all cases the offspring produced (a) _____ seeds, so roundness was said to be (b) _____ over wrinkledness. Because the second generation was produced by dissimilar parents, Mendel called it the (c) _____ generation. The parental generation is referred to as the (d) _____ and the offspring referred to as the (e) _____ generation. When the F_1 individuals are crossed, their progeny are usually designated as the (f) _____ generation.

4. When Mendel allowed members of the F_1 generation to fertilize themselves, the offspring were a mixture of round and wrinkled in a ratio of (a) _____ round to (b) _____ wrinkled.

5. To account for his results, Mendel made several assumptions. One was that when the (a) _____ are produced, (genetic) factors separate and are distributed as units to each (b) _____. This is called Mendel's law of (c) _____.

An individual with the genotype Rr will have the same (d) _____ as an individual with the genotype (e) _____. In order to determine whether an individual was RR or Rr, you would do a (f) _____. Or, you would cross the individual in a question with one that was homozygous (g) _____. If (h) ____% of the offspring had the recessive (i) _____ you could conclude that the one in question was heterozygous.

6. Mendel also cross-bred pea plants differing in two characteristics round seed coats and yellow cotyledons with those having wrinkled seed

coats and green cotyledons. All F dihybrid offspring were (a) _____-seeded and had (b) _____ cotyledons. Mendel next allowed F1 individuals to fertilize themselves and found that F2 offspring appeared with not two, but (c) _____, different phenotypes—some (d) _____-green, some (e) _____-yellow, plus those resembling the two P1 types—in a ratio of (f) _____. This indicated that the P1 (g) _____ for each trait were free to migrate independently. This fact came to be known as Mendel's law of (h) _____ _____.

7. Traits that range along a continum are said to vary in a (a) _____ way. This (b) _____ dominance gives rise to phenotypes that are intermediate to the parents. If several intermediate phenotypes are found, it is likely that (c) _____ than one pair of genes is controlling that trait. This is called the (d) _____ hypothesis or the theory of (e) _____ _____.

C. TESTING TERMS AND CONCEPTS

1. Offspring of a homozygous round-seed plant and a homozygous wrinkled-seed plant are termed the
   a) F_2 generation.
   b) P_1 generation.
   c) hybrid generation.
   d) independent generation.

2. The alternative form of a gene controlling a given characteristic is called
   a) a mutation.
   b) a recessive.
   c) an allele.
   d) a homozygote.

3. When gametes are being formed, the genetic factors for each characteristic separate and are distributed as units to each gamete. This is called Mendel's law of
   a) segregation.
   b) regeneration.
   c) redistribution.
   d) independent assortment.

4. The gamete-producing cells of the body are the
   a) somaplasm.
   b) germplasm.
   c) endoplasm.
   d) ectoplasm.

5. A light-green phenotype produced by a cross of a dark-green and a white phenotype would be an example of
   a) segregation.
   b) independent assortment.
   c) incomplete dominance.
   d) incomplete recessiveness.

6. In an attempt to <u>explain</u> the results of his cross-breeding experiments with garden peas, Mendel made a series of
   a) observations.
   b) assumptions.
   c) predictions.
   d) generalizations.

Questions 7 to 11 refer to the same situation.

7. Pollen from a plant with blue flowers is placed on the stamens of a plant with yellow flowers. All 154 offspring in the F_1 generation produced yellow flowers. The yellow-flower trait is
   a) dominant.
   b) recessive.
   c) neither, there is a lack of dominance.
   d) neither, this is a case of polygenic inheritance.

8. When two F_1 individuals were crossed, the offspring consisted of 127 yellow and 39 blue individuals. This is a(n)
   a) prediction.
   b) observation.
   c) deduction.
   d) hypothesis.

9. If an F_1 individual were crossed with a plant that produced blue flowers offspring would be predicted in a ratio of
   a) 3 blue: 1 yellow.
   b) 1 blue: 1 yellow.
   c) 1 blue: 3 yellow.
   d) all of an intermediate color.

10. When the cross described in question 9 was performed, the results were 65 blue and 59 yellow offspring. How do these data relate to your prediction?
    a) support the prediction
    b) refute the prediction
    c) assume the prediction
    d) deduce the prediction

11. If two plants with blue flowers were crossed, what ratio of color would you expect in the offspring?
    a) 1 blue: 1 yellow
    b) 3 blue: 1 yellow
    c) all blue
    d) all intermediate

12. If it was found that a certain kind of swamp-dwelling mouse adjusted to living in the desert and passed this ability on to its offspring, it would support
    a) Weismann's theory.
    b) Mendel's theory.
    c) Nilsson-Ehle's theory.
    d) Lamarck's theory.

13. An individual with the genotype Ww will produce gametes in which of the following proportions?
    a) 100% Ww
    b) 50% Ww: 25% Ww: 25% ww
    c) 50% Ww: 50% ww
    d) 50% W: 50% w

14. The correct answer to number 13 is based upon the law of
    a) segregation.
    b) dependent assortment.
    c) independent assortment.
    d) testcrosses.

15. A black dog is mated with a white dog and all the puppies are gray. All of the following terms have been used to describe this mode of inheritance except
    a) lack of dominance.
    b) codominance.
    c) quasi-dominance.
    d) incomplete dominance.

D. UNDERSTANDING AND APPLYING TERMS AND CONCEPTS

Decide whether the following are true (T) or false (F).

_____ 1. If a genetic factor passes through a generation but is not expressed in that generation, when it reappears in the next generation its expression is diminished.

_____ 2. A cross between the individuals (AAxaa) will produce a generation in which all individuals will have the same phenotype.

_____ 3. An F₁ generation produced by crossing parents differing in two traits is said to be a dihybrid.

_____ 4. Environment is sufficient to explain the full range of heights in humans.

_____ 5. There is ample evidence supporting the idea that traits acquired during a lifetime can be passed on to the offspring.

_____ 6. Geneticists make a prediction about the inheritance of a specific trait by using deductive reasoning.

7. One seldom gets results in the exact ratios that were predicted. Does this refute the theory underlying the study of genetics? _____ Why or why not? _____
   _____
   _____
   _____

8. Even today, genes remain hypothetical units; i.e., nobody has seen one. Why then are they accepted? _____

94

_____
_____
_____
_____

9. List all of the possible genotypes of the gametes produced
    by the following individuals:

a) AA _____
b) Aa _____
c) AABb _____
d) AaBb _____
e) AABBCc _____
f) AaBBCc _____

10. A rabbit breeder made the following observation: A male with
    long hair was crossed to a female with short hair, and all
    nine offspring had short hair.

a) Which phenotypic characteristic is dominant? _____
b) If the genotype of the male parent was <u>ll</u>, what was the genotype of the F_1s? _____
c) What kinds of gametes can be produced by the F_1s? _____
d) Set up and complete the following square for a cross between
   two F_1 individuals:

```
 _____
|        |        |
|        |        |
|        |        |
|_____|_____|
|        |        |
|        |        |
|        |        |
|_____|_____|
```

e) What percentage of the offspring will have short hair?
   _____

f) What percentage of the offspring will be homozygous dominant?
   _____

g) Suppose the breeder buys a rabbit with short hair. Tell him
   how to decide if his newly acquired animal is homozygous or
   heterozygous for short hair.

_____
_____
_____
_____

11. In the above rabbits, brown is dominant over gray. Our
    breeder has another pair of true-breeding rabbits: gray,
    short hair and brown, long hair.

a) What are the respective genotypes? _____

95

b) What kinds of gametes can be produced by the above?

c) What is/are the genotype(s) and phenotype(s) of the F_1 generation produced by mating the above pair? _____
_____

d) Set up and complete the following square for a cross between two F_1 offspring:

_____

_____

_____

e) What is the phenotypic ratio resulting from the above?
_____

12. A genetics student crossed a plant with red flowers to a plant with white flowers, and found that all the offspring produced pink flowers.

a) What is the mode of inheritance. _____
b) If the red plants have a genotype of RR, what will be the genotype of the white and pink plants? _____
c) Set up and complete the square for a cross between two pink-flowered plants:

d) What is the phenotypic and genotypic ratio for the F progeny?
_____
_____

ANSWERS TO CHAPTER EXERCISES

Reviewing Terms and Concepts

1. a) asexually b) environmental c) sexually d) parent

2. a) germplasm b) somaplasm

3. a) round b) dominant c) hybrid d) P_1 e) F_1 f) F_2

4. a) 3 b) 1

5. a) gametes b) gamete c) segregation d) phenotype e) RR
    f) testcross g) recessive h) 50 i) phenotype

6. a) round b) yellow c) four d) round e) wrinkled f) 9:3:3:1
    g) genes h) independent assortment

7. a) quantitative b) incomplete c) more d) multiple-factor
    e) polygenic inheritance

Testing Terms and Concepts

1.	c	6.	b	11.	c
2.	c	7.	a	12.	d
3.	a	8.	b	13.	d
4.	b	9.	b	14.	a
5.	c	10.	a	15.	c

Understanding and Applying Terms and Concepts

1. F 2. T 3. T 4. F 5. F 6. T 7. no; chance variation.
    This is the same reason that 100 tosses of a coin will seldom
    result in exactly 50 heads and 50 tails.

8. We accept genes because they account well for the observations,
    and we can make testable predictions based on them.

9. a) A b) A, a c) AB, Ab d) AB, ab, Ab, aB e) ABC, ABc
    f) ABC, ABc, aBC, aBc

10. a) short hair b) L1 c) 50% L, 50% 1 d)

	.5L	.51
.5L	.25LL	.25L1
.51	.25L1	.2511

    e) 75% f) 25% g) Do a testcross. Mate the unknown with a rabbit
    with long hair (11). If one-half have short hair and one-half
    have long hair then it is heterozygous. If all offspring have
    short hair then the unknown is homozygous dominant, i.e., LL
    for short hair.

11. a) bbLL, BB11 b) bL, B1 c) BbL1 only—all will be brown with
    short hair

d)

	¼BL	¼Bl	¼bL	¼bl
¼BL	1/16 BBLL	1/16 BBLl	1/16 BbLL	1/16 BbLl
¼Bl	1/16 BBLl	1/16 BBll	1/16 BbLl	1/16 Bbll
¼bL	1/16 BbLL	1/16 BbLl	1/16 bbLL	1/16 bbLl
¼bl	1/16 BbLl	1/16 Bbll	1/16 bbLl	1/16 bbll

e) 9 brown, short hair: 3 brown, long hair: 3 gray, short hair: 1 brown, long hair

12. a) lack of dominance b) white-rr, pink-Rr
    c)

	.5R	.5r
.5R	.25 RR	.25Rr
.5r	.25Rr	.25rr

d) 1 red (RR): 2 pink (Rr): 1 white (rr)

11

GENES ON CHROMOSOMES

I. REVIEWING THE CHAPTER

A. CHAPTER HIGHLIGHTS

11.1 Parallel Behavior of Genes and Chromosomes

Workers came to realize that certain of Mendel's rules of gene behavior could be explained if the genes were located on or in the chromosomes. Because there are so many more genes in an organism than chromosomes, each chromosome must contain many genes. The two alleles controlling a given trait are presumed to be located at corresponding places, called loci, on each of two homologous chromosomes.

11.2 Sex Determination

Chromosomes generally occur in homologous pairs, except for the sex chromosomes. All other chromosomes are termed autosomes. In humans, the female sex chromosomes are designated XX, in males, XY. In humans, although not in many insects, it is the presence of the Y chromosome that determines maleness. It was the discovery of the mechanism of sex determination in fruit flies that paved the way for testing the validity of the chromosome theory.

11.3 X-linkage

Another test involved the inheritance of a specific trait linked with the sex chromosomes. Such a trait is said to be X-linked. Examples are red-green color-blindness in humans and the condition known as hemophilia. The allele for these conditions is nearly always passed on by the mothers, who are heterozygous for the trait and hence do not express it in their phenotype. But the trait is passed on, via the X chromosome, to approximately one-half of their sons, who do express the trait in their phenotype.

11.4 Chromosome Abnormalities

In about 3% of human pregnancies, chromosomal abnormalities occur, often associated with severe deformities. Such abnormalities occur as (1) alterations in chromosome number or (2) alterations in the chromosome structure. In humans, most of the disorders of chromosome number arise as a result of a condition called nondisjunction, when the homologues fail to separate during meiosis. A condition resulting in severe mental retardation and caused by an extra number 21 chromosome, which is an autosome, is called Down's syndrome.

Abnormal numbers of sex chromosomes produce less severe deformities. XXY and XXXY karyotypes develop as males but are sterile. XYY males usually are tall and are found in relatively large numbers among prisoners and men in mental institutions. The fact that women with but a single X chromosome are viable supports the idea that only one X chromosome is functional in female cells. That would tend to account for the benignity of XXX, XXXX, and even XXXXX chromosome constitutions.

When a cell fails to divide after chromosome duplication, a condition known as polyploidy results. This occurs more frequently among plants than among animals. Such cells contain any multiple of the haploid number of chromosomes above two. Polyploidy in plants is often associated with larger size and greater vigor.

Structural alterations in chromosomes sometimes occur. When a chromosome breaks and part of it is lost, the condition is termed a deletion. When a break occurs and a chromosome segment is turned around, the condition is called an inversion. When a fragment of one chromosome attaches to a chromosome with which it is not homologous, the condition is called a translocation. Such alterations may result in changes in phenotype and also fertility problems. Many agents have been identified as causing chromosome breaks. Among them are X-rays and other types of ionizing radiation, plus a variety of chemical agents.

11.5 Linkage/11.6 Chromosome Maps

In some instances, two alleles inherited from one parent show a strong tendency to segregate together, a condition called linkage. During the process of crossing over, the original combination of alleles is broken up, with the result that linkage is disturbed. Gametes so formed are called recombinants. By analyzing linked genes on a given chromosome, it is possible to plot chromosome maps that show the sequence in which gene loci occur.

11.7 The Evidence of Creighton and McClintock/11.8 Assigning Linkage Groups to Chromosomes

Support for placing genes on chromosomes also came from the research of Creighton and McClintock. The worked with a variety of corn in which one chromosome had physical markers at each end. Knowing which genetic characteristics were on this chromosome and using the visual markers, they could distinguish the chromosome from its homologue with respect to the genotypes. Finding a situation such as this in humans is more unlikely, so a technique called somatic cell hybridization has been used to associate certain genes with specific chromosomes. Briefly, this involves the fusion of a human cell with a mouse cell. Since chromosomes are lost at each mitotic division, one waits until a biochemical marker is lost. It is then possible to associate the loss of a chromosome with the loss of a gene (trait).

11.9 Summary

Several types of evidence support the theory that genes are arranged linearly on chromosomes. The association of certain genes with the sex chromosomes has been among this evidence. Several types of abnormalities with respect to the physical appearance of chromosomes have been associated with changes in phenotypes.

B. KEY TERMS

linkage	trisomy
locus	nondisjunction
sex chromosomes	polyploidy
autosomes	inversion
homogametic	translocation
X-linkage	deletion
clone	linkage group
mosaic	crossing-over
karyotype	recombinants
Barr body	

II. MASTERING THE CHAPTER

A. LEARNING OBJECTIVES

When you have mastered the material in this chapter, you should be able to:

1. Define all key terms.

2. List four reasons why D. melanogaster has been such a successful subject for genetic study.

3. State the genetic conditions that determine maleness and femaleness in D. melanogaster.

4. Cite two examples of X-linked traits in humans, and state why males are more likely to inherit such traits than are females.

5. Cite the two major groups into which chromosomal abnormalities fall, and give two examples of each.

6. State the difference between chromosomal deletions, translocations, and inversions.

7. Cite one example of genetic linkage.

8. Explain how a testcross is used to study linkage.

9. State how the visual "markers" helped Creighton and McClintock relate genes to chromosomes.

10. Explain how somatic cell hybridization is used to study linkage.

B. REVIEWING TERMS AND CONCEPTS

1. Chromosomes generally occur in (a) _____ pairs, except for the (b) _____ chromosomes. All other chromosomes are termed (c) _____. In humans, the female sex chromosomes are designated (d) _____, in males, (e) _____. In humans it is the presence of the (f) _____ chromosome that determines maleness. It was the discovery of the mechanism of (g) _____ determination in fruit flies that first paved the way for testing the validity of the (h) _____ theory.

2. Another test involved the inheritance of one specific trait linked with the (a) _____ chromosomes. Such traits are said to be (b) _____-linked. Examples of such traits are red-green color-blindness in humans and the condition known as hemophilia. The (c) _____ for these conditions is nearly always passed on by the (d) _____, who are (e) _____-zygous for the trait and hence do not express it in their (f) _____. But the trait is passed on, via the (g) _____ chromosome, to approximately one-half of their (h) _____, who do express the trait in their (i) _____.

3. In about 3% of human pregnancies, chromosomal abnormalities occur, often associated with severe deformities. Such abnormalities occur as (1) alterations in chromosome (a) _____ or (2) altera-tions in chromosome _____. Most of the disorders of chromosome number arise as a result of a condition called (b) _____, when the homologues fail to separate during (c) _____. A con-dition resulting in severe mental retardation and caused by an extra number (d) _____ chromosome, which is an autosome, is called (e) _____ syndrome.

4. Abnormal numbers of (a) _____ chromosomes produce less severe deformities. XXY and XXXY karyotypes develop as males but are sterile. The fact that women with but a single (b) _____ chromosome are viable supports the idea that only one (c) _____ chromosome is (d) _____ in female cells.

5. When a cell fails to divide after chromosome duplication, a condition known as (a) _____ results. Such cells contain any multiple of the (b) _____ number of the chromosomes above two.

6. When a chromosome breaks and part of it is lost, the condition is termed a (a) _____. When a break occurs and a chromosome segment is turned around, the condition is called an (b) _____. When a fragment of one chromosome attaches to a chromosome with which it is not (c) _____, the condition is called a (d) _____.

7. In some instances, two alleles inherited from one parent show a strong tendency to segregate together, a condition called (a) _____. However, during (b) _____ _____, the original combination of alleles may be broken up, with the result that linkage is disturbed. Gametes so formed are called (c) _____.

8. By knowing the positions of gene (a) _____, one can construct a chromosome map. Genes belonging to a particular linkage group (b) _____ (will/will not) belong to another linkage group. The fact that the number of linkage groups equals the number of (c) _____ pairs also supports the theory that genes are located on (d) _____.

9. The association of a visually distinguishable chromosome and the (a) _____ of a certain trait also supports the theory that genes are located on chromosomes. A technique to assign genes to specific chromosomes in humans is called somatic cell (b) _____. The first step is to (c) _____ human cells with cells of another animal. The cells are then allowed to undergo (d) _____ while being monitored for a given human (e) _____, usually the activity of an (f) _____ unique to the human cells. As the cells divide, chromosomes are lost. When the activity of the enzyme is lost one looks to see which (g) _____ was lost. It can then be concluded that the (h) _____ for that enzyme was on the missing (i) _____.

C. TESTING TERMS AND CONCEPTS

1. Genetic defects leading to severe deformities may be produced by
   a) nondisjunction.
   b) Barr-body malfunction.
   c) autosomes.
   d) the X chromosome only.

2. In a woman with a genetic make-up of XXXXX, the number of Barr bodies present can be expected to be
   a) 1.
   b) 3.
   c) 4.
   d) 5.

3. The X and Y chromosomes are
   a) autosomes.
   b) sex chromosomes.
   c) found in fruit flies only.
   d) almost always linked.

4. When a segment of a chromosome breaks loose and becomes attached to a chromosome with which it is not homologous, the condition is termed
   a) a deletion.
   b) a translocation.
   c) an inversion.
   d) polyploidy.

5. You would expect an <u>XXY</u> genotype to be
   a) tall.
   b) grossly deformed.
   c) sterile.
   d) short.

6. In humans, the chromosome that determines maleness is the
   a) X.
   b) Y.
   c) XX pair.
   d) autosome.

7. Structural alterations in chromosomes can be caused by all of the
   following <u>except</u>
   a) X-rays.
   b) measles.
   c) chemicals.
   d) bacteria.

8. Chromosome maps show the
   a) relative number of genes on homologous chromosomes.
   b) relative spacing of chromosomes.
   c) sequence of gene loci on a given chromosome.
   d) sequence of gene functions.

9. Genes and chromosomes are similar in all of the following respects,
   <u>except</u>:
   a) they assort independently.
   b) they occur in pairs.
   c) they recombine at fertilization.
   d) they may show dominance.

10. If a certain plant has a diploid number of 20, and you find a
    triploid variant, how many chromosomes will the triploid have?
    a) 17
    b) 23
    c) 30
    d) 60

11. In reference to number 10, what is most likely to have happened to
    cause the triploid variant?
    a) One parent produced a diploid gamete, the other a normal gamete.
    b) One parent produced a triploid gamete, the other a gamete with
       no chromosomes.
    c) Each parent produced a gamete containing one-half of the tri-
       ploid number.
    d) Triploids are strictly "freaks of nature;" no one has a good
       explanation for them.

12. If a woman is a "carrier" for color-blindness, under what condition
    can she have a color-blind daugher?
    a) if her husband is also a "carrier"
    b) if her husband is color-blind
    c) Under any condition of her husband with respect to this trait
    d) Under no condition; i.e., she can not have a color-blind daughter

13. Evidence supporting the concept of crossing-over comes from all of the following, <u>except</u>:
    a) visual examination of chromosomes.
    b) genetic analysis, i.e., linkage studies.
    c) the Creighton-McClintock experiment.
    d) observation of the locations of the genes on the chromosomes.

14. Cell hybridization studies associate genes with chromosomes because
    a) the hybrids are superior to the parent.
    b) lost chromosomes correlate with lost gene functions.
    c) the hybrids have fused chromosomes and fused genes.
    d) fused genes are visualized on the hybrid chromosomes.

15. A person with Down's syndrome is said to be a trisomy 21. This means the person has
    a) 21 chromosomes.
    b) an extra number 21 chromosome.
    c) 21 number 3 chromosomes.
    d) a 3N chromosome number minus a number 21 chromosome.

D. UNDERSTANDING AND APPLYING TERMS AND CONCEPTS

Decide whether the following are true (T) or false (F).

_____ 1. The two alleles controlling a given trait are presumed to be located at corresponding loci on each of two homologous chromosomes.

_____ 2. Most of the disorders of chromosome number arise because of nondisjunction.

_____ 3. Down's snydrome results from a structural defect in chromosomes.

_____ 4. Both X chromosomes in females have been shown to be functional.

_____ 5. A Y chromosome would only be found in a sperm or sperm-producing cell.

_____ 6. A human male can not be a "carrier" of hemophilia.

_____ 7. Crossing-over occurs during meiosis II.

_____ 8. The closer two genes are on a chromosome, the higher the rate of crossing-over.

_____ 9. Gene mapping is done with the aid of autoradiography and a microscope.

_____ 10. Polyploid plants are often sought because of their desirable traits.

11. Here is an hypothetical chromosome. Draw the chromosome after the following events have happened.

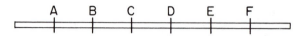

A   B   C   D   E   F

a) deletion of C-D

b) translocation with a

chromosome

R   S

c) inversion between C-E

12. Here are two hypothetical homologous chromosomes. Draw them in the process of crossing-over between the a and b genes. Darken the "recombinant."

a)

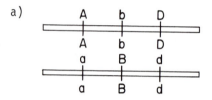

A   b   D

A   b   D

a   B   d

a   B   d

b) List the gametes produced by the original. _____

c) List the gametes produced after recombination. _____

d) If crossing-over occurred in 20% of the cells, what percentage of the gametes would be "recombinants"?

_____

13. Color-blindness is a sex-linked recessive trait in humans. A woman "carried" is married to a color-blind man. Draw the sex-chromosomes in their somatic cells and then in their gametes.

Woman   Man

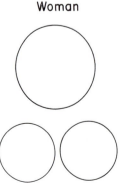   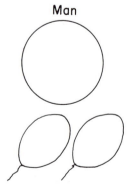

14. List the predicted frequency of the various phenotypes that could result from the above mating.

_____

_____

ANSWERS TO CHAPTER EXERCISES

Reviewing Terms and Concepts

1. a) homologous b) sex c) autosomes d) XX e) XY f) Y g) sex
   h) chromosome

2. a) sex b) X c) allele d) mothers e) hetero f) phenotype
   g) X h) sons i) phenotype

3. a) number, structure b) nondisjunction c) meiosis d) 21 e) Down's

4. a) sex b) X c) X d) functional

5. a) polyploidy b) haploid

6. a) deletion b) inversion c) homologous d) translocation

7. a) linkage b) crossing-over c) recombinants

8. a) loci b) will not c) homologous d) chromosomes

9. a) inheritance b) hybridization c) fuse d) mitosis e) trait
   f) enzyme g) chromosome h) gene i) chromosome

Testing Terms and Concepts

1. a	6. b	11. a
2. c	7. d	12. b
3. b	8. c	13. d
4. b	9. d	14. b
5. c	10. c	15. b

Understanding and Applying Terms and Concepts

1. T	5. F	9. F
2. T	6. T	10. T
3. F	7. F	
4. T	8. F	

11. a)

A B E F

b)

A B C D E F R S

(R-S could have been attached at other end, or in S R direction)

c)

12. a)

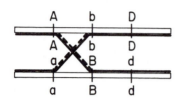

b) AbD, aBd

c) AbD, aBd, ABd, abD

d) 10%

13.

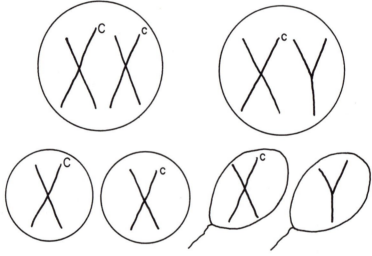

14. 25% "carrier" daughters ($\underline{X}^C\underline{X}^c$)

25% color-blind daughters ($\underline{X}^c\underline{X}^c$)

25% normal sons ($\underline{X}^C\underline{Y}$)

25% color-blind sons ($\underline{X}^c\underline{Y}$)

THE CHEMICAL NATURE OF GENES

I. REVIEWING THE CHAPTER

A. CHAPTER HIGHLIGHTS

12.1 DNA: The Substance of the Genes

The chromosomes of eukaryotes are made up of nucleic acids and
proteins. The nucleic acid is primarily DNA, which contains four
kinds of nucleotides. These are the purines adenine and guanine, and
the pyrimidines cytosine and thymine. One remarkable feature of DNA
is that whatever the amount of adenine present, the amount of thymine is
is equal to it; and whatever the amount of cytosine, the amount of
guanine is equal to it. Two important experiments linked DNA to the
genetic material of cells. In the first, bacteria with a particular
trait had that trait "transformed" by the addition of DNA from another
bacterial strain. The second series of experiments showed, with the
aid of radioactive isotopes, that it was the DNA portion of a virus
that entered the host cell. The viral DNA could then direct the host
to make more virus. In short, the experiments demonstrated that DNA
could carry and transfer information.

12.2 The Watson-Crick Model of DNA

In the Watson-Crick model of DNA, the backbone of the polymer
consists of two strands of alternating sugar and phosphate groups
twisted around each other something like a double spiral staircase.
This structure is called the double helix. Bridging the two strands
of the helix at regular intervals are paired nitrogen bases. This
pair is always composed of one purine linked to one pyrimidine since
if two pyrimidines formed one such pair, their linked length would not
span the distance between strands, and if two purines formed one such
pair, they would be too long. Adenine is always found linked with
thymine, never with cytosine. The guanine is always found linked with
cytosine, never with thymine. The reason for such pairing is that in
the former case hydrogen bonds linking the two nucleotides can form,
whereas in the second case they cannot.

12.3 DNA Replication

Chromosome duplication means DNA duplication, which occurs with
the "unzipping" of the double helix when hydrogen bonds linking the
base pairs are broken and the two halves of the molecule unwind. Once
exposed, the bases on each separate strand can pick up appropriate
nucleotides in the surrounding medium and reassemble a whole new
molecule identical to the original one. The newly acquired bases are
hooked together with the aid of an enzyme called DNA polymerase.

Because the two chains in a DNA molecule are complementary, the information in one is also coded in the other. This "complementary" relationship is analogous to prints and negatives in photography. The structure of the DNA molecule not only allows for accurate replication, but it is also suitable for damage repair. The latter is accomplished when an incorrect or damaged base is removed, and an appropriate one inserted.

12.4 Left-handed DNA

In addition to the normal DNA, which is twisted clockwise, a so-called left-handed DNA has recently been found. This newly discovered form not only turns counterclockwise but it is slightly elongated compared to the right-handed DNA. The role of left-handed DNA has not been clearly demonstrated, but it may be involved in controlling gene expression.

12.5 Mutations

Mutations, brought about by radiation or certain chemical mutagenic agents, involve changes in the sequence of bases. Such changes are called "point" mutations. Mutations occurring in somatic cells may kill the cell or make it cancerous. Mutations occurring in the sex cells are transmitted from parent to offspring (as in the case of hemophilia in Europe's royal families). Further, such transmitted mutations will then be present in every cell in the body.

Given the medical importance of genetic diseases, birth defects, and cancer, it is of great importance to understand the process(es) of mutation and to detect the agents causing mutations. Unfortunately, it is often many years before the exposure to a mutagen will be manifested in recognizable problem.

12.6 Summary

The double-stranded DNA molecule appears to be the major conveyor of genetic information in living organisms. Because its two strands are complementary it is provided with a mechanism for replication and repair. When the sequence of bases (A, T, C, G) is altered a harmful mutation may arise.

B. KEY TERMS

histones	nucleosome
transformation	left-handed DNA
bacteriophage	mutations
double helix	mutagenic agents
antiparallel	roentgen-rad-rem
DNA polymerase	carcinogenic

DNA ligase   teratogenic

semiconservative   revertant

II.   MASTERING THE CHAPTER

A.   LEARNING OBJECTIVES

When you have mastered the material in this chapter, you should be able to:

1.   Define all key terms.

2.   Name the basic hereditary material and list its components.

3.   List the four nucleotides in DNA.

4.   Diagram the basic structure of DNA, including as your labels, sugar group, phosphate group, pyrimidine, and purine.

5.   Cite experimental evidence supporting the theory that DNA is the genetic material.

6.   Outline an experiment that demonstrates the semi-conservative nature of DNA duplication.

7.   Cite two agents that cause point mutations.

8.   Explain how cells repair damaged DNA.

9.   Distinguish among the terms mutagenic, carcinogenic, and teratogenic.

B.   REVIEWING TERMS AND CONCEPTS

1.   The chromosomes of eukaryotes are made up of (a) _____ acids and (b) _____. The (c) _____ acid is primarily (d) _____, which contains four kinds of nucleotides. These are the purines (e)_____ and _____, and the pyrimidines (f) _____ and _____.

2.   One remarkable feature of DNA is that whatever the amount of adenine present, the amount of (a) _____ is equal to it; and whatever the amount of cytosine, the amount of (b) _____ is equal to it.

3.   Evidence that (a) _____ is the genetic material comes from several experiments. In bacteria, for example, a (b) _____ was transformed when the (c) _____ of one organism was given to another. Further evidence came from studies on a (d) _____, which consists of (e) _____ and a (f) _____ coat. In the experiment, it was observed that although only the (g) _____ was injected into the (h) _____ cell, the host was instructed to make a complete virus.

111

4. In the Watson-Crick model of DNA, the backbone of the polymer consists of two strands of alternating (a) _____ and _____ groups twisted around each other something like a double spiral staircase. This structure is called the double (b) _____. Bridging the two strands of the helix at regular intervals are paired nitrogen (c) _____. Always, one purine is linked to one pyrimidine, the combined length being just right. Adenine is always found linked with (d) _____, never with (e) _____. And guanine is always found linked with (f) _____, never with (g) _____. The reason for such pairing is that in the former case hydrogen (h) _____ linking the two (i) _____ can form, whereas in the second case they cannot.

5. Chromosome duplication means (a) _____ duplication, which occurs with the "unzipping" of the double helix when (b) _____ bonds linking the base pairs are broken and the two halves of the molecule unwind. Once exposed, the bases on each separate strand can pick up appropriate (c) _____ in the surrounding medium and reassemble a whole new molecule identical to the original one. The newly synthesized strand is said to be (d) _____ to the original. Another term used to describe this form of duplication is semi- (e) _____. The duplication and repair process(es) of DNA both involve enzymes, two of which are DNA (f) _____ and DNA _____.

6. Recent research has revealed a (a) _____-handed DNA. In contrast to most normally occurring DNA, it turns (b) _____. A definite role for the left-handed DNA (c) _____ (has/has not) been shown.

7. Mutations, brought about by (a) _____ or certain (b) _____ mutagenic agents, sometimes cause changes in the (c) _____ of bases. Such changes are called (d) _____ mutations. If a mutagenic substance alters the DNA in a somatic cell, it may lead to cancer and thus be called a carcinogen. When the alteration occurs in (e) _____ cells the damage can be passed on to the offspring. Agents that cause birth defects are called (f) _____. The Ames test is a well-known method for detecting potential chemical (g) _____.

C. TESTING TERMS AND CONCEPTS

1. The amount of cytosine in DNA is always
    a) different from the amount of guanine.
    b) the same as the amount of thymine.
    c) the same as the amount of guanine.
    d) the same as the amount of adenine.

2. Mutations can be induced by all of the following, except
    a) X-rays.
    b) gamma rays.
    c) ultra-violet rays.
    d) far-red rays.

3. The basic hereditary material is
    a) histone.
    b) protein.
    c) DNA.
    d) RNA.

4. Spanning the space between the double helix in DNA are
    a) a purine and a pyrimidine.
    b) two purines.
    c) two pyrimidines.
    d) a purine, pyrimidine and a histone.

5. Each DNA molecule contains how many copies of its information?
    a) one-half
    b) one
    c) two
    d) more than two

6. Histones are a class of
    a) DNAs.
    b) proteins.
    c) RNAs.
    d) nucleoproteins.

7. During DNA replication each exposed pyrimidine picks up
    a) a purine.
    b) another pyrimidine.
    c) one purine and one pyrimidine.
    d) one histone.

8. Changes in the sequence of bases in DNA are called
    a) reversions.
    b) revisions.
    c) point mutations.
    d) ligases.

9. Avery's experiments showed that the substance that transformed
    R-I cells into S-II cells was
    a) DNA.
    b) lipids.
    c) polysaccharides.
    d) proteins.

10. Which of the following is not a unit used for describing radiation
    dosage?
    a) roentgen
    b) rad
    c) rem
    d) rom

11. Mutations are all of the following, except:
    a) useful in research.
    b) always harmful.
    c) caused by certain chemicals.
    d) possibly obtained during medical X-rays.

113

12. Which enzyme "seals" the repaired portion of a repaired DNA molecule?
    a) DNA polymerase I
    b) DNA polymerase B
    c) DNA ligase
    d) all three working in unison

13. The two DNA strands run in opposite directions and are thus said to be
    a) antiparallel.
    b) complementary.
    c) anticomplementary.
    d) right-handed/left-handed.

14. The purines and pyrimidines in DNA are held together by
    a) hydrogen bonds.
    b) covalent bonds.
    c) ionic bonds.
    d) hydrophilic bonds.

D. UNDERSTANDING AND APPLYING TERMS AND CONCEPTS

Decide whether the following statements are true (T) or false (F).

____ 1. Adenine and thymidine are both pyrimidines.

____ 2. Proteins which make up DNA are called histones.

____ 3. A cell in which a genetic trait has been altered is said to be transformed.

____ 4. Deoxyribose is a 5-carbon sugar.

____ 5. The "rungs of the DNA ladder" are the deoxyribose and phosphate molecules.

6. Here is a segment of a DNA molecule:......A-T-T-A-C-G-G......

   a) Draw the complementary strand:

   _____

   b) Draw the two strands together, but assume that a chemical mutagen (M) has been substituted for the middle A after the complementary strand had been formed:

   _____

   c) Briefly describe how the mutagen might be removed and the strand repaired:

   _____

   _____

_____

_____

_____

7. Many tanning lotions contain sunscreens and are often advertised as preventing certain forms of cancer. What kinds of cancer might they prevent and how?

_____

_____

_____

8. It has been calculated that each of us will have acquired several unrepaired mutations by the time we are in our reproductive years. However, the incidence of mutant traits is far less than that calculated number. Why?

_____

_____

_____

_____

_____

9. After a person is past the "reproductive years" should they stop worrying about environmental mutagens? Why?

_____

_____

_____

10. Treatments for most bacterial infections have been found. Why have'nt scientists found an agent to kill viruses?

_____

_____

_____

_____

_____

115

11. Here is a partially "unzipped" piece of DNA. It is in a cell which contains a radioactive "tag" on the thymines (T*) along with normal As, Cs, and Gs. Show what will happen as replication occurs.

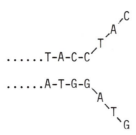

ANSWERS TO CHAPTER EXERCISES

Reviewing Terms and Concepts

1. a) nucleic b) protein c) nucleic d) DNA e) adenine, guanine
    f) cytosine, thymine

2. a) thymine b) guanine

3. a) DNA b) trait c) DNA d) virus (bacteriophage) e) DNA
    f) protein g) DNA h) host

4. a) sugar (deoxyribose), phosphate b) helix c) bases d) thymine
    e) cytosine f) cytosine g) thymine h) bonds i) nucleotides

5. a) DNA b) hydrogen c) nucleotides d) complementary e) conserva-
    tive f) polymerase, ligase

6. a) left b) counterclockwise c) has not

7. a) radiation b) chemical c) sequence d) "point" e) germ
    f) teratogens g) mutagens

Testing Terms and Concepts

1. c	6. b	11. b
2. d	7. a	12. c
3. c	8. c	13. a
4. a	9. a	14. a
5. c	10. d	

Understanding and Applying Terms and Concepts

1. F
2. F
3. T
4. T
5. F

6. a) ....T-A-A-T-G-C-C....

   b) ....A-T-T-M-C-G-G....

   ....T-A-A-T-G-C-C....

   c) The "M" would be enzymatically cleaved-out, the T would serve as a match for another A, and this A would be enzymatically reattached.

7. skin cancer-The sunscreens "filter" the harmful effects of UV rays and thus prevent DNA damage in skin cells.

8. Since most mutations alter a normal gene to a recessive one, our children are likely to receive the dominant allele from their other parent.

9. No. Many mutagens are also carcinogens.

10. Destroying the ability for viruses to be replicated by (by interfering with DNA synthesis) would simultaneously harm the host's DNA synthesis.

11. ....T-A-C-C $\diagup$ A $\diagup$ T-T* $\diagup$ A-G / C

   ....A-T-G-G $\diagdown$ T* $\diagdown$ A $\diagdown$ A-C / T-G

13

GENE EXPRESSION

I. REVIEWING THE CHAPTER

A. CHAPTER HIGHLIGHTS

13.1 The One Gene-One Enzyme Theory

Beadle and Tatum, working with Neurospora, developed the one gene-one enzyme theory of gene action: Each gene in an organism controls the production of a specific enzyme. It is these enzymes that in turn carry out all the metabolic activities of the organism, resulting in the development of a characteristic structure and function, that is, phenotype.

13.2 Inborn Errors of Metabolism

The one gene-one enzyme theory has shed light on several hereditary diseases, "inborn errors of metabolism." When certain genes are lacking, or are inoperative, certain symptoms appear. For example, people who manufacture exclusively a form of hemoglobin known as Hb^S suffer from the disease sickle-cell anemia. Victims of the disease inherit a defective gene from each parent, and the disease is almost always fatal. The defective hemoglobin is due to a single amino acid change in the beta chains of the hemoglobin molecule. Thus the one gene-one enzyme theory has been restated in terms of one gene-one polypeptide. Increasing evidence indicates that not only enzymes but all the proteins made by cells are the products of gene action.

13.3 Protein Synthesis

After it was determined that the information coded in genes is ultimately expressed as a sequence of amino acids in a protein the question became, how are proteins synthesized? The process begins with the formation of messenger RNA (mRNA) in the nucleus. The mRNA is produced on and takes information from the DNA (recall base pairing, but remember RNA contains uracil instead of thymine). Transcription is the name given to this step. Next, the mRNA migrates to the cytoplasm and attaches to ribosomes (chemically the ribosomes are made of proteins and special kinds of RNA, called ribosomal-RNA). Simultaneously, a third kind of RNA, transfer RNA, is being activated and attached to a specific amino acid. Now, the tRNAs, with their specific amino acids, pair-up (complementarity principle) with their "codon" on the mRNA. After the amino acids are placed in the correct position, peptide bonds are formed and the "chain grows." The actual production of the protein can be described in three phases: (1) initiation, (2) elongation, (3) termination. The process of converting the mRNA code into a particular amino acid sequence is called translation.

13.4   The Code

In understanding how genes operate a major problem was to determine the DNA (A, T, C, G) codes the information necessary to position the 20 amino acids. If two bases are employed, only 16 combinations are possible—four short. However, if three bases are used, 64 combinations can be made. As it turns out, the code is a "triplet" of bases and in some cases there is more than one triplet per amino acid. Some of these codons serve as punctuation in the genetic message. Mutations often arise when one base is substituted for another, for example, GAC⟶GGC. Such changes often result in major changes in phenotype since replacing one amino acid for another in a protein can greatly alter the functioning of the protein.

13.5   Multiple Alleles

Whenever a variety of mutant forms of a single gene are known, the expression multiple alleles is used. Although an individual can have only two alleles of a given gene, one on each chromosome, many alleles for a given gene can exist within a population of organisms. In fruit flies, for example, more than a dozen alleles at a single locus control eye color. In those cases where both the dominant and recessive allele produce their own effect on the phenotype, the situation is called codominance.

13.6   Verification of the Code

To clearly establish that the code was operative according to theory in living organisms, a correspondence between a nucleotide sequence and an amino acid sequence was necessary. This was accomplished by using a virus that used RNA rather than DNA for its genetic material. Through knowing the amino acid sequences for the proteins of the virus and correlating the base sequences in the RNA, it was possible to demonstrate the correctness of the relationships.

In addition to the universality of compounds involved in the genetic machinery, the code itself has been shown to be the same for organisms as diverse as bacteria, tobacco, and rabbits, to name a few. That is to say, a particular DNA codon for a given amino acid is the same in all organisms tested. An exception, however, has been the finding that mitochondrial DNA contains some codons that differ from their nuclear counterpart.

13.7   The Action of the Total Genome

The complete genetic library of an organism is called its genome. As yet, genetics has made only a beginning at relating the total organism to its genome, but three important principles now stand out: (1) Although a single gene usually produces but a single polypeptide, such a single polypeptide may have many effects. The creation of multiple effects by a single gene is called pleiotropy. (2) The expression of any trait is influenced by a large number of genes. Many genes contribute to the structure and arrangement of the petals on a flower, for example. (3) The action of genes is dependent also on the environment. The ability to manufacture chlorophyll, for example, is under gene control, but the plant must also be exposed to light.

13.8 Summary

The manner in which DNA encodes genetic material, and the way it is converted to a product has been established. An understanding of these processes helps account for normal characteristics, mutations, and malfunctions (genetic diseases).

B. KEY TERMS

spores	initiation
inborn errors of metabolism	initiator RNA
PKU	elongation
ribosomes	termination
ribosomal RNA	primary transcript
messenger RNA	wobble
transcription	silent mutations
RNA polymerase	multiple alleles
translation	codominance
codon	operon
transfer RNA	pleiotropy
anticodon loop	

II. MASTERING THE CHAPTER

A. LEARNING OBJECTIVES

When you have mastered the material in this chapter, you should be able to:

1. Define all key terms.

2. Cite Beadle and Tatum's evidence for their one gene-one enzyme theory.

3. Cite the biochemical defect in sickle-cell anemia.

4. Explain how PKU arises, genetically and physiologically.

5. List three different kinds of RNA and state their function in protein synthesis.

6. Explain what is meant by transcription and translation.

7. Cite reasons why a codon must be composed of more than two bases.

8. Give an example of "wobble."

9. Cite experimental evidence that shows the genetic code works according to theory.

10. List the three principles associated with the relationship of the total organism to its genome.

B. REVIEWING TERMS AND CONCEPTS

1. Beadle and Tatum developed the one (a) _____-one (b) _____ theory of gene action: Each (c) _____ in an organism controls the production of a specific (d) _____. It is these (e)_____ that then carry out all the (f) _____ activities of the organism.

2. When certain genes are lacking, or are inoperative, certain symptoms appear. For example, people who manufacture exclusively a form of hemoglobin known as (a) _____ suffer from the disease (b) _____ anemia. Victims of the disease inherit a defective gene from each (c) _____. The defective hemoglobin is due to a single (d) _____ acid change. Thus the one gene-one enzyme theory has been restated in terms of one gene-one (e) _____. Not only (f) _____ but all the proteins made by cells are the products of (g) _____ action.

3. Protein synthesis occurs on (a) _____. A type of RNA called (b) _____ RNA is synthesized in the (c) _____ by DNA in a process called (d) _____. The information coded in (e) _____ RNA is (f) _____ into a sequence of (g) _____ acids leading to protein synthesis. This process requires a third kind of RNA called (h) _____ RNA. The union of (i) _____ RNA and (j) _____ RNA requires the presence of ribosomes and occurs in the (k) _____.

4. Much work has been done to determine what sequence of the four (a) _____ in the messenger RNA molecule "codes" each of the (b) _____ amino acids. While pairs of bases can be arranged in only 16 different ways, (c) _____ of bases provide 64 possibilities called (d) _____. Only three of the 64 possible codons have been found not to code for any (e) _____ _____. Those three codons seem to serve as (f) _____ marks halting protein synthesis on the (g) _____.

5. Whenever several (a) _____ forms of a single gene are

known, the expression multiple (b) _____ is used. Although an individual can have only two (c) _____ of a given gene, one on each (d) _____, a population of organisms can have many (e) _____.

6. A correspondence between a (a) _____ sequence and an (b) _____ acid sequence was necessary to establish that the genetic (c) _____ operates in living organisms. This was accomplished by using a (d) _____ for which the (e) _____ sequence and _____ sequences were known. Since the genetic code appears to be (f) _____ (the same/different) in a wide variety of living organisms, it is said to be (g) _____.

7. The complete genetic library of an organism is called its (a) _____. As yet, genetics has made only a beginning at relating the total organism to its (b) _____. The creation of multiple effects by a single gene is called (c) _____. The expression of any (d) _____ is influenced by a large number of (e) _____. The action of genes is dependent also on environment. The ability to manufacture chlorophyll, for example, is under (f) _____ control, but the plant must also be exposed to (g) _____.

C. TESTING TERMS AND CONCEPTS

1. Victims of sickle-cell anemia suffer from a condition that is
    a) environmental.
    b) spread by mosquitoes.
    c) contagious.
    d) inherited.

2. The union of transfer RNA and messenger RNA requires the presence of
    a) lysosomes.
    b) ribosomes.
    c) DNA polymerase.
    d) nucleosomes.

3. The precise sequence of amino acids in protein is <u>controlled</u> by the base sequence in
    a) DNA.
    b) tRNA.
    c) enzymes.
    d) rRNA.

4. When each allele in a hybrid produces its own effect on the phenotype, the situation is called
    a) multiple alleles.
    b) codominance.
    c) multiple codons.
    d) parallel alleles.

5. The synthesis of a molecule of RNA complementary to a strand of DNA
    a) is called transcription.

b) is called translation.
c) requires DNA polymerase.
d) requires polysomes.

6. When a gene has more than one effect, the condition is referred to as
a) multiple alleles.
b) pleiotropy.
c) heterogeneity.
d) multiple codons.

7. The information transcribed and essential to the proper sequencing of amino acids in protein synthesis is carried by
a) transfer RNA.
b) messenger RNA.
c) ribosomal RNA.
d) DNA.

8. In eukaryotes, the transcription of DNA into messenger RNA occurs within the
a) cytoplasm.
b) nucleus.
c) polysomes.
d) Golgi apparatus.

9. Beadle and Tatum deduced that the information encoded in a gene finds ultimate expression in the
a) base pairs.
b) peptide bonds.
c) sequence of amino acids.
d) codon.

10. A DNA molecule can contain a tremendous amount of information because of
a) variety in sugar-phosphate combinations.
b) variety in base pair combinations.
c) variety in histone-DNA complexes.
d) variety of combination in the nucleotide sequence.

11. A given tRNA attaches to
a) only one specific amino acid.
b) any of the amino acids, but only one at a time.
c) as many as three different amino acids.
d) the growing peptide chain.

12. All of the following are true for polysomes, except
a) they are composed of several ribosomes.
b) they are the site of protein synthesis.
c) they perform their role within the nucleus.
d) they are made of RNA and protein.

13. The ultimate cause of sickle-cell anemia is
a) one DNA base that has been deleted.
b) one DNA that has been substituted for another.

c) one amino acid that has been substituted for another.
d) a translation error.

14. An anticodon loop would be found in
    a) DNA.
    b) mRNA.
    c) tRNA.
    d) rRNA.

15. Which of the following is <u>not</u> true regarding the action of genes?
    a) One gene may have several effects.
    b) Genes may influence the environment.
    c) Several genes may be involved with one trait.
    d) The environment may influence the functioning of a gene.

D. UNDERSTANDING AND APPLYING TERMS AND CONCEPTS

Decide whether the following statements are true (T) or false (F).

_____ 1. Two nontwins who are both homozygous for Hb^A will have
          different kinds of hemoblogin.

_____ 2. If a person is homozygous for a gene in which one base has
          been substituted for another, the person will necessarily
          be of a "mutant" phenotype.

_____ 3. The linking of two amino acids occurs without the expenditure
          of energy.

_____ 4. If a given gene consisted of 200 bases, there would be a
          maximum of 200 different mutations possible.

_____ 5. Humans are superior to other animals because of a different
          genetic code.

Use the following information for deciding whether 6 through 10 are
true or false.

> Recall that phenylalanine is an amino acid that one gets
> from their diet and that melanin is a skin pigment.
>
> $$Phenylalanine \xrightarrow[A]{Enzyme} Tyrosine \xrightarrow[B]{Enzyme} DOPA \xrightarrow[C]{Enzyme} Melanin$$

_____ 6. Several genes are involved in pigmentation.

_____ 7. A child that received a mutant gene coding for enzyme C
          from each parent would not be able to produce tyrosine.

_____ 8. Enzyme B is made of DNA.

_____ 9. A cell must have at least one correct copy of the gene that
          codes for enzyme A in order to convert phenylalanine to
          tyrosine.

_____ 10. If DOPA accumulates in a cell, you could conclude enzyme B is missing or defective.

11. Given—The following tRNAs are specific for the respective amino acids.

ACU-AA$_1$	ACC-AA$_5$
CGG-AA$_2$	GGG-AA$_6$
AUC-AA$_3$	AGC-AA$_7$
GAU-AA$_4$	UCC-AA$_8$

Here is a segment of DNA:

....A-T-C-A-C-C-G-A-T....

a) What is the sequence of bases in the complementary strand?

_____

b) Give the mRNA that would be transcribed from the given DNA.

_____

c) Arrange the tRNAs in the order they would take along the mRNA.

_____

d) After transcription and translation, what is the amino acid sequence?

_____

Suppose a mutation occurs to our DNA and the following results:

....A-T-C-A-G-C-G-A-T....

e) Give the new mRNA.

_____

f) Arrange the tRNAs as in c) for the mutant.

_____

g) What is the amino acid sequence in our "mutant" protein?

_____

ANSWERS TO CHAPTER EXERCISES

Reviewing Terms and Concepts

1. a) gene b) enzyme c) gene d) enzyme e) enzymes f) metabolic

2. a) Hb^S b) sickle-cell c) parent d) amino e) polypeptide
    f) enzymes g) gene

3. a) ribosomes b) messenger c) nucleus d) transcription
    e) messenger f) translated g) amino h) transfer i) transfer
    j) messenger k) cytoplasm

4. a) bases b) 20 c) triplets d) codons e) amino acids
    f) punctuation g) ribosomes

5. a) mutant b) alleles c) alleles d) chromosomes e) alleles

6. a) nucleotide b) amino c) code d) virus e) RNA, protein
    f) the same g) universal

7. a) genome b) genome c) pleiotropy d) trait e) genes f) gene
    g) light

Testing Terms and Concepts

1.	d	6.	b	11.	a
2.	b	7.	d	12.	c
3.	a	8.	b	13.	b
4.	b	9.	c	14.	c
5.	a	10.	d	15.	b

Understanding and Applying Terms and Concepts

1. F
2. F
3. F
4. F
5. F
6. T
7. F
8. F
9. T
10. F
11. a) T-A-G-T-G-G-C-T-A
    b) U-A-G-U-G-G-C-U-A
    c) A-U-C, A-C-C, G-A-U
    d) AA$_3$ —— AA$_5$ —— AA$_4$

    e) U-A-G-U-C-G-C-U-A
    f) A-U-C, A-G-C, G-A-U
    g) AA$_3$ —— AA$_7$ —— AA$_4$

THE ORGANIZATION OF
GENETIC INFORMATION

I. REVIEWING THE CHAPTER

A. CHAPTER HIGHLIGHTS

14.1 Reading Genes: DNA Sequencing

A fascinating and important practical aspect of molecular biology
begins with the determination of the gene's message. This involves
establishing the exact sequence of nucleotides in the DNA. Pioneer
research in this area employed a bacteriophage that had a single-
stranded DNA. The sequencing also required that the molecule be
broken into fragments of a known size. This aspect of the research
was facilitated by using the enzyme, "restriction endonuclease." The
fragments of DNA were then labeled with radioactive markers and, in
turn, separated by the use of a technique called electrophoresis. The
sequence of the nucleotides for the entire DNA was deciphered after
the sequence was determined for each of the fragments. Additional
support for the correctness of the triplet code hypothesis was obtained
when the nucleotide sequence correlated with the amino acid sequence
of the protein coded for by the DNA.

14.2 Overlapping Genes

Once again, the virus oX 174 has been a useful aid in understanding
how the genetic message is transcribed and translated into a protein.
By knowing the number of nucleotides in the single-stranded DNA as
well as the number of amino acids in the viral proteins, it was
determined that the former was not large enough to contain the information for the latter. This problem was explained by showing that
the same sequence of nucleotides could be "read" in different "frames."
In other words, by using overlapping sequences, the same nucleotides
can contain different messages.

14.3 Cloning Genes: Recombinant DNA

The task of sequencing and manipulating the DNA of a eukaryote
is much more complicated than it is for oX 174 for several reasons.
(1) There is more of it. (2) It is double stranded, thus having twice
as much information to decipher. (3) It is difficult to obtain large
quantities to work with. The techniques for overcoming these difficulties involves the "cloning" of "recombinant" DNA. The several
steps in this process are: (1) simultaneously cleaving the eukaryotic
and viral DNAs with restriction endonucleases, (2) attaching the
eukaryotic DNA to the virus with a ligase, (3) infecting a host cell

with the above product, and (4) allowing the virus and the attached DNA to replicate. After the host dies and releases the recombinant DNA, the task is to choose the plaque (zone of dead bacteria) that contains the desired gene. This problem is solved by using a radio-active mRNA specific for the correct product, and the principle of complementary pairing. That is, the desired DNA will be attached to the radioactive mRNA and thus recognized.

14.4 Split Genes

In contrast to viral and prokaryotic DNA, the genetic material in eukaryotes has sequences of nucleotides that are not transcribed and translated into known proteins. Thus the coding portion of the molecule is said to be "split." The code-containing portions of the molecule are called "exons" and the noncoding portions are called "intercons." It appears that the split genes are transcribed into "primary transcripts" (large mRNA molecules found in the nucleus), and then the introns are removed before the messenger is sent to the cytoplasm. When mRNA is allowed to base pair with its respective coding strand of DNA, bulges (R-loops) of intron DNA are formed. The bulges correspond to unpaired DNA.

14.5 Jumping Genes

Jumping genes refers to a situation in which two genes that are physically separated at one point in time are found together later. This phenomenon has been demonstrated with DNA from antibody synthesizing cells. In embryonic cells the genes for light-chain proteins are far removed from one another, but after embryonic development they are adjacent. When these genes are in nonantibody producing cells, the genes do not jump.

14.6 The Prospects of Genetic Engineering

Genetic engineering holds the promise of applying the knowledge of molecular biology to the solution of many agricultural and medical problems. In agriculture, one possible goal would be to insert the genes for nitrogen-fixation into eukaryotic crop plants in order to produce plants that could be grown without massive applications of fertilizer.

The medical applications of genetic engineering take two ap-proaches: (1) the insertion of eukaryotic genes into prokaryotes thereby allowing these microbes to produce large quantities of the gene pro-duct, and (2) the insertion of genes into eukaryotes, thus allowing them to manufacture their own gene products. The enzyme reverse transcriptase is providing a useful tool in both the theory and ap-plication of genetic engineering. Specifically, reverse transcriptase facilitates the synthesis of DNA from RNA. This allows for the pro-duction of a DNA that can be transcribed and translated by prokaryotes, the advantage being large-scale production of such things as hormones and vaccines. The second approach, inserting genes into eukaryotes, essentially means correcting rather than treating genetic disorders. For example, it should be possible to insert the gene for synthesizing

Hb^A into sickle-cell anemia patients, who are homozygous recessive for abnormal hemoglobin. However, four major obstacles need to be overcome: (1) identifying and purifying the desired genetic information, (2) inserting the molecules into receptive cells, (3) having the host cell replicate the new DNA, and (4) having the host cells express the information.

Potential dangers also are found in genetic engineering. For example, what if DNA from tumor-causing viruses became part of the genetic constitution of the common intestinal bacterium, and then escaped? The possibility of such an accident has created much controversy around some genetic engineering.

14.7 Summary

Recent research has provided the knowledge and techniques for isolating, deciphering, and manipulating genes. This chapter presents several important breakthroughs in this area.

B. KEY TERMS

bacteriophage—oX 174	lawn
restriction endonuclease	plaque
DNA polymerase	shotgun
electrophoresis	split genes
electrophoretogram	introns/exons
anticoding strand	primary transcripts
overlapping genes	reading frame
template	domains
cloning	jumping genes
recombinant DNA	reverse transcriptase
sticky ends	interferon

II. MASTERING THE CHAPTER

A. LEARNING OBJECTIVES

When you have mastered the material in this chapter, you should be able to:

1. Define all key terms.

2. Outline the technique for sequencing DNA.

3. Cite the role of the following enzymes in genetic engineering: endonuclease, DNA ligase, DNA polymerase, and reverse transcriptase.

4. Outline the main steps in cloning genes.

5. Describe overlapping genes, split genes, and jumping genes.

6. Cite two approaches taken in manipulating genes, and give examples of what kinds of problems can be treated in each case.

7. Cite a situation in which genetic engineering might "backfire."

B. REVIEWING TERMS AND CONCEPTS

1. In order to manipulate genes, the first step is determining the sequence of (a) _____ in the DNA. Early research on techniques was performed on a virus oX 174, because its (b) _____ was of relatively small size and (c) _____-stranded. In order to sequence the DNA, it was first broken into fragments with certain known properties. This was done with the aid of an enzyme called (d) _____ _____. The fragments were then "tagged" with (e) _____ ATP. After the sequence of nucleotides for each fragment was determined, the sequence for the entire (f) _____ could be established. By knowing the sequences of nucleotides and the sequences of (g) _____ _____ in the proteins coded by those genes, validity of the code has been (h) _____ (fully/ somewhat) supported. Because of the double strand nature of the (i) _____, only one strand is transcribed into (j) _____. It is called the (k) _____ strand.

2. The oX 174 bacteriophage consists of a protein coat and a core of (a) _____ and _____. When the single-stranded DNA (+) is injected into the (b) _____ cells, it serves as a template for the synthesis of a (c) _____ or (-) strand. The (-) strands then serve as templates for synthesis of (d) _____ and complementary (+) strands of (e) _____. The mRNA then directs the host to produce ten different (f) _____. Four of these proteins are for oX 174 coat and six play a role in the (g) _____ process. The curious finding is that there are not enough (h) _____ in the virus (i) _____ to code for all the (j) _____ _____ that constitute the viral proteins. The problem created has been overcome by the use of (k) _____-lapping genes. Thus, some of the nucleotides are used (1) _____ (more/less) than once for coding purposes.

3. Before eukaryotic genes can be studied or manipulated, they must be obtained in large quantities. To accomplish this, "recombinant" (a) _____ must be (b) _____. The (c) _____ DNA is attached to a (d) _____ with the aid of a (e) _____ endonuclease prior to infecting host cells. The host cells then replicate the (f) _____ and the attached eukaryotic DNA. After the bacterial hosts die, they leave a (g) _____ that contains replicate DNA. To choose the plaque with the appropriate DNA, radio-active (h) _____ for the chosen gene products is added. Any

130

(i) _____ that attaches to the mRNA will then be (j) _____ and thus paired with the desired DNA.

4. In eukaryotic cells there appear to be portions of (a) _____ that do not code for any known protein. Therefore, the coding portion is said to be (b) _____ by noncoding portions. The noncoding regions have been termed (c) _____, while those that do code are called (d) _____. This discovery helps account for the observation that (e) _____ can not be found in the nucleus. Rather, the nucleus contains (f) _____ transcripts, large (g) _____ molecules that contain both introns and exons. Before migrating from the nucleus the (h) _____ are removed.

5. Some genes that are physically separated during embryonic stages are found to be adjacent later. In such cases the term (a) _____ genes has been applied. Curiously, in cells that do not express those genes, the genes (b) _____ (do/do not) jump.

6. The field of genetic engineering has been made possible by several things, among them are (a) _____ nuclease, cloning (b) _____ DNA in prokaryotes, and (c) _____ the nucleotides in DNA. In the future we are likely to see eukaryotic (d) _____ inserted into prokaryotes. These microbes will then be "living factories" for gene (e) _____. Also, genes will be inserted into eukaryotes and thereby (f) _____ defective ones. An enzyme important in producing DNA from RNA is (g) _____ transcriptase.

C. TESTING TERMS AND CONCEPTS

Some or all of the following enzymes are used in manipulating genes. Use them to answer questions 1 through 5.

            a) DNA polymerase
            b) DNA ligase
            c) reverse transcriptase
            d) restriction endonuclease

1. Which enzyme cleaves DNA of specific locations?

2. Which enzyme is used to covalently link "sticky ends" together.

3. The process RNA ⟶ DNA is unique in its requirement for what enzyme?

4. Making complementary DNA strands requires what enzyme?

5. Which enzyme is crucial in sequencing DNA.

6. The first step in manipulating genes is
    a) inserting them into proper cells.
    b) translating them.
    c) transcribing them into mRNA.
    d) determining the nucleotide sequence in their DNA.

7. A situation in which more genetic information is required than there are available nucleotides is called
    a) split genes.
    b) jumping genes.
    c) overlapping genes.
    d) exon genes.

8. As used in this chapter, recombinant DNA refers to a situation in which
    a) eukaryotic DNA is added to a virus DNA.
    b) eukaryotic mRNA is added to its complementary DNA.
    c) two (+) DNA are joined.
    d) two exons are joined.

9. The production of large numbers of a DNA segment is referred to as
    a) cloning.
    b) splitting genes.
    c) reverse transcription.
    d) DNA polymerizing.

10. Which of the following is <u>not</u> a nucleotide sequence?
    a) codon
    b) exon
    c) intron
    d) interferon

11. Which of the following can change position on a chromosome during development?
    a) recombinant genes
    b) jumping genes
    c) split genes
    d) prokaryotic genes

12. How many copies of a eukaryotic gene product can be produced in a single microbe?
    a) one
    b) two
    c) four
    d) thousands

13. What is the genotype of a person most likely to benefit from gene insertion?
    a) $Hb_S^A \ Hb_Z^A$
    b) $Hb_S^S \ Hb_S^Z$
    c) $HB_C^A \ Hb_S^S$
    d) $Hb^S \ Hb^S$

14. Primary transcripts are
    a) the first formed proteins of recombinant DNA.
    b) fragments prepared for DNA sequencing.
    c) RNA molecules which contain exons and introns.
    d) prepared by reverse transcriptase.

15. The field of genetic engineering is
    a) purely theoretical.
    b) one of the oldest in biology.
    c) beginning to pay off in practical applications.
    d) totally free of hazards.

D. UNDERSTANDING AND APPLYING TERMS AND CONCEPTS

Decide whether the following statements are true (T) or false (F).

_____ 1. We should be able to produce a gene(s) that allows people
          to avoid the requirement for energy.

_____ 2. PKU could be "cured" by giving an affected baby the gene
          that contains the information to make the enzyme that con-
          verts phenylalanine to tyrosine.

_____ 3. By knowing an amino acid sequence for a protein, it is
          possible to determine what the DNA nucleotide sequence for
          that protein should be.

_____ 4. Once established on a chromosome, the position of a gene
          can never change.

_____ 5. Alterations of the genetic material can occur by mechanisms
          other than point mutations.

6. Before a gene "transplant" can be termed a success, it must
   be _____
   _____ in the host cell.

7. Assume the following is a coding strand of DNA and the
   triplets correspond to the given amino acids.

   ...A - T - C - G - C - T - A - G - C - T...

   $AGC = AA_1$, $GCT = AA_2$, $CTA = AA_3$, $ATC = AA_4$, $CAT = AA_5$, $TCG = AA_6$

   a) What is the amino acid sequence if the message is read
      beginning at the first A? _____
      _____

8. A genetic engineer wants a certain recombinant DNA molecule.
   The engineer does not know which plaque has the desired one,
   but does have a radioactive mRNA for it. It is:
   ...*U - A - U - C - C - G - A - G - U...

   Which plaque will be chosen?

   a) A - U - A - G - G - C - U - C - A
   b) T - U - T - G - G - C - U - C - T
   c) A - T - A - G - G - C - T - C - A
   d) T - U - T - G - G - C - U - C - T

9. In talking about the potentials of genetic engineering, people often (and not altogether jokingly) refer to "designer genes." To what are they referring? _____
_____
_____

ANSWERS TO CHAPTER EXERCISES

Reviewing Terms and Concepts

1. a) nucleotides b) DNA c) single d) restriction endonuclease
    e) radioactive f) DNA g) amino acids h) fully i) DNA j) mRNA
    k) coding

2. a) protein, DNA b) host c) complementary d) mRNA e) DNA
    f) proteins g) replication h) nucleotides i) DNA j) amino
    acids k) over l) more

3. a) DNA b) cloned c) recombinant d) virus e) restriction f) virus
    g) plaque h) mRNA i) DNA j) complementary

4. a) DNA b) split c) introns d) exons e) mRNA f) primary g) RNA
    h) introns

5. a) jumping b) do not

6. a) restriction b) recombinant c) sequencing d) genes e) pro-
    ducts f) replace g) reverse

Testing Terms and Concepts

1. d	6. d	11. b
2. b	7. c	12. d
3. c	8. a	13. d
4. a	9. a	14. c
5. d	10. d	15. c

Understanding and Applying Terms and Concepts

1. F 2. T 3. T 4. F 5. T
6. inserted, replicated, transcribed, and translated
7. a) $AA_4 - AA_2 - AA_1$ b) $AA_6 - AA_3 - AA_2$
8. c
9. the potential to design and make genes for a given purpose, and
   to insert them where we want them.

15

THE REGULATION OF GENE EXPRESSION

I. REVIEWING THE CHAPTER

A. CHAPTER HIGHLIGHTS

15.1 Modulation of Gene Activity

All cells have the ability to respond to certain signals that
reach them from their environment. E. coli cells contain all the
genetic information they need to metabolize, grow, and reproduce. However, the genes are expressed only when their products (enzymes) are
required.

In order for certain genes to be expressed, they must be stimulated
by certain substances. The term modulation is used to describe the
turning off and on of genetic information in a cell in response to
changes in its environment.

Each enzyme in prokaryotes is encoded by a separate gene, called
a structural gene. These genes are turned on and off by regulator
genes, but not necessarily directly. Associated with structural genes
of a prokaryote may be a smaller gene called the operator, which in
combination with its structural genes is called the operon. The
regulator genes seem to act on the operator gene, which then acts on
the structural gene. The regulators perform their task by producing a
so-called repressor, which appears to combine with an environmental
signal—a corepressor. It is this combination of repressor and corepressor that determines how the regulator gene will function.

15.2 Modulation in Eukaryotes

Like prokaryotes, eukaryotes contain more genetic information than
they use at any one time and are subject to modulation. There are
examples among eukaryotes of hormones serving as powerful modulators
of gene expression. And it now appears that at least some of the effects of hormones are brought about by their influence on gene transcription. For example, in mammalian females the hormones estrogen and
progesterone have been shown to have important physiological effects.
In addition, it has been shown that they are bound in "target" tissues
to special "receptor" proteins. The receptor-hormone complex then appears to somehow regulate gene activity by attaching to one of the
nonhistone proteins of the chromatin.

15.3 The Giant Chromosomes and Differential Gene Action

Differential gene action refers to the observation that not all genes function in all cells all the time. Studies on giant chromosomes have been used to demonstrate this phenomenon. These chromosomes have visual bands (thought to be gene loci) that "puff" in specific cells under certain conditions. By treating organisms with hormones and radioactive uracil (found in RNA), it is possible to show that the puffs represent regions of RNA synthesis. Thus, there is evidence that the genetic information in certain cells is responding to an environmental signal to produce a specific effect.

15.4 Translation Controls

Eukaryotes, in contrast to prokaryotes, are able to produce stable messengers, which can remain in the cell for long periods. So the opportunity exists to control the rate of gene expression by controlling the rate of messenger translation. There is some evidence that modulation in eukaryotes can involve differential messenger translation in addition to differential gene transcription.

15.5 Summary

Mechanisms have been described that help explain how cells, through specific receptors, can respond to specific environmental signals. However, what determines which receptors are present in particular cells remains a challenge for further research.

B. KEY TERMS

induce	corepressor
repress	progesterone
permease	estrogen
beta-galactosidase	actinomycin D
modulation	receptor
structural genes	histone, nonhistone
regulator genes	ecdysome
repressor	differential gene action
operator	"puffs"
operon	

II. MASTERING THE CHAPTER

A. LEARNING OBJECTIVES

When you have mastered the material in this chapter, you should be able to:

1. Define all key terms.

2. Distinguish between a structural gene, an operon, and a regulator gene by citing the function of each.

3. Distinguish between transcription and translation.

4. Cite one example each of stability and instability of messenger RNA, and the consequence of that stability or instability.

5. Explain the Jacob-Monod hypothesis of gene control.

6. Show how a given hormone can influence a specific event in a specific tissue.

7. Cite the evidence that nonhistone proteins are involved in gene regulation.

8. Explain how the use of giant chromosomes has been employed to demonstrate differential gene action.

B. REVIEWING TERMS AND CONCEPTS

1. All cells have the ability to respond to certain signals that reach them from their (a) _____. Cells contain all the (b) _____ information they need to metabolize, grow, and reproduce.

2. In order for certain genes to be expressed, they must be stimulated by certain substances. The term (a) _____ is used to describe the turning on and off of genetic information in a cell in response to changes in its environment.

3. In certain metabolic pathways in E. coli each enzyme is encoded by a separate gene, called a (a) _____ gene. These genes are turned on and off by (b) _____ genes, but not necessarily directly. Associated with structural genes of a prokaryote may be a smaller gene called the (c) _____, the combination being called the (d) _____. The regulator genes seem to act on the (e) _____ gene, which then acts on the (f) _____ genes.

4. Like prokaryotes, eukaryotes contain more genetic information than they use at any one time and are subject to modulation. There are examples among eukaryotes of (a) _____ serving as powerful modulators of gene (b) _____. And it now appears that at least some of the effects of hormones are brought about by their influence on gene (c) _____.

5. Not all genes function all the time. This is called (a) _____ gene action. The use of giant chromosomes with their visual (b) _____, which are thought to be (c) _____ loci, has helped us to understand this phenomenon. For example, when exposed to a certain hormone, the chromosome will form a (d) "_____," this region has been shown to be very active in (e) _____ synthesis. Hence, there is evidence that the (f) _____

material in the chromosome responds to an (g) _____ stimulus.

6. Eukaryotes, in contrast to prokaryotes, are able to produce stable (a) _____, which can remain in the cell for long periods. So the opportunity exists to control the rate of gene (b) _____ by controlling the rate of messenger (c) _____. There is some evidence that (d) _____ in eukaryotes can involve differential messenger translation in addition to differential gene (e) _____.

C. TESTING TERMS AND CONCEPTS

1. Modulation of genes in eukaryotes may involve all of the following, except:
    a) differential messenger translation.
    b) differential gene transcription.
    c) response to hormones.
    d) response to giant chromosomes.

2. Protein synthesis is a direct example of
    a) translation
    b) transcription
    c) "puffing"
    d) modulation

3. When estrogens enter a target cell, they become tightly bound to
    a) the nucleus
    b) the nuclear membrane
    c) protein receptors
    d) regulators

4. In eukaryotes, messenger RNA is produced during
    a) transcription
    b) translation
    c) selective translation
    d) transformation

5. In eukaryotes, gene transcription occurs in the
    a) mitochondria
    b) cytoplasm
    c) nucleus
    d) Golgi apparatus

6. Those cells capable of responding to hormonal signals are
    a) prokaryotes
    b) eukaryotes
    c) target cells
    d) receptor cells

7. The stimulatory effect of estrogens on cells of the endometrium is inhibited by
    a) transfer RNA
    b) ribosomal RNA
    c) ecdysone
    d) actinomycin D

138

8. Repressor substances inhibiting the production of a given enzyme
    are produced through the action of
    a) structural genes
    b) regulator genes
    c) operons
    d) modulator genes

9. "Puffs" refer to
    a) enlarged chromosomes.
    b) outbursts of DNA synthesis.
    c) "swollen" regions along the chromosomes.
    d) clusters of ribosomes.

10. The environmental signal that acts in conjunction with the product
    of the regulator gene is called a(n)
    a) corepressor.
    b) operon.
    c) modulator.
    d) inhibitor.

11. Chromatin consists of all the following, <u>except</u>:
    a) DNA.
    b) saturated lipids.
    c) histones.
    d) nonhistone proteins.

12. Of the chromatin constituents, which one appears to establish which
    genes will be transcribed?
    a) DNA
    b) RNA
    c) histones
    d) nonhistone proteins

13. In a series of developmental stages, "puffing" has been shown to
    be
    a) random.
    b) independent.
    c) sequential.
    d) nonfunctional.

14. In the modulation process, one environmental signal
    a) will affect only one gene.
    b) may affect several genes at once.
    c) must be a hormone.
    d) may not have more than one effect on a cell.

15. The visualization of RNA synthesis around "puffs" has been shown
    with the aid of the technique known as
    a) centrifugation.
    b) electrophoresis.
    c) electronmicroscopy.
    d) autoradiography.

  D. UNDERSTANDING AND APPLYING TERMS AND CONCEPTS

Decide whether the following statements are true (T) or false (F).

_____ 1. E. coli has the ability to synthesize new genes to cope with new environmental conditions.

_____ 2. Eukaryotic cells contain more genetic information than they use at any one time.

_____ 3. Because hormones are transported in the blood, they will affect every cell in the organism.

_____ 4. Modulation of gene activity has not been demonstrated in plants.

_____ 5. In a multicellular organism, each cell does not express its total genetic "library."

6. Modify statement 1 to make it true. _____
_____
_____
_____

7. Defend your answer to statement 2 by using puberty as an example. _____
_____
_____
_____
_____
_____

8. Using liver cells and brain cells as examples, defend the correct answer to statement 5. _____
_____
_____
_____
_____
_____

9. Pretend that you are following a radioactive molecule of estrogen from the time it enters the cell until it produces its effect within the cell. List the various organelles and components it would come in contact with on its trip. _____
_____
_____
_____
_____

10. Once at its destination, how does estrogen probably function?
_____
_____
_____

11. How could you test your answer to number 10? _____
_____

ANSWERS TO CHAPTER EXERCISES

Reviewing Terms and Concepts

1. a) environment b) genetic

2. a) modulation

3. a) structural b) regulator c) operator d) operon e) operator
    f) structural

4. a) hormones b) expression c) transcription

5. a) differential b) bands c) gene d) puff e) RNA f) genetic
    g) environmental

6. a) messengers b) expression c) translation d) modulation
    e) transcription

Testing Terms and Concepts

1. d	6. c	11. b
2. a	7.	12. d
3. c	8. b	13. c
4. a	9. c	14. b
5. d	10. a	15. d

Understanding and Applying Terms and Concepts

1. F 2. T 3. F 4. F 5. T

6. E. coli has the ability to "turn genes on or off" in order to cope
    with new environmental conditions.

7. Humans contain the genetic information for adult sex characteristics
    throughout childhood, but don't express them until the time of
    puberty.

8. The genetic material in the nucleus of both liver and brain cells
    is the same. However, it is obvious that liver and brain cells
    are quite different in function and therefore different in which
    genes are being expressed.

9. cross cell membrane, enter cytoplasm, attach to a specific receptor
    protein, migrate to nucleus, bind to chromatin

10. As a complex, along with its receptor and a nonhistone protein,
    the estrogen "turns on or off" certain genes.

11. Look for RNA synthesis in the presence and absence of actinomycin
    D.

REPRODUCTION IN PLANTS

I. REVIEWING THE CHAPTER

A. CHAPTER HIGHLIGHTS

16.1 Alternation of Generations

Sexual reproduction involves the two processes of fertilization and meiosis. In the former, the nuclei of two gametes fuse, raising the chromosome number from haploid to diploid. In the latter process, the chromosome number is reduced again from the diploid to the haploid condition.

In plants, fertilization and meiosis divide the life of the organism into two distinct generations: (1) the gametophyte generation begins with spores produced by meiosis. A spore and all cells derived from it are haploid in genetic constitution. Gametes are produced by this generation. When two gametes fuse, the sporophyte generation begins. This generation thus starts with a zygote and it contains the diploid number of chromosomes. Eventually, however, certain cells undergo meiosis and form spores that start the gametophyte generation again.

16.2 The Problems to Be Solved

Plants have had to solve two problems of reproducing sexually on land: (1) bringing about the union of two delicate gametes in a pro- tected environment without the aid of locomotion and (2) dispersal of new generations to locations suitable for healthy growth. Different solutions to these problems have been employed by different kinds of plants.

16.3 Mosses

In mosses, the gametophyte generation is the one commonly seen. Male gametophytes support the male reproductory organs, called the antheridia, which contain sperm. A female moss plant contains the female reproductory organs, called archegonia, each of which contains a single egg. In spring, rains facilitate the transport of swimming sperm from the male plants to the archegonia of the female plants. Mitotic division of the resulting zygotes gives rise to the sporophyte generation. The sporophyte, which is dependent on the gametophyte, is diploid. Meiosis occurs in the sporanigium and yields haploid spores. Upon germination, the spores develop into a protonema and the game- tophytic generation begins again.

16.4 Ferns

In ferns, it is the sporophyte generation (diploid) that we commonly see. Meiosis produces spores which germinate in favorable circumstances and each grows into a structure called a prothallus, the haploid gametophyte generation. Each plant of this generation contains both antheridia and archegonia. Sperm released from the antheridia swim to an archegonium, usually on anotherprothallus, because the two kinds of sex organs generally do not mature simultaneously on a single prothallus. This circumstance results in cross-fertilization. Fertilizaton takes place within the archegonium and the new sporophyte generation is initiated. Note both the gametophyte and sporophyte are capable of independent existence.

16.5 Gymnosperms Angiosperms

The gymnosperms, including pine and spruce trees, produce microspores, which germinate to form the male gametophyte generation, and megaspores, which develop into the female gametophyte generation. Gymnosperm spores, in contrast to a single spore of mosses and ferns, cannot be agents of dispersal. The only place the windblown pollen grain (a four-celled structure which developed from the microspore) of a gymnosperm can germinate is on a female reproductive organ (or cone) of the same or a different plant. In gymnosperms, the function of dispersal is taken over by the seed, which can be carried by the wind. In the gymnosperms the gametophyte is now dependent on the sporophyte for nourishment, and has become little more than a reproductive mechanism.

16.6 The Flower and its Pollination

In the angiosperms, microspores and megaspores are produced in flowers. The microspores are produced in the stamens, and the megaspores are produced in the pistils. The pistil contains the ovary, in which the mature female gametophyte generation is produced. The stamen produces the (male) pollen grains. Wind or animals are usually the agents of transport of the pollen grains to the pistils, a process called pollination.

On reaching the female organ of a flower, a pollen grain germinates into a pollen tube, which contains two nuclei—the tube nucleus and the generative nucleus. As the tube starts to grow down into the ovary, the generative nucleus divides by mitosis and forms two sperm nuclei. The pollen tube enters the ovule through the micropyle and one sperm nucleus fuses with an egg and forms a zygote ($2n$). The other sperm nucleus fuses with the two polar nuclei to form an endosperm cell ($3n$).

16.7 The Seed

When the zygote nucleus and the endosperm nucleus divide by mitosis, a seed is formed. The seed consists of: (1) a plumule, consisting of two embryonic leaves and a terminal bud; (2) structures that will develop into stem and root; (3) cotyledons, which store food used by germinating seed. The seed is a dormant embryo sporophyte with stored food and protective coats. Its two functions are dispersal of the species and maintenance of the species over periods of unfavorable climatic conditions.

16.8 Seed Dispersal: The Fruit/16.9 Germination

A fruit is a development of the ovary (sometimes along with other flower parts as well) of a plant. It contains the seeds and so aids in the dispersal of the species. Germination is the resumption of growth of new individuals in a plant species. Proper temperature, moisture, and oxygen are necessary for seed germination to occur. But what is proper for one species is not necessarily proper for another. A period of dormancy also is required for the germination of many seeds For example, a period of cold is required for apple seeds to germinate.

16.10 Asexual Reproduction in Plants

Asexual reproduction is an alternative form of reproduction in most plants and has its own particular advantages and disadvantages. Stems are used most frequently for asexual propagation, although virtually all plant organs have been exploited for this purpose. This form of reproduction lacks the advantage of providing new genetic combinations.

16.11 Summary

The plant kingdom exhibits many mechanisms for sexual reproduction. In the primitive plants, the gametophyte is often dominant, but through a transition to angiosperms, it is the sporophyte that predominates. However, in all cases meiosis and fertilization bring about new genetic combinations, as well as reproducing the species. Asexual reproduction only provides the latter.

B. KEY TERMS

sexual reproduction	germination
asexual reproduction	tube cell
alternation of generations	generative cell
gametophyte	stamens
sporophyte	pistil
antheridia	imperfect/perfect flowers
archegonia	monoecious
sporangium	dioecious
protonema	cotyledon

leaves (fronds)	plumule
rhizome	hypocotyl
microspores (microsporangium)	radicle
megaspores (megasporangium)	germination
cones	apomixis
endosperm	clone

II. MASTERING THE CHAPTER

A. LEARNING OBJECTIVES

When you have mastered the material in this chapter, you should be able to:

1. Define all key terms.

2. Cite the two problems to be solved by sexually reproducing land plants.

3. Distinguish between the gametophyte and sporophyte generations in plants.

4. List the three main structures of the mature sporophyte generation of mosses, and briefly describe the function of each. Repeat for ferns.

5. Distinguish between a gymnosperm and an angiosperm, giving an example of each.

6. Compare the dispersal structure of ferns, for instance, with that of a spruce tree.

7. Cite the two most "important" cells in the female gametophyte of flowering plants and describe the function of each.

8. Describe what happens to a pollen grain from the time it reaches the stigma of flower until a seed begins to form.

9. List the three main components of a seed.

10. List the two chief functions of seeds.

11. List the three—and sometimes four—conditions necessary for seed germination.

12. List seven organs used in asexual reproduction.

B. REVIEWING TERMS AND CONCEPTS

1. Sexual reproduction involves the two processes of of (a) _____ and (b) _____. In the former, the nuclei of two (c) _____ fuse, raising the chromosome number from (d) _____ to (e) _____. In the latter process, the chromosome number is reduced again from the (f) _____ to the (g) _____ condition.

2. In plants, fertilization and meiosis divide the life of the organism into (a) _____ distinct generations: (1) The (b) _____ generation begins with a spore produced by meiosis. The spore and all cells derived from it are (c) _____ in genetic consititution. Gametes are produced by this generation. When two gametes fuse, the (d) _____ generation begins. This generation thus starts with a (e) _____ and it contains the (f) _____ number of chromosomes.

3. In mosses, the (a) _____ generation is the one commonly seen. The male gametophyte supports the (b) _____ reproductory organs, called antheridia and containing (c) _____. A female moss plant contains the female reproductory organs, called (d) _____ and containing a single (e) _____ each.

4. In ferns, it is the (a) _____ generation that we commonly see. Windblown spores germinate in favorable conditions and each grows into a structure called a (b) _____, the (c) _____ generation. Each plant of this generation contains both (d) _____ and _____. (e) _____ released from the antheridia swim to an (f) _____, usually on another prothallus, because the two kinds of sex organs generally do not mature simultaneously on a single (g) _____. This circumstance results in cross- (h) _____. Fertilization takes place within the (i) _____ and the new (j) _____ generation is initiated.

5. The gymnosperms, including pine and spruce trees, produce (a) _____, which germinate to form the male gametophyte generation, and (b) _____, which develop into the female gametophyte generation.

6. In the angiosperms, microspores are produced in (a) _____. The microspores are produced in the (b) _____, and the megaspores are produced in the (c) _____. The pistil contains the (d) _____ in the ovules of which the mature female gametophyte generation is produced. The (e) _____ produces the (male) pollen grains. Wind or animals usually are the agents of transport of the pollen grains to the pistils, a process called (f) _____.

7. On reaching the female organ of a flower, a pollen (a) _____ germinates into a pollen (b) _____, which contains (c) _____ nuclei—the (d) _____ nucleus and the _____ nucleus.

8. A seed consists of: (1) a (a) _____, consisting of two embryonic leaves and a terminal bud; (2) structures that will develop into stem and (b) _____; (3) (c) _____, which

store food used by the germinating seed.

9. A (a) _____ is a development of the ovary of a plant and sometimes accessory structures. It contains the seeds and so aids in the dispersal of the species.

10. (a) _____ is the resumption of growth of new individuals in a plant species. Proper temperature, moisture, and oxygen are necessary for seed (b) _____ to occur. But what is proper for one species is not necessarily proper for another. A period of (c) _____ also is required for the germination of many seeds.

11. Asexual reproduction is not a primitive form or substitute for (a) _____ reproduction, but rather an (b) _____ form. (c) _____ are the most widely used organ for asexual re-production. Some special names for stems that are used for this pur-pose are (d) _____ if above ground, and (e) _____, _____, _____, and _____ if below ground. Another form of asexual reproduction, (f) _____, occurs when a diploid egg develops without being (g) _____. Cuttings or (h) _____ are commonly used methods to propagate popular nursery products.

C. TESTING TERMS AND CONCEPTS

1. In gymnosperms, the mature male gametophyte generation is formed from the
    a) prothallus.
    b) megaspores.
    c) microspores.
    d) endosperm.

2. In flowering plants, the megaspores are produced in the
    a) stamen.
    b) pistil.
    c) cotyledons.
    d) hypocotyl.

3. The male reproductory organ in mosses is the
    a) antheridium.
    b) archegonium.
    c) sporophyte.
    d) microspore.

4. In mosses, the gametophyte generation originates in the
    a) protonema.
    b) archegonia.
    c) sporangium.
    d) prothallus.

5. Fruits are produced by
    a) gymnosperms.
    b) angiosperms.
    c) mosses.
    d) ferns.

6. In gymnosperms, dispersal is a function of the
    a) seed.
    b) endosperm.
    c) microspore.
    d) rhizome.

7. Food is stored in a seed in the
    a) hypocotyl.
    b) plumule.
    c) cotyledons.
    d) radicle.

8. The haploid generation in plants is the
    a) sporophyte.
    b) heterophyte.
    c) haplophyte.
    d) gametophyte.

9. The endosperm of flowering plants is produced by the cell containing the
    a) polar nuclei.
    b) egg.
    c) sperm.
    d) cotyledon.

10. In ferns, fertilization takes place within the
    a) prothallus.
    b) archegonium.
    c) foot.
    d) protonema.

11. The megaspores in gymnosperms form the
    a) female gametophyte generation.
    b) male sporophyte generation.
    c) female sporophyte generation.
    d) male gametophyte generation.

12. Seed germination in some instances may require all of the following, except
    a) a period of dormancy.
    b) exposure to light.
    c) freezing and thawing.
    d) temperature shock greater than 40°C.

13. In which group of plants is the sporophyte dependent on the gametophyte for nourishment?
    a) mosses
    b) ferns
    c) gymnosperms
    d) angiosperms

14. Which of the following is not an organ used for asexual reproduction?
    a) corn
    b) tuber

c) petal
d) bulb

15. Grafting is a
    a) type of dormancy.
    b) form of sexual reproduction.
    c) naturally occurring event.
    d) means of asexual propagation.

D. UNDERSTANDING AND APPLYING TERMS

Decide whether the following statements are true (T) or false (F).

_____ 1. The fusion of two gametes gives rise to the gametophyte
         generation.

_____ 2. The prothallus is an independent, autotrophic plant.

_____ 3. In flowering plants, the gametophyte is the dominant genera-
         tion.

_____ 4. Most of our grain crops are angiosperms.

_____ 5. In tropical plants there is no need for plants to produce
         seeds since there is no winter (non-growing season).

      6. a) What is an imperfect flower? _____
            _____
            _____

         b) If the imperfect flowers are on separate plants, the con-
            dition is called _____.

         c) This condition ensures _____ -pollination.

     The following represent typical life cycles for several kinds of
plants. Fill in, where appropriate, the following letters or words:
(N) or (2N), M_itosis___ or M_eiosis___. [dependent] or [independent].

7. Moss

Sporophyte (a) ( )   (b) M_____   Spores (c) ( ) Gametophyte (d) ( )
(j) [   ]   (e) [   ]
                              Eggs (g) ( )
      Zygote (i) ( )   (f) M_____
                           Sperm (h) ( )

8. Fern

Sporophyte (a) ( )   (b) M_____   Spores (c) ( )   Prothallus (d) ( )
(k) [   ]

                eggs (h) ( )
  Zygote (j)( )   (g) M_____   Gametophyte (e) ( )
                sperm (i) ( )   (f) [   ]

149

9. Gymnosperm

Sporophyte (a) ( )

(m) [   ]

male cone (b) M_____ microspores (d) ( )

female cone (c) M_____ megaspores (e) ( )

male gametophyte (f) ( )
(h) [   ]

seed (l) ( )   (k) M_____   zygote (j) ( )

female gametophyte (g) ( )
(i) [   ]

10. Angiosperm

Sporophyte (a) ( )
(o) [   ]

(n) M_____

anther microspore mother cell (b) ( )   (d) M_____

pistil megaspore mother cell (c) ( ) (e) M_____

generative cell (j) ( )   male gametophyte (f) ( )

embryo (m) ( )   zygote (l) ( )

egg cell (k) ( )   female gametophyte (g) ( )
(i) [   ]

ANSWERS TO CHAPTER EXERCISES

Reviewing Terms and Concepts

1. a) fertilization b) meiosis c) gametes d) haploid e) diploid
   f) diploid g) haploid

2. a) two b) gametophyte c) haploid d) sporophyte e) zygote
   f) diploid

3. a) gametophyte b) male c) sperm d) archegonia e) egg

4. a) sporophyte b) prothallus c) gametophyte d) antheridia,
   archegonia e) Sperm f) archegonium g) prothallus h) fertilization i) archegonium j) sporophyte

5. a) microspores b) megaspores

6. a) flowers b) stamens c) pistils d) ovary e) stamen f) pollination

7. a) grain b) tube c) two d) tube, generative

8. a) plumule b) root c) cotyledons

9. fruit

10. a) Germination b) germination c) dormancy

11. a) sexual b) alternative c) Stems d) stolons e) rhizomes,
    bulbs, corms, tubers f) apomixis g) fertilized h) graftings

Testing Terms and Concepts

1. c	6. a	11. a
2. b	7. c	12. d
3. a	8. d	13. a
4. a	9. a	14. c
5. b	10. b	15. d

Understanding and Applying Terms and Concepts

1. F 2. T 3. F 4. T 5. F
6. a) A flower that has either stamens or pistils, but not both.
    b) dioecious c) cross

7. a) 2N b) meiosis c) N d) N e) independent f) mitosis g) N
    h) N i) 2N j) dependent

8. a) 2N b) meiosis c) N d) N e) N f) independent g) mitosis
    h) N i) N j) 2N k) independent

9. a) 2N b) meiosis c) meiosis d) N e) N f) N g) N h) dependent
    i) dependent j) 2N k) mitosis l) 2N m) independent

10. a) 2N b) 2N c) 2N d) meiosis e) meiosis f) N g) N
    h) dependent i) dependent j) N k) N l) 2N m) 2N n) mitosis
    o) independent

17

REPRODUCTION IN ANIMALS

I. REVIEWING THE CHAPTER

A. CHAPTER HIGHLIGHTS

17.1 Asexual Reproduction in Animals

Asexual reproduction is much less common in animals than in plants.
Some worms are capable of breaking into several fragments, each of
which can grow into a new individual. In others, buds develop as
growths on the body of the parent.

17.2 The Formation of Gametes

Animals, unlike plants, do not exhibit an alternation of diploid
and haploid generations. Fertilization is still preceded by meiosis,
but the products of meiosis are the gametes themselves. In all animals,
distinctly different gametes, called heterogametes, are produced. One
gamete, the sperm, is motile and small, and one gamete, the egg, is
filled with food reserves. A sperm cell, produced in the testes, consists of a head, two centrioles, and a tail. An egg, produced in an
ovary, is the sperm's counterpart for the reproductive process.

17.3 Bringing the Gametes Together

Among aquatic animals, fertilization occurs directly in the
water after each parent releases its gametes, usually in close proxi-
mity. When life forms migrated to the land, problems of reproduction
arose. How to bring the gametes together? Special reproductive organs
evolved permitting direct introduction of sperm from the male into the
female in a process called copulation.

17.4 Fertilization

The process of fertilization begins when a sperm becomes attached
to the egg, and is followed by a reorganization of egg cytoplasm. In
many cases, the completion of the second meiotic division does not
occur until after fertilization. The final event occurs when the
enlarged "pronuclei" of both gametes fuse to form a diploid nucleus.

17.5 Care of the Young

An important relationship exists between the number of gametes,
especially eggs, produced by different animals and the care given the
embryos during development. For a population to continue at a stable
size, an average of two offspring should reach maturity for every two

parents.

The eggs of most aquatic animals are relatively small. In a
large number of species, the organism that hatches from the egg is not
a miniature replica of the adult but a larval stage quite different in
appearance. Usually the larvae are free-swimming and feed on microscopic plant and animal life called plankton. In so doing, they acquire
additional food materials for further growth. After a period of growth,
the larvae undergo metamorphosis and the body structure is reorganized
on the plan of the adult.

Terrestrial animals all need some means of protecting the developing
embryo from the drying action of the air. In some organisms, such as
birds, insects, and reptiles, a hard protective coating is provided
around the egg to assure a moist internal environment. In such cases,
fertilization generally is achieved within the body of the female, thus
assuring sperm an environment conducive to mobility toward the egg.

Birds, reptiles, and mammals all produce extraembryonic membranes
consisting of: (1) a yolk sac, which connects the embryo with its major
source of food; (2) the amnion, which encloses the embryo in a fluid-
containing cavity; (3) the chorion, which lines the inner surface of
the egg shell and functions in gase exchange; and (4) the allantois,
which serves as a reservoir for the metabolic wastes excreted by the
embryo and later functions in gas exchange. Mammals that produce little
yolk retain the embryo within the reproductive tract where the extra
embryonic membranes penetrate the walls of the uterus.

17.6 The Sex Organs of the Male

The sexual apparatus of the human male has two major reproductive
functions—the production of sperm cells and the delivery of these cells
to the reproductive tract of the female. In males, the testes produce
sperm and testosterone, the chief male sex hormone, which is responsible
for the development of the secondary sex characteristics of males. Two
pituitary hormones, luteinizing hormone (LH) and follicle stimulating
hormone (FSH), also play important roles in sperm production. These
two, in turn, are influenced by releasing hormones which are products
of the hypothalamus.

17.7 The Sex Organs of the Female

The human female reproduction physiology is considerably more
complex than that of the male. She must not only manufacture sex cells
(eggs) but be equipped to: (1) receive sperm from the male, (2) provide
the right conditions for fertilization to occur, and (3) nourish the
developing baby. Also, for fertilization to be effective, careful
coordination is required, i.e., the uterus must be at the proper stage
to receive the fertilized egg. To accomplish this synchrony of ovarian
and uterine events, estrogen and progesterone (both produced in the
ovary) LH and FSH (both produced in the pituitary), and gonadotropin
releasing hormone (an hypothalamus product) are required.

17.8 Copulation and Fertilization

For fertilization to occur, sperm must be deposited in the vagina fairly near to the time of ovulation, which is when the egg is released and migrates down into the Fallopian tube. The sperm are contained in a mixture of fluids which are important in the fertilization process. These fluids are produced by the seminal vesicles and prostate gland.

17.9 Pregnancy and Birth

Fertilization occurs and embryonic development begins when the fertilized egg is still within the Fallopian tube. After implantation development continues in two major divisions of cells and tissues: (1) the embryo proper, which ultimately becomes the baby, and (2) the various extraembryonic membranes, which play a number of vital roles.

During the first two months of pregnancy the basic structure of the baby, now called the embryo, is being formed. This involves cell division, cell migration, and the development of cells into the types found in the adult organism. After this period, all the systems of the baby, now called the fetus, have been established in a rudimentary way. If during the first five months or so of pregnancy, progesterone secretion should be stopped, uterine contractions begin and the embryo or fetus is aborted.

Labor begins with hormonally induced uterine contractions. Labor opens the cervix, ruptures the amnion, and eventually expels the baby. At this time the umbilical cord can be severed and the infant begins breathing, which entails a major switchover in the circulatory system. Within two or three days following birth, the mother's breasts begin to secrete milk.

17.10 Reproductive Engineering: The Prospects

Present and foreseeable techniques put humans on the edge of controlling human and animal reproductive processes. It is currently possible to (1) freeze semen for later use, (2) transplant animal embryos to "foster mothers," and (3) use amniocentesis to determine the sex or genetic/chromosomal abnormalities of a fetus. Epitomizing these feats in humans was the birth of a child conceived by "test tube fertilization" of an egg that was subsequently transplanted into the mother's uterus. In the future it may be possible to "clone" higher animals and humans just as we have done with some lower ones.

17.11 Summary

Sexual reproduction in animals involves gamete production and fertilization. These processes are controlled by elaborate hormone mechanisms. Many methods for ensuring the survival of some of the zygotes have been exploited. Some species simply produce thousands of free-floating offspring; others have developed elaborate methods for retaining only one or two potential offspring within the mother.

B. KEY TERMS

fragmentation	allantois
budding	testosterone
proglottids	luteinizing hormone
parthenogenesis	follicle stimulating hormone
heterogametes	releasing hormones
spermatogonium	estrogen
spermatocyte (primary and secondary)	progesterone
spermatids	negative feedback
sperm	menopause
oogonium	menstruation
oocyte (primary and secondary)	corpus luteum
egg	seminal vesicles
polar body	prostate gland
testes/ovaries	implantation
animal/vegetal pole	extra embryonic membranes
copulation	amnion
pronucleus	placenta
gray crescent	umbilical cord
follicle	oxytocin
yolk sac	prostaglandins
amnion	in vivo/in vitro
chorion	amniocentesis

II. MASTERING THE CHAPTER

A. LEARNING OBJECTIVES

When you have mastered the material in this chapter, you should be able to:

1. Define all key terms.

2. Beginning with spermatogonia, list the steps leading to the formation of haploid sperm cells.

3. Distinguish between the production of oogonia in a human female and in a female frog.

4. Distinguish between internal and external fertilization, and cite an example of each.

5. Cite six events occurring between sperm and egg during fertilization.

6. Cite adaptations (or lack thereof) in at least four different species for care of the newly born young.

7. Cite and give the functions of the four extraembryonic membranes of the shelled egg.

8. State the two functions of the testes.

9. List the four things a successfully reproducing human female must be capable of doing.

10. State the functions of FSH and LH in both males and females.

11. State three functions of the hormone progesterone.

12. Cite the two major divisions of cells and tissues produced in the blastocyst.

13. Cite at least three functions of the placenta.

14. Cite at least four methods of contraception, and describe how each works.

B. REVIEWING TERMS AND CONCEPTS

1. Asexual (a) _____ is found in animals, but occurs (b) _____ (more/less) frequently than in plants. Some worms do this through (c) _____, whereas others form (d) _____, which are outgrowths of the parent. When an egg develops without being fertilized the process is called (e) _____.

2. Animals, unlike plants, do not exhibit alternation of (a) _____ and _____ generations. Fertilization is still preceded by meiosis, but the products of meiosis are the (b) _____ themselves. In all animals, distinctly different gametes, called (c) _____, are produced. One gamete, the (e) _____, is filled with food reserves.

3. A sperm cell, produced in the (a) _____, consists of a (b) _____, two (c) _____, and a (d) _____. An egg, produced in an (e) _____, is the sperm's counterpart.

4. The process of (a) _____ begins when a sperm becomes

156

attached to the egg. The sperm (b) _____ eventually fuses with the egg. The final event in fertilization occurs when the sperm (c) _____ initiates the formation of a (d) _____ upon which first the sperm (e) _____ and then the egg (f) _____ become arranged. At this point a zygote with the (g) _____ number of chromosomes has been formed.

5. Birds, reptiles, and mammals all produce extraembryonic membranes consisting of: (1) a (a) _____ sac, which connects the embryo with its major source of (b) _____; (2) the (c) _____, which encloses the embryo in a fluid-containing cavity; (3) the (d) _____, which lines the inner surface of the egg shell and functions in (e) _____ _____; and (4) the (f) _____, which serves as a reservoir for metabolic wastes.

6. In males, the (a) _____ produce sperm and (b) _____, the chief male sex hormone, which is responsible for the development of the (c) _____ sex characteristics of males.

7. For fertilization to occur, (a) _____ must be deposited in the (b) _____ fairly near to the time of (c) _____, which is when the egg is released and migrates down into the (d) _____ tube. The uterus must be in the proper stage for (e) _____ to occur. The sequencing of ovulation and uterus preparation is under the control of the two important ovarian hormone, (f) _____ and _____. These are influenced by two pituitary hormones called (g) _____ and _____.

8. During the first (a) _____ months of pregnancy the basic structure of the baby, now called the (b) _____, is being formed. This involves cell (c) _____, cell (d) _____, and the development of cells into the types found in the adult organism. After this period, all the systems of the baby, now called the (e) _____, have been established in a rudimentary way. If during the first (f) _____ months or so of pregnancy, (g) _____ secretion should be stopped, uterine (h) _____ begin and the embryo or fetus is (i) _____.

9. The manipulation of reproductive processes has taken or will take several directions. One current practice is to freeze (a) _____ for future use. Test tube (b) _____ has also been accomplished. Prenatal diagnoses, or (c) _____ as it is called, also serves to give people control over the (d) _____ (kinds/numbers) of children they can have. It also appears possible to transplant (e) _____ from one animal into many receptive cells and thus (f) _____ the "donor."

C. TESTING TERMS AND CONCEPTS

1. Each primary spermatocyte gives rise to
   a) two sperm cells.
   b) four sperm cells.
   c) eight sperm cells.
   d) an indefinite number of sperm cells.

157

2. The extraembryonic membrane that serves as a reservoir for metaboli wastes is the
a) chorion.
b) yolk.
c) amnion.
d) allantois.

3. The uterus is inhibited from contractions by the hormone
a) progesterone.
b) testosterone.
c) estrogen.
d) oxytocin.

4. The extraembryonic membrane that connects the embryo with its food source is the
a) amnion.
b) allantois.
c) chorion.
d) yolk sac.

5. Primary oocytes in human females are
a) haploid.
b) diploid.
c) polyploid.
d) semidiploid.

6. The eggs of mammals generally lack a
a) yolk.
b) shell.
c) chorion.
d) follicle.

7. The corpus luteum produces the hormone
a) testosterone.
b) estrogen.
c) progesterone.
d) luteinizing hormone.

8. In humans, fertilization normally occurs in the
a) follicle.
b) Fallopian tube.
c) uterus.
d) cervix.

9. The placenta secretes
a) estrogen.
b) progesterone.
c) LH.
d) FSH.

10. Growth of an unfertilized egg into a new organism is called
a) amniocentesis.
b) ovulation.
c) parthenogenesis.
d) nonfertilization reproduction.

11. In a human female, meiosis II is usually completed in the
    a) ovary.
    b) follicle.
    c) Fallopian tube.
    d) uterus.

12. The cell(s) of a potential egg that does not develop is called a(n)
    a) gray crescent.
    b) allantois.
    c) blastocyst.
    d) polar body.

13. Animals that go through several quite different appearing stages
    between zygote and adulthood are said to undergo
    a) parthenogenesis.
    b) orthogenesis.
    c) implantation.
    d) metamorphosis.

14. Semen contains all of the following, except:
    a) sperm.
    b) at least one sugar.
    c) oxytocin.
    d) secretions of the prostate gland.

15. A major problem of sexual reproduction on land was
    a) keeping gametes moist.
    b) finding a mate.
    c) preventing predators from eating the eggs.
    d) the lack of a placenta.

D. UNDERSTANDING AND APPLYING TERMS AND CONCEPTS

Decide whether the following statements are true (T) or false (F).

_____ 1. Under normal circumstances there is no mixing of maternal
         and fetal blood.

_____ 2. A woman should be most concerned about teratogens during her
         last two months of pregnancy.

_____ 3. Progesterone and estrogen are steroid hormones—thus they are
         classified as lipids.

_____ 4. Follicle stimulating hormone is produced in females only.

_____ 5. Meiosis (specifically the fate of chromosomes) is essentially
         the same in both plants and animals.

6. When is conception most likely to occur in humans? _____

7. Castration prevents the development of many adult characteristics in males. Why? _____

159

8. What would happen if the ovaries were removed from a human female before puberty? _____
_____

9. Explain how you could determine if a woman was pregnant with a Down's syndrome male. _____
_____
_____
_____

10. How could the cloning of humans be used to help shed light on the "nature versus nuture" (heredity versus environment) debate over personality and behavior? _____
_____
_____
_____

11. Place in correct order the major tissues and organs through which a human sperm passes. Begin with the spermatogonium and finish when it fertilizes an egg. _____
_____
_____
_____

12. Place the correct hormones in the blanks.

PITUITARY

initiates the
completion of
egg development

(a)_____

causes follicle to
release egg

(d)_____

OVARY

initiates uterine
wall thickening

(b)_____

keeps uterus ready
to receive ferti-
lized egg

(e)_____

UTERUS

causes lowering
of FSH and LH

(c)_____

13. The action of estrogen is an example of _____ feedback control.

ANSWERS TO CHAPTER EXERCISES

Reviewing Terms and Concepts

1. a) reproduction b) less c) fragmentation d) buds e) parentheno-genesis

2. a) diploid, haploid b) gametes c) heterogametes d) sperm e) egg

3. a) testes b) head c) centrioles d) tail e) ovary

4. a) fertilization b) nucleus c) centriole d) spindle e) chromo-somes f) chromosomes g) diploid

5. a) yolk b) food c) amnion d) chorion e) gas exchange f) allantois

6. a) testes b) testosterone c) secondary

7. a) sperm b) vagina c) ovulation d) Fallopian e) implantation f) estrogen, progesterone g) FSH, LH

8. a) two b) embryo c) division d) migration e) fetus f) five g) progesterone h) contractions i) aborted

9. a) sperm (semen) b) fertilization c) amniocentesis d) kinds e) nuclei f) clone

Testing Terms and Concepts

1. b	6. b	11. c
2. d	7. c	12. d
3. a	8. b	13. d
4. d	9. b	14. c
5. b	10. c	15. a

Understanding and Applying Terms and Concepts

1. T 2. F 3. T 4. F 5. T

6. about 14 days after the onset of menstruation.

7. The testes produce testosterone. The testosterone is responsible for secondary sex characteristics.

8. She would not develop the normal secondary sex characteristics— breast development, hair patterns, etc.

9. (1) remove some amniotic fluid (it contains cells of the fetus) (2) do a karyotype (3) look for an extra number 21 chromosome

(4) look for a <u>Y</u> chromosome

10. The clones would be genetically identical; thus the hereditary factor or component would be "controlled."

11. seminiferous tubule, epididymis, vas deferens, penis (urethra), vagina, cervix, uterus, Fallopian tube.

12. a) FSH b) estrogen c) estrogen d) LH e) progesterone

13. negative

18

EARLY DEVELOPMENT

I. REVIEWING THE CHAPTER

A. CHAPTER HIGHLIGHTS

18.1 Stages in the Development of the Adult/18.2 Cleavage/18.3
      Morphogenesis/18.4 Differentiation

     Fertilization starts a series of elaborate and well-organized
changes that eventually give rise to a new adult. We use the term
development to describe these changes, which occur in four major
stages: (1) During cleavage, the zygote nucleus undergoes mitotic
divisions and daughter nuclei usually become partitioned off in
separate cells. (2) During morphogenesis, the many new cells produced
by cleavage continue dividing but also move about and organize themselves into distinct layers and masses. (3) During differentiation,
the cells of the developing embryo begin to take on specialized
structures and functions characteristic of the adult. (4) During
growth, the organism becomes larger by continued cell division and/or
enlargement. Growth depends on the intake of more matter and energy
than is needed simply to maintain the organism's normal functions.

     Although there are great variations in the details of these
stages, some generalities can be made: (1) Cleavage provides a stock-
pile of cells for later use and establishes the normal relationship
between the nucleus and cytoplasm. (2) Morphogenesis involves the
positioning of cells as well as their taking on specialized shapes and
internal organization for the tasks they are to perform. (3) Differentiation marks the "point of no return," i.e., after this point
cells are committed to be of a certain type.

18.5 Evidence that Differentiating Cells Retain the Entire Genome

     Differentiating cells all contain the entire genome. There are
several lines of evidence that this is so. For example, the DNA content of virtually all differentiated cells in the same organism is the
same; and differentiated cells contain the same number of chromosomes.
In one instance, a fully differentiated carrot root cell was observed
to divide repeatedly and give rise to an entire new plant. Identical
twins originate from the same zygote after one or two cleavages when
the zygote separates into two parts. Identical twins always are of
the same sex and are identical in their body chemistry. The finding
that some genes "jump" may help explain how some of the genome becomes
permanently "turned off."

18.6 Cytoplasmic Factors Affecting Gene Expression During Differentiation

Since differentiating cells all contain the entire genome, how can we account for some cells differentiating in one way, and others in a completely different way? We are faced with the problem of differential gene activity. Certain experiments beginning with those of Spemann demonstrated that certain properties of the cytoplasm influence what a nucleus can or cannot do. Genes, then, may be selectively activated by a particular cytoplasmic environment. The "puffing" of giant chromosomes during differentiation also provides evidence that gene expression is influenced by cytoplasmic factors.

18.7 Extracellular Factors Affecting Gene Expression During Differentiation

Spemann also was first to discover that the pattern of development of cells is influenced by the activities of other cells. Cells from the dorsal lip of the blastopore, when transplanted to the ventral side of a normal gastrula, resulted in the formation there of a second notochord, neural groove, and ultimately a second head. The process is termed induction and the chemical agents responsible for it are termed inducers. Most inducers are macromolecules but some may be small, such as vitamin A and certain steroids.

The differentiation of cells seems also to be influenced by inhibitory substances from adjacent cells. Fully differentiated cells have been observed to inhibit undifferentiated cells from differentiating into cells like the differentiated ones. Differentiated tissues secrete substances, called chalones, that inhibit mitosis of the cells os that tissue. It also is becoming increasingly clear that the course of embryonic development depends not just on the induction or inhibition of some cells by other cells, but on interactions occurring between both groups of cells.

The ability of two like cells to recognize each other as they migrate from one place to another in the developing embryo depends on the presence of specific sticky substances on their surfaces. Cancer cells lack this sticky substance and so do not recognize other cells or stop their migration and cell division. Thus they escape from the controls that regulate the normal cells of the body. In normal cells, it now begins to look as though the reacting differentiating tissue contains all the information required to specialize and simply needs some fairly unspecific influence to enable it to do so.

18.8 The Reversibility of Differentiation

The question regarding the reversibility of differentiation can not be given a clear-cut answer. In plants, examples can be cited where the dedifferentiation of cells can be followed by complete new individuals arising from them. In animals, the healing of a wound indicates dedifferentiation is followed by differentiation. However, it is not clear if dedifferentiated cells can give rise to cell types other than their own kind.

18.9 Summary

All the cells of an adult come from a single fertilized egg, and contain the same genome. During the many stages of development it appears that various signals stimulate and/or inhibit different portions of the genome in the different cells.

B. KEY TERMS

cleavage	mesoderm
morphogenesis	ectoderm
differentiation	coelom
growth	notochord
embryo	induction
blastocoel	organizer
blastopore	chalone
blastula	dedifferentiation
gastrulation	somites
endoderm	

II. MASTERING THE CHAPTER

A. LEARNING OBJECTIVES

When you have mastered the material in this chapter, you should be able to:

1. Define all key terms.

2. List the four major stages in development, and describe briefly what happens at each stage.

3. Cite the two chief things accomplished by cleavage in the developing embryo.

4. Distinguish between the vegetal-pole and animal-pole cells, citing at least two body parts deriving from each.

5. List two organs or tissues deriving from each of the three germ layers.

6. Cite two lines of evidence indicating that each differentiated cell contains the entire genome of the zygote.

7. Cite two examples of the cytoplasm exerting influence what a nucleus can or cannot do.

8. Cite evidence on which the theory of induction is based.

B. REVIEWING TERMS AND CONCEPTS

1. Fertilization starts a series of elaborate and well-organized changes that eventually give rise to a new adult. We use the term (a) _____ to describe these changes, which occur in four major stages: (1) During (b) _____, the zygote (c) _____ undergoes mitotic divisions. (2) During (d) _____, the many new cells produced by cleavage continue dividing, move about, and organize themselves into distinct layers and masses. (3) During (e) _____, the cells of the developing embryo begin to take on specialized (f) _____ and _____ characteristic of the adult. (4) During (g) _____, the organism becomes larger by continued cell division and/or enlargement.

2. Differentiating cells all contain the entire (a) _____. There are several lines of evidence that this is so. For example, the (b) _____ content of virtually all differentiated cells of the same organism is the same; and differentiated cells contain the same number of (c) _____.

3. Certain experiments demonstrated that certain properties of the (a) _____ influence what a nucleus can or cannot do. Genes, then, may be selectively activated by a particular (b) _____ environment. The pattern of development of cells also is influenced by the activities of other cells. Cells from the dorsal lip of the (c) _____, when transplanted to the ventral side of a normal gastrula, resulted in the formation there of a second (d) _____, neural groove, and ultimately a second (e) _____. The process is termed (f) _____ and the chemical agents responsible for it are termed (g) _____.

4. The differentiation of cells seems also to be influenced by inhibitory substances from (a) _____ cells. Differentiated tissues secrete substances, called (b) _____, that inhibit (c) _____ of the cells of that tissue. It also is becoming increasingly clear that the course of (d) _____ development depends not just on the induction or inhibition of some cells by other cells, but on (e) _____ occurring between both groups of cells. In normal cells, it now begins to look as though the reacting differentiating tissue contains (f) _____ (all/some of) the information required to specialize and simply needs some fairly unspecific influence to enable it to do so.

C. TESTING TERMS AND CONCEPTS

1. Each differentiated cell of the body retains
   a) a selected, partial genome present in the zygote.
   b) the full genome.
   c) a genome that differs from one individual to the next.
   d) a genome specific for a particular tissue.

2. The cells of the developing embryo begin to take on specialized structure and function during
   a) morphogenesis.
   b) growth.
   c) differentiation.
   d) cleavage.

3. The ability of two like cells to recognize each other depends on the presence of specific substances
   a) on their surfaces.
   b) in their nuclei.
   c) in their cytoplasm.
   d) called chalones.

4. Organized development of an organism proceeds by the process of
   a) reduction.
   b) induction.
   c) preduction.
   d) transduction.

5. A fully differentiated plant cell has the potential of producing
   a) an entire new individual.
   b) undifferentiated cells.
   c) only more cells like itself.
   d) no more cells.

6. Fully differentiated cells produce inhibitory substances called
   a) chalones.
   b) somites.
   c) polytene.
   d) coelom.

7. The zygote nucleus starts a series of mitotic divisions during
   a) morphogenesis.
   b) cleavage.
   c) differentiation.
   d) blastopore formation.

8. While passing through the first three stages of development, an animal is called a(n)
   a) zygote.
   b) embryo.
   c) fetus.
   d) blastula.

9. The amount of DNA in a woman's liver cell is
   a) the same as that found in the liver cells of all mammals.
   b) half as much as in one of her eggs.
   c) proportional to the size of that particular cell.
   d) the same as in one of her brain cells.

10. Several lines of evidence suggest that during development
    a) genes are lost from various tissues.
    b) chromosomes are lost from various tissues.

c) genes are repressed in various tissues.
d) chromosomes lose their identity.

11. The observation that killed cells could still induce properties of the organizer suggests that the inducer is
a) chemical.
b) electrical.
c) neither chemical nor electrical.
d) both chemical and electrical.

12. During which stage do cells of the developing embryo begin to take on specialized roles?
a) cleavage
b) morphogenesis
c) differentiation
d) induction

13. Genetically speaking, how a nucleus behaves may be determined by the
a) genes only.
b) whole organism.
c) influence of outside stimuli only.
d) cytoplasm.

14. Chromosomes and the genes on them can be derepressed by
a) hybrid derepression.
b) specialized differentiation.
c) hybrid differentiation.
d) somatic cell hydridization.

15. Development has been studied by all of the following techniques, except:
a) grafting.
b) microscopy.
c) biochemical.
d) electrolysis.

D. UNDERSTANDING AND APPLYING TERMS AND CONCEPTS

Use this key for 1 through 3.

        a) endoderm
        b) mesoderm
        c) ectoderm

1. The nervous system comes from: _____

2. Muscles arise from: _____

3. Hair arises from: _____

Decide whether 4 through 7 are true (T) or false (F).

_____ 4. The yolk of a chicken egg is the nucleus.

168

_____ 5. The "white" of a chicken egg is the cytoplasm.

_____ 6. By transplanting nuclei from one kind of cell to another, we learn nothing about cytoplasmic influences.

_____ 7. If you transplant a diploid nucleus from an apple tree to a nucleus-free cell of a pear tree, you would get an apple tree (assume that it grows and develops).

8. Why are organ transplants more successful between identical twins than fraternal twins? _____
_____
_____
_____

9. If a person loses a limb, is it theoretically possible to have a new one grow back? _____ Why? _____
_____
_____
_____

10. What is a reasonable explanation for the observation that human limbs do not grow back? _____
_____
_____

ANSWERS TO CHAPTER EXERCISES

Reviewing Terms and Concepts

1. a) development b) cleavage c) nucleus d) morphogenesis e) differentiation f) structures, functions g) growth

2. a) genome b) DNA c) chromosomes

3. a) cytoplasm b) cytoplasmic c) blastopore d) notochord e) head f) induction g) inducers

4. a) adjacent b) chalones c) mitosis d) embryonic e) interactions f) all

Testing Terms and Concepts

1. b	6. a	11. a
2. c	7. b	12. c
3. c	8. b	13. d
4. b	9. d	14. d
5. a	10. c	15. d

Understanding and Applying Terms and Concepts

1. c 2. b 3. c

4. F 5. F 6. F 7. F

8. Identical twins begin as one fertilized egg and therefore are genet-
    ically identical. This reduces the rejection problems that can
    arise from "foreign tissue."

9. Yes. The person would still have cells with nuclei that are the
    same as those in the original limb. In other words, the information
    to make the limb still exists.

10. After differentiation is complete, the genes that control that
    specific task are permanently turned off.

19

LATER DEVELOPMENT

I. REVIEWING THE CHAPTER

A. CHAPTER HIGHLIGHTS

19.1 Growth

Development includes growth. Most growth among animals takes place after the completion of morphogenesis and differentiation. For growth to occur, the rate of synthesis of complex molecules—proteins, for example—must exceed the rate of their breakdown. Growth thus involves gaining more material from the environment than is given back to it in the form of metabolic wastes. It is also more than the simple accumulation of matter, rather it is the production of components of the organism from the raw materials. Growth can be the increase in cell number, size, or both.

Several periods are distinguishable in growth. The first period is the lag period and is characterized by little or no actual growth. Next comes the exponential period, one of rapidly increasing growth. Finally there is a period of decelerating growth, a slowing process in which growth eventually ceases altogether. In plants, growth occurs in the localized areas of meristem tissue. In animals, growth occurs through all of the tissues and organs of the organism.

In humans, the growing regions of the bones become fully "ossified" during the late teens and early twenties. The cartilaginous matrix of these regions becomes replaced with a bony matrix and further skeletal growth ceases. In some animals, particularly those that go through metamorphosis, growth may not occur in all stages.

19.2 Regeneration

All organisms appear to have the ability to repair damaged parts of themselves through a process called regeneration; but various organisms have this ability to varying degrees. Generally, plants have the greatest powers of regeneration. Sponges, hydra, and planarian worms have the ability to regenerate an entire new organism from a part of themselves. Lobsters and salamanders can regenerate a missing leg, and lizards a lost tail.

Mature, differentiated cells of a given tissue synthesize and secrete an inhibitory substance called a chalone that prevents mitosis of the young cells of the same tissue. In the early stages of regeneration, there are no mature cells and thus there is no inhibition of cell division. As regenerated cells differentiate, chalone

171

production starts and gradually stops regenerative growth. Within a single organism, at least among the vertebrates, there seems to be a progressive loss of regenerative ability with increasing age.

19.3 The Biology of Cancer

Cancer is any unchecked proliferation of cells. It is an example of failures to control normal morphogenesis and differentiation. When the abnormal cells move to other parts of the body and begin to grow, the process is called metastasis. Some cancer cells also appear to be dedifferentiated, that is, they behave in some respects like embryonic cells. Cancer occurs most frequently in tissues where there is rapid mitosis. Cancers do not appear to spread to other cells, but rather new cancer cells are clones of the original cell. Solid tumors can cause the diversion of blood supplies so that they are nourished like normal tissue.

19.4 What Causes Cancer?

The causation of cancer should be considered in two parts: (1) What are the changes from normal to cancerous growth? (2) What triggers the change? The genetic component of cancer is reflected by the fact that cancer cells pass on the "trait" to descendant cells. The potential to develop certain forms of cancer also appears to be hereditary. Carcinogenic chemicals are thought to affect DNA since about 90% of them also cause mutations. Certain viruses and radiation (particularly X-rays and UV rays) have been strongly linked to cancer. There is also evidence that carcinogenesis is a two-stage process, that is, after it is initiated, a second agent (promoter) is required to complete the process. It is often years between exposure to an agent and the manifestation of the cancer. This leads to great difficulties in understanding its causation, and determining if certain agents are causative.

19.5 Aging: The Facts

Aging can be defined as the progressive deterioration of the structures and functions of a mature organism, the process leading to death. Among the bacteria, aging does not occur. These cells may be killed, but they do not die by aging. When they reach full growth, they divide and growth begins again. All normal mammalian cells do age and die, although some that are not normal can be propagated indefinitely. About 50 cell doublings seem to be the limit for normal mammalian cells. A number of organisms fail to show signs of aging, although most eventually die. Woody perennials, for example, die because of disease or accident, not because of aging. The symptoms of aging in mammals are decreased muscular strength, decreased lung capacity, decreased pumping of blood from the heart, decreased urine formation in the kidney, and decreased metabolic rate, for example.

19.6 Aging: The Theories

One theory about aging says that rapid differentiation and growth, accompanied by a high rate of metabolism, are responsible for aging.

According to another theory, called the "clinker" theory, each cell unavoidably accumulates poisonous wastes during its lifetime. The accumulation gradually reduces the ability of the cell to function, and the cell thus ages. Many biologists feel that even under the best environmental conditions, organisms will age at a rate determined by the nature of their genes. And most accept the idea that aging results from an interaction of both hereditary and environmental factors.

19.7 Death

The outcome of aging is death. Although other organs may fail first, we generally mark death in humans at the time the heart stops beating. But it may soon be necessary to reexamine our definition of death because of life support systems capable of taking over the function of the heart, the lungs, etc. Many feel that irrevocable loss of brain activity is death of the organism even if not all of its constituent cells.

19.8 Summary

Reproduction, heredity, morphogenesis, differentiation, growth, aging, and abnormalities of the above processes all shape a central nature—they are controlled by genes and the way the genes are modulated by interactions with the environment.

B. KEY TERMS

growth	polarity
lag period	metaastasis
exponential period	promoter
decelerating growth	carcinogenic
imaginal disks	Delaney Clause
regeneration	oncogenic virus
chalone	

II. MASTERING THE CHAPTER

A. LEARNING OBJECTIVES

When you have mastered the material in this chapter, you should be able to:

1. Define all key terms.

2. State the two ways in which growth occurs on the cellular level.

3. List the three periods of growth of an organism, and state what is characteristic of each period.

4. Cite examples of regeneration in four different species, listing those with greater regenerative powers first.

5. Cite two ways in which the processes of embryonic development and regeneration are similar.

6. List five symptoms of aging in humans.

7. State the essence of two theories that relate aging to living.

8. Describe the role in aging played by collagen.

B. REVIEWING TERMS AND CONCEPTS

1. Most growth among animals takes place after the completion of morphogenesis and (a) _____. For growth to occur, the rate of (b) _____ of complex molecules—proteins, for example— must exceed the rate of their (c) _____.

2. Several periods are distinguishable in growth. The first period is the (a) _____ period and is characterized by little or no actual growth. Next comes the (b) _____ period, one of (c) _____ increasing growth. Finally there is a period of (d) _____ growth, a slowing process in which growth eventually ceases altogether. In plants, growth occurs in the localized areas of (e) _____ tissue. In animals, growth occurs through all of the (f) _____ and _____ of the organism.

3. All organisms appear to have the ability to repair damaged parts of themselves through a process called (a) _____ _____. Sponges, hydra, and planarian worms have the ability to regenerate an entire new (b) _____ from a part of themselves.

4. Mature, differentiated cells of a given tissue synthesize and secrete an inhibitory substance called a (a) _____ that pre- vents (b) _____ of the young cells of the same tissue. Within a single organism, at least among the vertebrates, there seems to be a progressive (c) _____ (increase/decrease) in re- generative ability with increasing age.

5. Cancer is the unchecked (a) _____ of cells. Cancers start in one location, the (b) _____ tumor, and often migrate throughout the body in a process called (c) _____. Some cancers de- (d) _____ and revert back to an (e) _____ state. Most frequently, cancer occurs in cells undergoing rapid (f) _____. Cancer (g) _____ (does/does not) appear to "infect" other cells, and therefore when it spreads it does so as a (h) _____.

6. By whatever means cancer is caused, the transition from a normal cell occurs at the level of a (a) _____. Evidence for this comes from the finding that certain cancers result from (b) _____ disorders. Furthermore, most (c) _____ agents act at the DNA level, and also cause (d) _____. The agents associated with carcinogenesis are (e) _____, _____,

and _____. Some tumors may not develop unless given a (f) _____. This has led to a (g) _____-stage theory. One of the difficulties in studying cancer causation is the (h) _____ period of time between exposure to an agent the clinical symptoms. Furthermore, the studies are often statistical because of the (i) _____ exposure doses.

7. (a) _____ can be defined as the progressive deterioration of the structures and (b) _____ of a mature organism, the process leading to death. Among the bacteria, aging (c) _____ (does/does not) occur. These cells may be killed, but they do not die by aging. All normal mammalian cells do age and (d) _____, although some that are not normal can be propagated (e) _____. About (f) _____ doublings seem to be the limit for normal mammalian cells. A number of organisms fail to show signs of aging, although most eventually die. Woody perennials, for example, die because of (g) _____ or (h) _____, not because of aging.

8. The symptoms of aging in mammals are decreased (a) _____ strength, decreased (b) _____ capacity, decreased (c) _____ of blood from the heart, decreased (d) _____ formation in the kidney, and decreased (e) _____ rate, for example.

9. There are several theories on aging. One states that the more (a) _____ (rapidly/slowly) an organism lives, the sooner it ages. The (b) _____ theory relates aging to the unavoidable accumulation of (c) _____ during life. It is also held that aging may result from damage due to (d) _____, particularly X-rays and cosmic rays. Another theory is that aging is intrinsic, that is, it is due to the nature of the (e) _____. Most biologists feel, however, that aging is a result of both (f) _____ and _____ influences.

C. TESTING TERMS AND CONCEPTS

1. Growth occurs in localized areas in
   a) viruses.
   b) bacteria.
   c) animals.
   d) plants.

2. The action of chalones is
   a) species-specific.
   b) tissue-specific.
   c) organ-specific.
   d) host-specific.

3. Death can best be described as loss of
   a) heart function.
   b) brain function.
   c) lung function.
   d) metabolic function.

4. Growth occurs as a result of an increase in the

a) number of cells.
b) size of cells.
c) both a and b.
d) neither a nor b.

5. Regeneration is most extensive among
    a) reptiles.
    b) mammals.
    c) bacteria.
    d) plants.

6. A seed preparatory to germination is in that period of growth termed
    a) exponential.
    b) lag.
    c) decelerating.
    d) accelerated.

7. All of the following stop growing at a certain size, <u>except</u>:
    a) cows.
    b) frogs.
    c) fish.
    d) birds.

8. Regeneration is similar to
    a) embryonic development.
    b) cancer.
    c) lag phase growth.
    d) declining growth.

9. When cancer cells begin to grow in a new location the term to describe the event is
    a) nodal.
    b) chalone-resistant growth.
    c) benign.
    d) metastasis.

10. One reason for considering cancer cells to be dedifferentiated is that they
    a) respond well to chalone(s).
    b) express embryonic states (conditions).
    c) eventually revert back to differentiated cells.
    d) have abnormal-looking nuclei.

11. In order to survive, a solid tumor must do all of the following, <u>except</u>:
    a) reproduce.
    b) obtain nutrition.
    c) exchange gases.
    d) excrete.

12. Carcinogenic agents most likely interact directly with, and therefore alter
    a) proteins.

b) DNA.
c) mRNA.
d) histones.

13. There is general agreement among biologists that aging
    a) is strictly a result of heredity.
    b) is solely the result of environmental "insults."
    c) can be accounted for by the "clinker" theory.
    d) has both genetic and environmental components.

14. In vitro, normal mammalian cells
    a) do not divide.
    b) divide at no determinable rate.
    c) divide only a limited number of times.
    d) divide indefinitely.

15. In a human, aging occurs at all of the following levels, except:
    a) organ.
    b) tissue.
    c) atomic.
    d) protein.

D. UNDERSTANDING AND APPLYING TERMS AND CONCEPTS

Decide whether statements 1 through 7 are true (T) or false (F).

_____ 1. The larval stage of development of a moth is an example of incomplete development.

_____ 2. Within a single organism, at least among vertebrates, there also seems to be a progressive loss of regenerative ability with increasing age.

_____ 3. An embryo is more likely to grow by increase in cell number than cell size.

_____ 4. From infant to adult, proportions as well as size change.

_____ 5. Humans are incapable of any form of regeneration other than wound healing.

_____ 6. Annual plants appear to be "programmed" to die after a certain period of time.

_____ 7. In some cases, death of some cells is necessary for development of others.

8. How does "pinching" terminal buds of many plants cause them to be "fuller"? _____
_____
_____

9. During the healing process, why doesn't the chalone inhibit cell division and hence prevent the regeneration of tissue?

177

_____
_____
_____
_____

10. If cells normally grow only "amongst their own kind," what is likely to be happening to allow metastasis to occur?
_____
_____

11. Cancers of muscle are rare. Why? _____
_____

12. How does xeroderma pigmentosum exemplify the combined effects of heredity and environment in cancer causation? _____
_____
_____

13. Assume identical twins are both exposed to equal amounts of a chemical that is thought to cause cancer. Five years later one twin develops cancer and the other does not. Give a possible explanation. _____
_____
_____
_____

ANSWERS TO CHAPTER EXERCISES

Reviewing Terms and Concepts

1. a) differentiation b) synthesis c) breakdown

2. a) lag b) exponential c) rapidly d) decelerating e) meristem
   f) tissues, organs

3. a) regeneration b) organism

4. a) chalone b) mitosis c) decrease

5. a) proliferation b) primary c) metastasis d) differentiate
   e) embryonic f) mitosis g) does not h) clone

6. a) gene b) inherited c) carcinogenic d) mutations e) chemicals,
   viruses, radiation f) promoter g) two h) long i) low

7. a) Aging b) functions c) does not d) die e) indefinitely f) 50
   g) disease h) accident

8. a) muscle b) lung c) pumping d) urine e) metabolic

9. a) rapidly b) clinker c) wastes d) radiation e) genes
   f) heredity, environmental

Testing Terms and Concepts

1. d	6. b	11. a
2. b	7. c	12. b
3. b	8. a	13. d
4. c	9. d	14. c
5. d	10. b	15. c

Understanding and Applying Terms and Concepts

1. F   2. T   3. T   4. T   5. F   6. T   7. T

8. Growth is diverted to lateral buds which, in turn, form branches.

9. The regenerating cells are immature and do not produce chalone.

10. The cancer cells are unable to respond to signals from the tissue they are invading.

11. Muscle cells are nondividing, and therefore less likely to become cancerous.

12. Ultraviolet light causes damage to DNA. Normal people repair the damage. Those who are genetically deficient can not make the repairs.

13. The twin with cancer was subsequently exposed to a "promoter."

20

HETEROTROPHIC NUTRITION

I. REVIEWING THE CHAPTER

A. CHAPTER HIGHLIGHTS

20.1 Introduction/20.2 Requirements of Heterotrophic Nutrition

Multicellular organisms expend large amounts of energy in maintaining constant amounts of nutrients in their extracellular fluid. Nutrition that involves dependence upon preformed organic molecules is called heterotrophic nutrition, and the organisms using it are called heterotrophs. The organic molecules that serve as a source of material and energy are sugars, amino acids, fatty acids, and glycerol.

20.3 Intracellular Digestion/20.4 Extracellular Digestion

In some heterotrophic organisms, the first steps in the breakdown of food—called digestion—are intracellular, occurring after the solid material has been engulfed by a cell. In other organisms, digestion is extracellular, occurring outside the cellular system before the broken-down food products are absorbed by the cells.

20.5 A Filter Feeder: The Clam

Barnacles, clams, and certain other aquatic organisms are filter feeders. They move a stream of water through themselves and filter out tiny organisms and organic particles, then ingest them into the alimentary canal for eventual digestion and absorption. The tubular digestive system of the clam is an efficient one because it permits a one-way flow of food materials. Incoming food need not mix with food that has already been partially processed, as it does in a flatworm or jellyfish.

20.6 Active Food Seekers: The Grasshopper and the Honeybee

In the grasshopper, food first passes from the mouth through the esophagus into the crop, which is a temporary storage organ, then into the gizzard, where the food is ground up, then into the stomach, where chemical digestion takes place, and lastly into the intestine, where absorption occurs. The wastes are stored temporarily in the rectum before being egested through the anus.

20.7 Ingestion/20.8 The Stomach

The human digestive system is similar to that of the grasshopper. Mechanical digestion begins with chewing in the mouth. While still in the mouth, food is acted on by saliva, a lubricant also containing a starch-digesting enzyme called amylase. The mucin-secreting glands of the esophagus further lubricate the food, and peristalsis pushes the food along. Next, the food reaches the stomach, where gastric juices and mechanical mixing further work on the material. Hydrochloric acid denatures proteins and breaks down connective tissue. Pepsin hydrolyzes proteins, yielding short polypeptide chains. Very little absorption occurs here.

20.9 Pancreas

Now in liquid form, the food next passes into the small intestine, the first part of which is called the duodenum. Under the influence of the duodenal hormones secretin and cholecystokinin, the pancreas releases sodium bicarbonate, to neutralize the acidity of the mixture, and a variety of digestive enzymes, which hydrolyze starches, fats, proteins, and nucleic acids.

20.10 The Small Intestine

Small food molecules, such as disaccharides, peptides, fatty acids, and monoglycerides, are absorbed into the microvilli and villi that line the inner surface of the small intestine. Fats are resynthesized and move into the lymphatic system. Carbohydrates and proteins are further digested, and amino acids and monosaccharides are absorbed directly into the bloodstream.

20.11 The Liver

The liver secretes a substance called bile, which is important in the digestion of fats, and which is stored between meals in the gall bladder. The liver also acts as a gatekeeper between the digestive system and the rest of the body. All of the blood from the small intestine is collected by the hepatic portal system and fed into the liver before passing into the general circulation. Nonnutritive materials, including drugs and poisons, are removed and destroyed. Excess nutritive materials are removed and stored.

20.12 The Large Intestine

What is left of the food next passes into the large intestine, which contains numerous bacteria. The chief function of the large intestine is reabsorption of water. The remaining residue is discharged into the rectum and out of the body through the anus.

B. KEY TERMS

heterotrophic	cholecystokinin
saprophytes	retum
peristalsis	nuclease
duodenum	villi

carboxypeptidase   disaccharidase

aminopeptidase   hepatic portal system

bile   food vacuole

digestion   ingestion

amylase   proboscis

pepsin   gastrin

sodium bicarbonate   lipase

trypsin   chymotrypsin

homeostasis   microvilli

egestion   secretin

II. MASTERING THE CHAPTER

A. LEARNING OBJECTIVES

When you have mastered the material in this chapter, you should be able to:

1. Define all key terms.

2. List those organic molecules humans must take in ready-made from the environment and those that we are able to synthesize.

3. Distinguish between intracellular and extracellular digestion and list organisms that rely on each.

4. Distinguish between mechanical and chemical breakdown of food.

5. List the major organs of the human digestive system and, in general terms, tell what happens in each.

6. List the three kinds of cells associated with the stomach's gastric glands and cite the function of each.

7. List the six components of the pancreatic fluid and cite the function of each.

8. List three hormones produced by digestive organs and describe their effects.

9. Discuss the importance of the villi and microvilli.

10. Cite three means by which the end products of digestion enter the bloodstream.

11. List three functions of the liver.

12. Distinguish between the functions of the small and large intestines.

B. REVIEWING TERMS AND CONCEPTS

1. Nutrition that involves dependence upon preformed organic molecules is called (a) _____ nutrition, and the organisms using it are called (b) _____. The organic molecules that serve as a source of material and energy are (c) _____, _____, (d) _____ acids, and _____ acids.

2. In some heterotrophic organisms, the first step in the break-down of food—called (a) _____—is (b) _____-cellular, occurring after the solid material has been engulfed by a cell. In other organisms, digestion is (c) _____-cellular, occurring "outside" the cellular system before the broken-down food products are (d) _____ by the cells.

3. The human digestive system is similar to that of the grass-hopper. (a) _____ breakdown begins with chewing in the mouth. While still in the mouth, food is acted on by (b) _____, a lubricant also containing a starch-digesting enzyme called (c) _____. The food is swallowed through the pharynx and into the (d) _____. Next, the food reaches the (e) _____, where _____ juices and mechanical mixing further work on the material.

4. Now in liquid form, the food passes into the (a) _____ intestine, the first part of which is called the (b) _____. Substances produced by the (c) _____—such as sodium bicarb-onate, lipase, and proteases—neutralize acidity and further break down the food. The secretion of these pancreatic juices is under the control of (d) _____.

5. The liver secretes a substance called (a) _____, which is important in the digestion of (b) _____, and which is stored between meals in the (c) _____ bladder. The liver also acts as a gatekeeper between the digestive system and the rest of the body. The blood from the small intestine flows through the (d) _____ system into the liver, and there (e) _____ materials are removed and destroyed and excess (f) _____ are removed and stored.

6. The absorption of the end products of digestion is the function (a) _____ and _____, which line the inner surface of the (b) _____ intestine. (c) _____ of carbohydrates and protein comes to an end when what were originally macromolecules have been converted into (d) _____ acids and (e) _____ ready for passage into the bloodstream.

7. What is left of the food next passes into the (a) _____ intestine, which contains numerous bacteria. The chief function of the (b) _____ intestine is (c) _____ of water. The

remaining residue is discharged into the (d) _____ and out of
the body through the (e) _____.

C. TESTING TERMS AND CONCEPTS

1. Nutrition involving dependence on preformed organic molecules is
   called
   a) autotrophic.
   b) heterotrophic.
   c) intracellular.
   d) extracellular.

2. Mechanical breakdown of food in the grasshopper is the function of
   the
   a) crop.
   b) gizzard.
   c) caeca.
   d) food vacuole.

3. An organism that does not use extracellular digestion is the
   a) human.
   b) clam.
   c) common bread mold.
   d) amoeba.

4. Which answer is not true? An adult human must have in his diet
   certain specific
   a) amino acids.
   b) vitamins.
   c) sugars.
   d) unsaturated fats.

5. The chemical responsible for digestion in the mouth is called
   a) mucin.
   b) lipase.
   c) protease.
   d) amylase.

6. Food is propelled along the esophagus and other parts of the
   digestive tract by a rhythmic wave of muscular contraction called
   a) active transport.
   b) swallowing.
   c) peritonitis.
   d) peristalsis.

7. The "chief" cells of the stomach secrete
   a) pepsin.
   b) pepsinogen.
   c) hydrochloric acid.
   d) gastrin.

8. Secretion of pancreatic juice is under the control of
   a) secretin and cholecystokinin.
   b) trypsin and chymotrypsin.

c) lipase and protease.
d) gastrin.

9. The pH of the material in the duodenum is raised to about 8 by
a) $NaHCO_3$.
b) lipase.
c) trypsin.
d) cholecystokinin.

10. The liver is a digestive organ in that it is responsible for
emulsifying fats. Where does this process take place?
a) in the small intestine
b) in the large intestine
c) in the hepatic portal system
d) in the liver itself

11. The hormone that stimulates the release of bile from the gall
bladder is
a) gastrin.
b) chymotrypsin.
c) secretin.
d) cholecystokinin.

12. Absorption of digested food occurs in the
a) mouth.
b) stomach.
c) small intestine.
d) large intestine.

13. Ingested fats are hydrolyzed into fatty acids and monoglycerides
by
a) carboxypeptidase.
b) lipase.
c) HCl.
d) bile.

14. The end products of protein and carbohydrate digestion are
a) disaccharides and peptides.
b) monosaccharides and amino acids.
c) fatty acids and monoglycerides.
d) nucleotides and monosaccharides.

15. If your digestive tract was somehow stripped of its microvilli,
what process would decrease the most?
a) movement of food down the small intestine
b) digestion of proteins, starch, and fats
c) absorption of water, sugars, and amino acids
d) deamination of amino acids and synthesis of glycogen

D. UNDERSTANDING AND APPLYING TERMS AND CONCEPTS

1. a) Which is the more common strategy in higher animals, intracellular or extracellular digestion of food? _____

b) List four specific animals that rely on this strategy.

_____ , _____ , _____ , _____

c) What is the particular value of that strategy, making it so common? _____

_____

2. a) Name two organisms that rely on only chemical breakdown of food.

_____

b) Name two organisms that rely on mechanical and chemical break-down. _____

3. Match the vitamin to its deficiency disease:

_____ a) anemia   A. nicotinic acid

_____ b) pellagra   B. thiamine

_____ c) slow blood clot formation   C. riboflavin

_____ d) rickets   D. B_{12} and folic acid

_____ e) scurvy   E. D

_____ f) damage to tongue and   F. C
         eyes
                                             G. K
_____ g) beriberi
                                             H. A
_____ h) night blindness

4. Deficiencies in thiamine, riboflavin, and nicotinic acid all cause widespread damage in the body. Why? _____

_____

_____

5. Match the terms in Column B with the descriptions in Column A. Terms may be used once, more than once, or not at all.

Column A   Column B

a) Starch is eventually converted   A. pancreas
   into the sugar _____
                                             B. parietal cells
b) Bile is helpful in the diges-
   tion of _____   C. reabsorption of water

c) This hormone is produced by   D. absorption of food
   the stomach when food enters
   it _____   E. cardiac sphincter

d) Where the first stage of   F. pyloric sphincter
   mechanical digestion occurs

   _____

186

e) Upper valve of the
stomach _____

f) This hormone stimulates
action of the pancreas

_____

g) The only digestive enzyme
secreted by the stomach

_____

h) The chief function of
the large intestine is

_____

i) Several juices enter the
duodenum from the _____

j) Food enters the esophagus
from the _____

k) This material is produced in
the pancreas and raises the
pH of material in the duodenum

_____

l) Gastric juices are produced
in the walls of this organ

_____

m) This substance contains
mucins _____

n) These enzymes hydrolyze
ingested nucleic acids

_____

o) Food is moved along the
alimentary canal by muscular
action called _____

p) The first 10 inches of the
small intestine compose the

_____

q) HCl is secreted into the
stomach by the _____

r) Increase absorptive surface
area _____

s) These synthesize and secrete
pepsinogen _____

G. villi

H. glucose

I. pepsinogen

J. mouth

K. stomach

L. $NaHCO_3$

M. saliva

N. peristalsis

O. fats

P. protein

Q. gastrin

R. pharynx

S. secretin

T. starches

U. esophagus

V. nucleases

W. duodenum

X. "chief" cells

t) Salivary amylase works
   on _____

u) Saliva action occurs here
   _____

6. Suppose you swallow a paper clip and a piece of paper. Why are
   you unable to absorb the nutrients potentially available in these
   items?

   a) paper clip_____
   b) paper_____

7. Number the following events in the correct sequence. (1 = first,
   14 = last)

____ a) removal of water from contents of digestive tract

____ b) action of bile

____ c) salivation

____ d) first instance of peristalsis

__1_ e) imagine the taste of a delicious grilled hamburger

____ f) initial digestion of starch

____ g) initial digestion of small fat droplets

____ h) initial digestion of protein molecules

____ i) initial mechanical breakdown of food

____ j) ingestion

____ k) food enters stomach

____ l) pH of food drops markedly

____ m) pH of food rises markedly

____ n) absorption of monoglycerides into body

ANSWERS TO CHAPTER EXERCISES

Reviewing Terms and Concepts

1. a) heterotrophic b) heterotrophs c) sugars, glycerol d) amino,
   fatty

2. a) digestion b) intra- c) extra- d) absorbed

3. a) Mechanical b) saliva c) amylase d) esophagus e) stomach
   f) gastric

4. a) small b) duodenum c) pancreas d) hormones

5. a) bile b) fats c) gall d) hepatic portal e) nonnutritive
   f) nutrients

6. a) villi, microvilli b) small c) Digestion d) amino e) monosaccharides

7. a) large b) large c) reabsorption d) rectum, e) anus

Testing Terms and Concepts

1.	b	6.	d	11.	d
2.	b	7.	b	12.	c
3.	d	8.	a	13.	b
4.	c	9.	a	14.	b
5.	d	10.	a	15.	c

Understanding and Applying Terms and Concepts

1. a) extracellular b) molds, clams, grasshoppers, humans
   c) It allows an organism to feed on foods larger than its cells.

2. a) amoebas, molds, b) grasshoppers, humans

3. a) D b) A c) G d) E e) F g) C h) B i) H

4. All are involved in the widespread reactions of cellular
   respiration.

5. a) H   d) J   g) I   j) R   m) M   p) W   s) X
   b) O   e) E   h) C   k) L   n) V   q) B   t) T
   c) Q   f) S   i) A   l) M   o) N   r) G   u) J

6. a) Mechanical digestion cannot even begin to reduce a paperclip
      to pieces that could be absorbed.

   b) Mechanical digestion will reduce the paper to cellulose mole-
      cules, but humans have no enzyme that can reduce that molecule
      further, and cellulose is to large to be absorbed.

7. a) 14 b) 11 c) 2 d) 6 e) 1 f) 5 g) 12
   h) 9 i) 4 j) 3 k) 7 l) 8 m) 10 n) 13

GAS EXCHANGE IN PLANTS AND ANIMALS

I. REVIEWING THE CHAPTER

A. CHAPTER HIGHLIGHTS

21.1 Gas Exchange in Aquatic Organisms

To carry on respiration, cells need to acquire oxygen and dispose of carbon dioxide. For photosynthesis, green plant cells need to acquire carbon dioxide and dispose of oxygen. Gas exchange with the environment, then, is important to all cells.

The amoeba takes in oxygen and expels carbon dioxide by simple diffusion. It can do this because it has an adequate surface-to-volume ratio, as does the planarian worm. But in larger aquatic animals, cells deep within the body cannot exchange gases directly with the environment. Gills and and a circulating blood supply make an indirect exchange of gases possible.

21.2 Water Versus Air

Terrestrial organisms can collect oxygen more easily than aquatic organisms can, because air contains at least 21 times more oxygen than water does. However, terrestrial organisms do have the problem of preventing their gas exchange surfaces from drying out.

21.3 Gas Exchange in Roots and Stems

In roots and stems, living cells are arranged in thin layers, near the surface of the plant, and usually in contact with air passageways. Tiny openings called lenticles allow air to pass through otherwise airtight bark. Thus, no specialized gas exchange organ is needed.

21.4 Gas Exchange in the Leaf

In leaves, openings called stomata regulate gas exchange and water loss, by opening and closing in response to internal and external environmental conditions. Light stimulates the opening of stomata by causing an active accumulation of potassium ions within the stomatal guard cells. Turgor increases, the guard cells bend, and the stoma is enlarged. Carbon dioxide can now be absorbed by the leaf and used in photosynthesis. During the night, the stomata close and so reduce water loss by transpiration. Even during the day, if transpiration causes a greater loss of water than can be replaced by the root system, the guard cells will lose their turgor and close the stomata.

21.5 Tracheal Breathing

Most animals have special gas exchange organs enclosed within the body for protection from physical damage and from desiccation. Insects, for example, take in air through holes in their exoskeleton called spiracles, the air then circulating through an elaborate network of tubes called tracheae. In small, inactive insects, oxygen passage through the tracheae is by diffusion. Larger insects ventilate their tracheae by a muscular pumping action.

21.6 Lung Breathers

The gas exchange organ in terrestrial vertebrates is the lung. In the frog, the lungs are simple sacs, which are supplemented by a moist and highly vascular skin. Lungs are more elaborate and contain a greater absorptive surface area in reptiles, birds, and mammals.

21.7 Mechanism of Breathing in Humans/21.8 The Pathway of Air

In humans, oxygen in the air enters the lungs from the mouth or nostrils. It is forced into the lungs through the action of the muscular diaphragm and external intercostal muscles. Once through the pharynx, air passes through the glottis and into the trachea, which branches into a right and left bronchus. The two bronchi branch many times into ever smaller bronchioles, which terminate at the lung alveoli, where gas exchange actually takes place. Oxygen diffuses into the blood through the alveoli capillaries. It is then attached to hemoglobin of the red blood cells and transported to all functioning cells of the body. Carbon dioxide diffuses out of the cells and is carried to the alveoli capillaries where it diffuses into the lungs for discharge through exhalation.

21.9 Control of Breathing

Oxygen deprivation plays a very minor role in regulating the rate at which ventilation of the lungs occurs. Carbon dioxide is the operative agent, stimulating the medulla oblongata. When the carbon dioxide level of the blood reaching the medulla exceeds a certain level, the medulla responds by increasing the number and rate of nerve impulses controlling the diaphragm. The bronchioles also respond to high levels of carbon dioxide by dilating and so facilitating the passage of a greater volume of air to and from the lungs. The medulla, then, is an important homeostatic device of the body.

21.10 Air Pollution and Health

Urban air pollution and cigarette smoking place severe stresses on our respiratory systems. These stresses can overwhelm the cleaning mechanisms there. If one analyzes death rates from all causes among United States males, the rate is directly proportional to: (a) number of cigarettes smoked per day, (b) depth of inhalation, and (c) number of years since smoking began.

B. KEY TERMS

gills	lenticels
stomata	transpiration
spiracles	lungs
air sacs	hemoglobin
vital capacity	pleural membranes
bronchioles	bronchus
emphysema	pneumonia
guard cells	skin
tracheae	nostrils
diaphragm	nasal cavities
glottis	nasopharynx
alveoli	precancerous
medulla oblongata	

II. MASTERING THE CHAPTER

A. LEARNING OBJECTIVES

When you have mastered the material in this chapter, you should be able to:

1. Define all key terms.

2. Explain how an aquatic microorganism, such as the ameoba, acquires oxygen and rids itself of carbon dioxide, and how ocean and lake water acquire oxygen.

3. Cite the three distinct parts of the gas-exchange system in a lobster.

4. Cite the chief advantage, in the context of a gas-exchange system, that a terrestrial mammal, for instance, has over a fish.

5. Cite four reasons why plants generally do not require specialized organs for gas exchange.

6. Describe the conditions that regulate the opening and closing of leaf stomata.

7. Distinguish between a tracheal and lung breathing system.

8. State why the oxygen demands of a small, active bird are so much greater than those of a reptile or amphibian.

9. Describe the action of the rib cage and that of the diaphragm as you breathe.

10. List the various structures through which oxygen passes from the time it is inhaled until it enters a liver cell, for instance.

11. Discuss the important role of the alveolus in the process of gas exchange.

12. Distinguish between pneumonia and emphysema.

13. Cite the substance that regulates the rate of breathing and state how such regulation is achieved.

14. Cite the sources of the two major categories of air pollution, and describe the defenses of the respiratory system against impurities in the air.

B. REVIEWING TERMS AND CONCEPTS

1. To carry on respiration, cells need to acquire (a) _____ and dispose of (b) _____ _____. For photosynthesis, green plant cells need to acquire (c) _____ _____ and dispose of (d) _____. Gas exchange with the environment, then, is important to all cells.

2. The amoeba takes in oxygen and expels carbon dioxide by simple (a) _____. It can do this because it has an adequate (b) _____ - _____ - _____ ratio, as does the planarian worm. However, larger animals require specialized (c) _____ organs. Among mollusks, shrimp, and fishes, the gas exchange organs are (d) _____.

3. Terrestrial organisms can collect (a) _____ more easily than can aquatic organisms, because air contains at least (b) _____ times more oxygen than does water. However, terrestrial organisms must somehow prevent their gas exchange surfaces from (c) _____ _____.

4. In roots and stems tiny openings called (a) _____ allow air to pass through the bark to the living cells of the plant. No specialized (b) _____ _____ organ is required.

5. In leaves, those leaf openings called (a) _____ regulate gas exchange. (b) _____ stimulates the opening of the stomata by stimulating an active accumulation of potassium by stomatal (c) _____ _____. The stomatal (d) _____ cells respond by building up turgor, thus opening the stomata. Darkness and dry conditions result in the (e) _____ of the stomata. Action of the stomata regulates a plant's water loss through (f) _____.

6. The gas exchange organs of terrestrial animals are enclosed within the body to protect them from (a) _____ _____ and from (b) _____ .

7. Air is supplied to the cells of an insect's body by small openings along the side of the body, called (a) _____ , leading to tubes, called (b) _____ .

8. Terrestrial vertebrates accomplish gas exchange by means of (a) _____ . In humans, oxygen is forced into the lungs through the action of the muscular (b) _____ . Once through the pharynx, air passes through the (c) _____ and into the trachea, which branches into a right and left (d) _____ . The two bronchi branch many times into ever-smaller (e) _____ , which terminate at the lung (f) _____ . Oxygen diffuses into the blood through the (g) _____ capillaries. It is then attached to (h) _____ of the (i) _____ blood cells and transported to all functioning cells of the body. (j) _____ _____ diffuses out of the cells and is carried to the alveoli capillaries, where it diffuses into the lungs for discharge through exhalation.

9. (a) _____ deprivation plays a very minor role in regulating the rate at which ventilation of the lungs occurs. (b) _____ _____ is the operative agent, stimulating the (c) _____ _____ of the brain. When the (d) _____ _____ level of the blood reaching the medulla exceeds a certain level, the medulla responds by increasing the number and rate of nerve impulses controlling the (e) _____ .

C. TESTING TERMS AND CONCEPTS

1. Oxygen enters an amoeba through
    a) active transport.
    b) osmosis.
    c) endocytosis.
    d) diffusion.

2. In its gas-exchange system, a fish is most like a
    a) whale.
    b) clam.
    c) planarian worm.
    d) frog.

3. Whether or not gas exchange between an organism and its environment is accomplished by diffusion alone depends on
    a) the efficiency of its gas-exchange organ.
    b) its surface-to-volume ratio.
    c) the size of its nucleus.
    d) whether the organism is aquatic or terrestrial.

4. Plants do not accomplish gas exchange through their
    a) lenticels.
    b) roots.
    c) stomata.
    d) vascular system.

5. During a typical day, a leaf's guard cells <u>do</u> <u>not</u>
    a) absorb water from surrounding cells.
    b) absorb and fix energy from the sun.
    c) absorb oxygen from the atmosphere.
    d) absorb carbon dioxide from the atmosphere.

6. The one-way flow of air in the grasshopper is accomplished by a type of
    a) simple diffusion.
    b) simple osmosis.
    c) lung breathing.
    d) tracheal breathing.

7. In which of the following organisms is a circulatory system used to distribute oxygen to the body's cells?
    a) grasshopper
    b) tree
    c) planarian worm
    d) earthworm

8. Which is <u>not</u> a gas exchange surface of a terrestrial animal?
    a) frog skin
    b) bird air sac
    c) grasshopper trachea
    d) human alveolus

9. A sparrow's oxygen demands are greater than those of a frog because the sparrow is
    a) homeothermic.
    b) dependent on air sacs.
    c) dry-skinned.
    d) more active during the day than during the night.

10. During human inhalation, the rib cage
    a) is an active agent.
    b) is a passive agent.
    c) is antagonistic to the diaphragm.
    d) decreases in volume.

11. Which of the following is true?
    a) Gas exchange occurs only across the walls of the alveoli.
    b) Gas exchange occurs across the walls of the alveoli, bronchiols, and bronchi.
    c) The absorption of oxygen is aided by active transport.
    d) The absorption of oxygen and the release of carbon dioxide are both aided by active transport.

12. Where does one get the power to blow up a toy balloon?
    a) from the natural elacticity of lung tissue
    b) from contraction of the diaphragm
    c) from contraction of the internal intercostals
    d) from contraction of the external intercostals

13. The rate at which ventilation of the lungs occurs is usually

regulated by
a) oxygen concentration in the blood.
b) heart rate.
c) respiratory center in the lung.
d) carbon dioxide concentration in the blood.

14. Not included among the structures that help prevent dust from damaging lung tissues are
a) hairs in the nasal passageways.
b) mucus in the bronchioles.
c) cilia in the alveoli.
d) phagocytic cells in the alveoli.

15. The disease of the lungs most likely to cause death directly is
a) chronic bronchitis.
b) emphysema.
c) asthma.
d) pneumonia.

D. UNDERSTANDING AND APPLYING TERMS AND CONCEPTS

1. The amoeba, planarian, and earthworm have no specialized gas-exchange organs. Give two reasons why all fish, reptiles, and birds require gills or lungs.
a)_____
b)_____
_____

2. Decreasing the oxygen supply in an area would interfere with the respiration of the animals living there. Would such a change be more serious in a pond or a forest?
a)_____ Why? b)_____
Which would have the greater effect in the forest, decreasing the oxygen supply or increasing the carbon dioxide level by the same amount? c)_____ Why? d)_____
_____

3. Gills and lungs differ conspicuously, gills protruding out into a watery environment and lungs folding into the body in air. They do have two very important structural similarities, however. These are:
a)_____
b)_____

4. Plants and insects are also surprisingly similar in the structure of their respiratory systems. Describe two general similarities.
a)_____
b)_____

5. Check the following changes that would significantly decrease the amount of oxygen available to the organism involved.

_____a) waterlogging the soil in which an earthworm is living.

_____ b) waterlogging the soil in which a tree is growing.
_____ c) removing the plants from a fish's pond.
_____ d) moving a mouse to an atmosphere containing twice the normal
carbon dioxide.
_____ e) restricting the muscular movement of a grasshopper.
_____ f) moving a frog from a moist to a dry atomosphere.
_____ g) restricting the rib movements of a frog.
_____ h) restricting the mouth and neck movements in a lizard.
_____ i) decreasing the blood supply to a bird's air sacs.
_____ j) decreasing the number of alveoli in a human while holding their
total volume the same.

6. In what sequence would a molecule of oxygen pass through the
following structures on its way to a body cell? Arrange the following in order (1 = first, 12 = last).

  1 a) nostril
_____ b) glottis
_____ c) alvelolus
_____ d) red blood cell
_____ e) trachea
_____ f) nasal passage
_____ g) larynx
_____ h) bronchiole
_____ i) nasopharynx
_____ j) oral pharynx
_____ k) bronchus
_____ l) capillary

7. List five ways smoking cigarettes interferes with breathing.
  a)_____
  b)_____
  c)_____
  d)_____
  e)_____

ANSWERS TO CHAPTER EXERCISES

Reviewing Terms and Concepts

1. a) oxygen b) carbon dioxide c) carbon dioxide d) oxygen

2. a) diffusion b) surface-to-volume c) respiratory d) gills

3. a) oxygen b) 21 c) drying out

4. a) lenticels b) gas exchange

5. a) stomata b) Light c) guard cells d) guard e) closing
   f) transpiration

6. a) physical damage b) desiccation

7. a) spiracles b) tracheae

8. a) lungs b) diaphragm c) glottis d) bronchus e) bronchioles
   f) alveoli g) alveoli h) hemoglobin i) red j) Carbon dioxide

9. a) Oxygen b) Carbon dioxide c) medulla oblongata d) carbon
   dioxide e) diaphragm

Testing Terms and Concepts

1. d	6. d	11. a
2. b	7. d	12. c
3. b	8. b	13. d
4. d	9. a	14. c
5. c	10. a	15. d

Understanding and Applying Terms and Concepts

1. a) Fish, reptiles, and birds are larger. Each has a greater volume
   per unit surface area. b) Fish, reptiles, and birds are more
   active and have a greater oxygen demand.

2. a) pond b) Water contains much less oxygen than air, so for
   aquatic organisms, the supply is only slightly greater than the
   demand. Terrestrial animals have a considerable excess. c)
   increasing the carbon dioxide level d) Again, oxygen levels in
   air are greater than our needs. Our bodies (spedifically, the
   medulla) are much more sensitive to carbon dioxide levels. The
   specific effect would be to greatly increase ventilation rates.

3. a) Both gills and lungs are highly vascular structures, so oxygen
   can be absorbed and transported throughout the body. b) Both
   gills and lungs are finely divided structures, gills into filaments and lungs into alveoli. This elaboration of these organs
   gives them the large surface area necessary for high absorption
   rates.

4. a) Plants have openings for air called lenticels and stomata; insects have openings called spiracles. b) These openings lead
   to air passageways among loosely packed parenchyma cells in
   plants and to similar passageways called tracheae in insects.

5. a, b, c, e, f, j

6. a) 1 b) 5 c) 10 d) 12 e) 7 f) 2 g) 6 h) 9 i) 3 j) 4
   k) 8 1) 11

7. a) There is less oxygen in cigarette smoke and more carbon dioxide
   and carbon monoxide. b) The smoke irritates and destroys the
   alveoli. c) It stimulates mucus production, which decreases
   lung capacity. d) It inhibits ciliary action, which decreases
   the cleansing of the lungs. e) Smoke causes cancers to develop,
   which destroys lung tissue.

22

THE TRANSPORT OF MATERIALS
IN THE VASCULAR PLANTS

I. REVIEWING THE CHAPTER

A. CHAPTER HIGHLIGHTS

22.1 Importance

The need for a transport system in land plants stems from the plant's need to get water and minerals from the ground and to receive an adequate supply of light from leaf display high off the ground. Water and minerals must be transported up to the leaves, and food must be transported down to the roots. The transport of materials in plants is called translocation and takes place in a system of tissues called vascular bundles.

22.2 Xylem/22.3 Phloem

The xylem vessels are composed of dead cells, arranged end-to-end in tubelike fashion, and are responsible for the transport of water and minerals up from the roots to all other parts of the plant. Xylem tracheids function similarly. The conducting elements of phloem are sieve tubes, composed of living cells, also arranged end-to-end, and are responsible for the transport of sugars and hormones throughout the plant.

Xylem and phloem are arranged differently in the root, stem, and leaf and in different individual plants and different species. The two major subdivisions of the flowering plants, the monocots and dicots, differ conspicuously in the organization of their vascular tissue.

22.4 The Organization of the Root

Roots consist of a variety of tissues and organs. At the root tip is a meristem, where root growth occurs by mitosis. The new cells produced first elongate and then soon differentiate into specialized structures. These include root hairs, which take in water from the soil; the cortex, a food storage area; the pericycle, from which secondary roots develop; and the xylem and phloem tissues, responsible for transport. In older roots, there is another meristem, the cambium, between the xylem and phloem. It produces new xylem and phloem and so increases the diameter of the root.

22.5 The Woody Dicot Stem/22.6 The Herbaceous Dicot Stem/22.7 The Monocot Stem

The outermost layer of a woody stem consists of dead cork cells impregnated with waterproof suberin. Oxygen and carbon dioxide are exchanged with the air through openings called lenticels. Beneath the cork is a layer of cortex for food storage, a meristematic cork cambium that replaces cork lost by weathering, alternating bundles of phloem and food-storing parenchyma, and another meristematic cambium that produces new xylem and phloem each year as happens in the root. All these tissues make up the bark of the plant. It is the xylem that makes up the woody part of a tree. In addition to providing transport, the xylem also provides structural support for woody plants. The innermost core of a young woody stem is the pith, another food storage tissue.

In a herbacious dicot, there is less xylem and no bark. Pith is the dominant tissue. The vascular bundles are arranged in a circle toward the outside of the stem. In a monocot, vascular bundles are scattered almost uniformly throughout the pith.

22.8 The Leaf Veins

The vascular bundles of the leaves are direct extensions of the vascular bundles of the stem. They not only carry materials to and from the leaf, but they provide support to the soft tissues of the leaf as well.

22.9 The Pathway

Water and minerals enter the plant through the root hairs and pass through and between the cells of the cortex. However, to enter the vascular tissue, materials must pass through the cytoplasm of the cells of the endodermis. Thus, the endodermis acts as a kind of gatekeeper to the rest of the plant. Once in the xylem, water and dissolved ions move up and laterally across the plant, eventually entering the leaves through the veins.

22.10 The Magnitude of the Flow/22.11 Factors Affecting the Rate of Transpiration

Of all the water moved through a plant, only 1 to 2% is used in photosynthesis, the rest evaporating from the leaves in a process called transpiration. A number of factors affect the rate of transpiration. These include light, temperature, humidity, wind, and the availability of soil water.

22.12 Root Pressure/22.13 The Dixon-Joly Theory/22.14 Evidence for the Theory

Two theories have been proposed to explain the upward transport of water in the xylem. One, called the root pressure theory, cites the difference in water concentration of the soil water and the sap in the xylem ducts. This concentration gradient results in an inflow of water by osmosis, a buildup of pressure within the xylem, and a movement of water up the plant. While root pressure may account for transport in some small plants, i.e. the tomato, it cannot account for transport in most plants and large trees. According to the Dixon-Joly theory, the transpiration of water from the leaf exerts a pull on the water in the xylem ducts and so draws more water into the leaf. Such a process depends upon two fundamental properties of water: cohesion, which is the sticking of water molecules to one another, and adhesion, which is the sticking of water to other materials, such as the walls of the xylem vessels. These two forces are so strong that columns of water can be pulled to the tops of the tallest trees without breaking or pulling away from the sides of the vessels.

22.15 The Pathway/22.16 Mechanism of Food Transport

Food manufactured in the leaves enters the phloem and is transported both up and down the plant. It then may be used as an energy source or stored as starch in the cortex. The mechanism by which sugars and other molecules are translocated through the phloem is not yet understood. It seems to depend in part on the metabolic activity of the phloem cells. Decreased temperature and lack of oxygen both depress translocation in the phloem. One theory suggests that sap moves through sieve tubes by bulk flow. The phloem, especially in the leaves where sugars are manufactured, is hypertonic, and osmosis develops a pressure that pushes the sap down and up the plant. Another theory suggests that each sieve tube element somehow individually moves its contents along.

B. KEY TERMS

translocation	root hairs
vessels	pericycle
tracheids	transpiration
plasmodesmata	adhesion
differentiation	xylem
endodermis	sieve tube
heartwood	monocot
vascular bundles	cortex
phloem	sapwood
dicot	cohesion

II. MASTERING THE CHAPTER

A. LEARNING OBJECTIVES

When you have mastered the material in this chapter, you should be able to:

1. Define all key terms.

2. Explain why it is so important that land plants have a vascular system.

3. Distinguish between xylem vessels and xylem tracheids and describe their taxonomic distribution.

4. Distinguish in function and structure between xylem vessels and phloem sieve tubes, listing at least two functions of each.

5. Distinguish between dicots and monocots, and cite two examples of each.

6. Trace the path taken by a molecule of water from the time it enters a plant until the time it leaves the plant through transpiration.

7. Cite the five chief factors affecting the rate of transpiration.

8. Cite the two theories accounting for the rise of liquids in plants and state why one is favored over the other.

9. Trace the biochemical and physical pathways of a product that originated in the leaf but ends up as molecules of starch stored in a root.

10. Cite three factors affecting the rate of food translocation in a plant.

11. Explain the mechanism involved in the "pressure flow" theory of phloem transport.

B. REVIEWING TERMS AND CONCEPTS

1. The need for a transport system in land plants stems from the plant's need to get (a) _____ and _____ from the ground, and to receive an adequate supply of (b) _____ from leaf display high off the ground. The transport of materials in plants is called (c) _____ and takes place in a system of ducts called (d) _____ bundles.

2. The (a) _____ vessels are composed of dead cells, arranged end-to-end in tubelike fashion, and are responsible for the transport of (b) _____ and (c) _____ up from the (c) _____ to all other parts of the plant.

3. The conducting elements of phloem are (a) _____ _____, composed of (b) _____ (living/dead) cells, also arranged end-to-end, and are responsible chiefly for the transport of (c) _____ manufactured in the (d) _____ and carried throughout the plant.

4. Roots consist of a variety of tissues. At the root tip is a (a) _____, where root growth occurs by (b) _____. The new cells produced first elongate and then (c) _____ into specialized structures. These include root (d) _____, which take in water from the soil; the (e) _____, a food storage area; the (f)

_____, from which secondary roots develop; and the (g)
_____ and _____ tissues, responsible for transport. In older roots another meristem called (h) _____ lies
between the (i) _____ and _____ and produces new vascular tissue.

5. The outermost layer of a woody stem consists of dead (a)
_____ cells impregnated with waterproof (b) _____.
As these cells wear off, new cork is produced by a meristem called (c)
_____ _____. A still deeper component of the bark is
(d) _____, which transports a sugary sap. Wood is really
(e) _____ tissue.

6. An herbaceous dicot stem has less (a) _____ and more
(b) _____ than a woody dicot. A monocot stem is also largely (c)
_____ and has (d) _____ _____ scattered
throughout its interior.

7. In leaves, the vascular bundles are the (a) _____.
They transport materials to and from the leaf and provide (b) _____
_____ to the soft tissues of the leaf.

8. Of all the water moved through a plant, only (a) _____ % is
used in photosynthesis, the rest evaporating from the leaves in a process called (b) _____. Two theories have been proposed to explain the upward transport of water in the (c) _____. One,
called the (d) _____ pressure theory, relies on the osmosis of
water from the (e) _____ (hypertonic/hypotonic) soil water
to the (f) _____ (hypertonic/hypotonic) root contents.
According to the Dixon-Joly theory, water transpiring from the (g)
_____ exerts a pull on the water in the (h) _____
ducts and so draws more water into the leaf. Such a process must depend on that property of water known as (i) _____ and the fact
that water in the xylem is under (j) _____ rather than pressure.

9. Food manufactured in the (a) _____ enters the (b)
_____ and is transported both up and down the plant. It then
may be used as an (c) _____ source or stored as (d) _____
in the cortex. The mechanism by which sugars and other molecules are
translocated through the phloem is not yet understood. It seems to
depend in part on the (e) _____ activity of the phloem cells.
Decreased (f) _____ and lack of (g) _____ both depress
translocation in the phloem.

C. TESTING TERMS AND CONCEPTS

1. Why is a vascular system so important to a land plant?
    a) The leaves are often a long distance from one another.
    b) The leaves are often a long distance from the roots.
    c) Trunks and branches have a larger diameter than seaweeds and
       other algae.
    d) Land plants are more complex and have a greater variety of
       tissues.

2. The cells with the greatest diameter and therefore the greatest ability to carry water and minerals through the plant are
   a) tracheids.
   b) companion cells.
   c) sieve tube elements.
   d) vessel elements.

3. Cytoplasm is lost at maturity in the development of
   a) sieve tube cells.
   b) companion cells.
   c) plasmodesmata.
   d) xylem vessel cells.

4. In ferns and conifers, the only water transport ducts are
   a) sieve tubes.
   b) pits.
   c) vessels.
   d) tracheids.

5. Water enters a plant through the
   a) root meristem.
   b) root cap.
   c) root hairs.
   d) root cortex.

6. Sugars from the stem and root phloem are stored by the
   a) tracheids.
   b) epidermis.
   c) cortex.
   d) sieve tubes.

7. The innermost tissue in a dicot root is the
   a) xylem.
   b) pericycle.
   c) cortex.
   d) pith.

8. If you removed the cambium from a section of root or stem, that part of the plant would be
   a) unable to grow.
   b) unable to produce branches or secondary roots.
   c) considerably thinner.
   d) considerably less rigid.

9. The herbaceous dicot stem lacks
   a) an epidermis.
   b) rings of woody xylem.
   c) functioning cortex cells.
   d) an active cambium.

10. In a three-year-old woody dicot stem, which is the most abundant tissue?
    a) phloem
    b) pith

        c) cortex
        d) xylem

11.   Rate of transpiration increases with
        a) an increase in humidity.
        b) the onset of night.
        c) a decrease in soil water.
        d) an increase in wind.

12.   In large trees, transpiration is great because
        a) there is such a great volume of wood in the trunk and branches.
        b) whenever photosynthesis takes place, transpiration greatly
            increases.
        c) large, mature vessel elements have greatly perforated end walls.
        d) as a tree grows, bark becomes old, split, and less waterproof.

13.   Of the theories advanced to account for the upward movement of
        water through the xylem, the one most favored is the
        a) root-pressure theory.
        b) active transport theory.
        c) transpiration-pull-cohesion theory.
        d) single-cell-pump theory.

14.   Poisoning the metabolic processes of a plant would interfere least
        with the
        a) push of food up from the roots early in the spring.
        b) pull of water up the plant by the leaves during midsummer.
        c) movement of minerals into the xylem of the root system.
        d) production of new xylem by the cambium.

15.   The Dixon-Joly theory provides an explanation for the ability of
        certain vascular plants, such as the mangrove, to
        a) transport foods through the phloem.
        b) live in salt water.
        c) carry on photosynthesis.
        d) push water long distances against the force of gravity.

   D. UNDERSTANDING AND APPLYING TERMS AND CONCEPTS

   1. What are two similarities between the cells that make up xylem and
        those that make up phloem?
        a)_____
        b)_____
        What is the major difference between these tissues?
        c)_____

   2. In what order would a water molecule pass through these tissues
        on its way to the mesophyll of a leaf? Number the tissues and
        place a zero in front of any tissue not on a direct route to the
        leaf mesophyll.
   ____a) root xylem   ____h) stem phloem
   ____b) root epidermis   ____i) stem cambium
   ____c) root cambium   ____j) root cortex
   ____d) leaf mesophyll   ____k) endodermis

_____ e) leaf xylem   _____ l) pericycle
_____ f) leaf phloem   _____ m) pith ray
_____ g) stem xylem

3. Match the terms in Column B with the descriptions in Column A. All terms will not be used.

Column A

a) Lilies, palms, orchids, and grasses are examples of this major subdivision of flowering plants: _____

b) These structures arise from individual, cylindrical, dead cells oriented end-to-end:
   _____

c) Secondary roots develop from this tissue: _____

d) This tissue serves as an area of food storage in plants: _____

e) The rate of transpiration is affected by: _____
   and _____

f) These structures are in the cork and facilitate gas exchange: _____

g) The process by which plants lose water from the leaves: _____

h) This tissue is composed of living, cylindrical cells arranged end-to-end, and transports food and hormones:
   _____

i) All elements needed by a plant are taken up in the form of: _____

j) That portion of the xylem contributing primarily support to a tree is called the: _____

Column B

A. phloem

B. inorganic molecules or ions

C. cortex

D. pericycle

E. organic nutrients

F. humidity

G. vessels

H. monocots

I. lenticels

J. Light

K. dicots

L. sapwood

M. heartwood

N. transpiration

4. An increase in light intensity greatly increases transpiration, even if there is <u>no</u> change in internal or external temperature. Why? _____

_____

5. Many leaves and herbaceous stems are fuzzy with tiny hairlike structures. List two ways this could decrease transpiration.
   a)_____
   b)_____

6. Over-fertilizing some plants seems to affect the transpiration rate. What effect would you expect? (a)_____
   Why? (b)_____

ANSWERS TO CHAPTER EXERCISES

Reviewing Terms and Concepts

1. a) water, minerals b) light c) translocation d) vascular

2. a) xylem b) water, minerals c) roots

3. a) sieve tubes b) living c) sugars d) leaves

4. a) meristem b) mitosis c) differentiate d) hairs e) cortex
   f) pericycle g) xylem, phloem h) cambium i) xylem, phloem

5. a) cork b) suberin c) cork cambium d) phloem e) xylem

6. a) xylem b) pith c) pith d) vascular bundles

7. a) veins b) structural support

8. a) 1 to 2 b) transpiration c) xylem d) root e) hypotonic
   f) hypertonic g) leaf h) xylem i) cohension j) tension

9. a) leaves b) phloem c) energy d) starch e) metabolic
   f) temperature g) oxygen

Testing Terms and Concepts

1. b	6. c	11. d
2. d	7. a	12. b
3. d	8. a	13. c
4. d	9. b	14. b
5. c	10. d	15. b

Understanding and Applying Terms and Concepts

1. a) Both are long, large-diameter cells. b) They have end walls that are perforated or completely degenerated. c) Phloem is alive and plays an active role in transporting materials. Xylem is dead and passive.

2. a) 6 b) 1 c) 5 d) 9 e) 8 f) 0 g) 7 h) 0 i) 0 j) 2 k) 3
   l) 4 m) 0

3. a) H b) G c) D d) C e) F, J f) I g) N h) A i) B j) M

4. Stomata open to admit carbon dioxide and thus let water vapor out.

5. a) Hairs slow air movements near the surface of the leaf and allow
      the local humidity to rise.
   b) Hairs shade the leaf surface and reduce the temperature of the
      leaf.

6. a) Transpiration should decrease.
   b) Fertilizer will lower the water gradient from the soil into the
      root and thus reduce the root pressure.

23

ANIMAL CIRCULATORY SYSTEMS

I. REVIEWING THE CHAPTER

A. CHAPTER HIGHLIGHTS

23.1 Simple Transport Mechanisms

Diffusion, active transport, and cytoplasmic streaming supply
microorganisms with food and other materials. Higher organisms have
specialized circulatory systems.

23.2 A "Closed" System: The Earthworm/23.3 An "Open" System: The
Grasshopper/23.4 The Squid

A generalized circulatory system, such as that in the earthworm,
includes (1) a fluid to carry dissolved materials, (2) a system of
vessels to carry the fluids, (3) a pump to move the fluids, and (4)
specialized organs to carry out exchanges between the fluid and the
environment. When blood is contained entirely within the system of
vessels, the system is said to be a closed system. When the blood is
confined in vessels only part of the time, as in insects, the system
is said to be open. The closed circulatory system of the squid
possesses three separate hearts. One pumps the blood to all internal
organs and tissues while the other two pump blood from the organs and
tissues to the gills.

23.5 Single Pump: The Fish/23.6 Three Chambers: The Frog and the
Lizard

Modern fishes have a single heart-pump, a muscular ventricle that
contracts and forces the blood to and then out of the capillary network
of the gills to the rest of the body. It returns to the heart by
entering a collecting chamber called the atrium. The amphibian heart
is more efficient because it has three chambers, two atria and a ventricle. The right atrium receives oxygen-deficient blood from the body
tissues while the left atrium receives oxygen-rich blood from the lungs.
Although both kinds of blood tend to mix in the ventricle, the mixing
is held to a minimum by the presence of narrow chambers. One important
advantage amphibians have over the fishes is that blood supplying gas-
exchange organs and body tissues is under full pressure. Reptiles have
a further advantage, a septum in the ventricle containing an opening
which closes when the ventricle contracts, thus preventing mixing of
the two kinds of blood.

23.7 Four Chambers: Birds and Mammals/23.8 The Heart

Mammals and birds have a four-chambered heart. The right atrium receives oxygen-deficient blood from the body, and the right ventricle pumps this blood forcibly to the lungs, where it gives off carbon dioxide and picks up a fresh supply of oxygen. This oxygenated blood then returns to the left atrium, passes into the left ventricle, and is pumped out forcibly to all other organs and tissues. The efficiency of having two separate circulatory systems makes possible the high rate of cellular respiration on which the homeothermic mammals and birds depend.

23.9   The Systemic Blood Vessels

Blood leaving the heart through the aorta moves because of the force exerted by the contraction of the left ventricle. The surge of blood at each contraction can be detected as the pulse beat. Even when the heart is relaxed, a state called diastole, there is pressure in the arterial system. When the heart contracts, in the state called systole, the pressure increases.

23.10   The Capillaries/23.11   Return of Blood to the Heart

The pressure of arterial blood is largely dissipated when the blood enters the capillaries. These are tiny, thin-walled vessels through which exchanges of materials between the blood and the tissue take place. From the capillaries, blood flows into the veins for return to the heart. Since it is under low pressure on leaving the capillaries, blood depends on the squeezing effect of active muscles to be moved along. Valves prevent backward movement of the blood.

23.12   The Blood Cells/23.13   The Plasma

The medium of transport in the circulatory system is blood. It carries oxygen, carbon dioxide, glucose, amino acids, metabolic wastes such as urea, ions of various salts, and hormones. Blood also distributes heat throughout the body and guards the body against infective disease agents. Blood is a liquid tissue. It contains red blood cells, which are manufactured in the bone marrow and are about 90% hemoglobin. Oxygen in the lung capillaries attaches to the hemoglobin and so is carried to cells throughout the body. White blood cells are less numerous than the red ones, their chief function being to protect the body from infection. Among the white blood cells are the neutrophils and monocytes, which combat foreign agents by the process of endocytosis. The lymphocytes give rise to the antibodies and so combat disease. Platelets are cell fragments, also produced in the bone marrow. They are very important in the process of blood clotting. The fluid in which the blood cells are suspended is called the plasma. Mostly water, the plasma carries food molecules to our cells, removes metabolic wastes, and carries vitamins and hormones. The plasma also contains fibrinogen, an essential component of the clotting process.

23.14   Oxygen Transport/23.15   Carbon Dioxide Transport

Hemoglobin plays several roles in the transport of gasses through the body. It aggressively attracts and binds oxygen in lung tissues, and it readily releases that oxygen in the body tissues. Hemoglobin is also involved, in two different ways, in transporting most of the carbon dioxide from the body tissues to the lungs for excretion. Without this latter action, the pH of the blood would be fatally lowered.

23.16 Exchanges Between The Blood and the Cells/23.17 The Lymphatic System

A substantial amount of plasma entering a capillary bed passes into the tissue space, some of it not entering the venule end of the capillary. That remaining fluid is picked up by tiny vessels of the lymphatic system and is called lymph. The lymph is ultimately returned to the blood circulation. When lymph begins to accumulate in the tissues and distend them, the condition is called edema.

23.18 The Heart/23.19 Auxiliary Control of the Heart

The stimulus that maintains the heart's rhythmic beat is self-contained and originates in a specialized region of the right atrium called the pacemaker. It is an electrical stimulus and can be recorded in an electrocardiogram. Without action of the pacemaker, the ventricles continue pumping, but more slowly, and there is danger that their action will become disorganized and random, which leads to death through a condition known as ventricular fibrillation. Persons subject to this condition may be helped by an artificial pacemaker. Two auxiliary nervous control centers of heartbeat are located in the medulla. During stress, the accelerator nerves cause an increase in the rate and strength of the heartbeat. During unnecessary or excessive heart activity, the vagus nerves of the medulla transmit impulses that cause a slowing of the heartbeat.

23.20 Peripheral Control of Circulation

In times of danger or other stress, the arterioles supplying blood to the skeletal muscles will be dilated while the bore of the vessels supplying the digestive organs will be decreased. This action results from nervous stimulation and also stimulation by a hormone called adrenaline. Many other chemical substances also help regulate blood flow. In mammals, the kidney is responsible for monitoring blood pressure and, if the pressure drops, it releases an enzyme called renin. This initiates a series of reactions in blood components, producing an end substance that causes the muscular walls of the arterioles to contract. This closes down capillary beds and by bringing the volume of the functioning blood vessels back into balance with the volume of blood, restores normal blood pressure.

23.21 The Transport of Heat

The circulatory system carries heat through the body, as well as nutrients, from sites of heat production or collection to sites of dissipation. This action is important in both homeotherms (animals with a relatively constant body temperature) and poikilotherms (variable body temperature).

Homeotherms exploit a variety of mechanisms to conserve heat, including increased metabolism, increased muscular activity, decreased blood flow to the skin and to extremities, and extraction of heat from blood flowing to extremities by a mechanism called a countercurrent exchanger. To get rid of excess heat, homeotherms can increase blood flow to the surface and to extremities, and they can evaporate water by sweating or panting.

23.22 Clotting of Blood

When blood vessels are cut, blood loss must be prevented. This is accomplished by a series of reactions among clotting factors that give rise to a substance called fibrin, which gradually forms a mesh in which the blood cells become embedded, and hence bleeding gradually stops.

B. KEY TERMS

hemoglobin	muscle pump
atrium	white blood cells
pericardium	lymphocytes
pulmonary system	eosinophils
angina pectoris	platelets
torr	hemocoel
anemia	septum
monocytes	pulmonary artery
antigens	coronary arteries
leukemia	pulse
capillaries	red blood cells
arteriole	pulmonary vein
venule	aorta
ventricle	coronary system
tricuspid valve	lymph node
bicuspid valve	neutrophils
coronary occlusion	antibodies
homeothermic	basophils

plasma	poikilothermic
albumin	endothermic
interstitial fluid	edema
thoracic duct	diastole
systole	thrombin
renin	serum
fibrinogen	lymph
globulins	pacemaker
lymphatic system	vagus
countercurrent exchanger	ectothermic

II. MASTERING THE CHAPTER

A. LEARNING OBJECTIVES

When you have mastered the material in this chapter, you should be able to:

1. Define all key terms.

2. List the four components of the earthworm's closed circulatory system.

3. Distinguish between a closed and an open circulatory system, describe the advantages and limitations of each, and list animals in which each can be found.

4. Distinguish between a double and a single (heart) pumping system, and cite an organism with each, and explain why the double pump evolved.

5. Describe the improvements in the hearts of the frog, lizard, and bird, over that of the fish, and explain the effects of these improvements on blood flow.

6. Draw a diagram showing the human circulatory system, including details of the heart, with arrows indicating direction of blood flow.

7. Discuss the function and importance of the capillaries.

8. List six substances for which blood serves as the medium of transport.

9. List the three types of "formed" elements in the blood and cite the chief functions of each.

10. Name and discuss the functions of the liquid part of the blood.

11. Describe the conditions that permit hemoglobin both to pick up oxygen in the lungs and to release it in the body tissues.

12. Describe the two mechanisms enabling hemoglobin to transport carbon dioxide from the tissues to the lungs.

13. Explain why interstitial fluid leaves and reenters the capillary network.

14. State how lymph in the tissue space reenters the blood system.

15. List three causes of edema.

16. Describe the basic intrinsic control of the heartbeat and the extrinsic control exercised by the medulla of the brain.

17. List four mechanisms that regulate blood flow in the periphery of the body, away from the heart.

18. State the circulatory condition that results in shock, and how recovery is made possible.

19. Discuss the role of the circulatory system in maintaining a constant temperature in homeotherms.

20. Explain the roles of platelets, thrombin, and fibrin in the formation of a blood clot.

21. Summarize the three major functions of the circulatory system and describe two specific examples of each.

B. REVIEWING TERMS AND CONCEPTS

1. Diffusion, (a) _____ transport, and (b) _____ streaming supply microorganisms with food and other materials. Higher organisms have specialized (c) _____ systems for this purpose.

2. A generalized circulatory system, such as that in the earthworm, includes (1) a fluid to carry dissolved materials, (2) a system of (a) _____ to carry the fluids, (3) a (b) _____ to move the fluids, and (4) specialized organs to carry out (c) _____ between the fluid and environment. When blood is contained entirely within the system of vessels, the system is said to be a (d) _____ system. In insects, the system is said to be (e) _____ because part of the time the blood is not confined to the vessels. The closed circulatory system of the squid is maintained by (f) _____ separate hearts.

3. Modern fishes have a (a) _____ (single/double) heart-pump,

214

a muscular (b) _____ that contracts and forces the blood to
and then out of the capillary network of the (c) _____ to the rest
of the body. Blood returns to the heart and enters a collecting
chamber called the (d) _____. The amphibian heart is more efficient because it has three chambers, two (c) _____ and a (f)
_____. The right (g) _____ receives oxygen-deficient blood
from (h) _____ _____ by way of (i) _____, while
the left (j) _____ receives oxygen-rich blood from the (k)
_____.

   4. Mammals and birds have a (a) _____-chambered heart. The
right (b) _____ receives oxygen (c) _____ blood from the
body, and the right (d) _____ pumps this blood forcibly to the
(e) _____, where it gives off (f) _____ _____
and picks up a fresh supply of (g) _____. This blood then returns
to the left (h) _____, passes into the (i) _____, and is
pumped out forcibly to all other organs and tissues.

   5. The pressure of arterial blood is largely dissipated when the
blood enters the (a) _____. These are tiny, thin-walled vessels
through which exchanges of materials between the (b) _____ and the
(c) _____ take place. From the (d) _____, blood flows into
the veins for return to the (e) _____. Since it is under (f)
_____ (low/high) pressure on leaving the capillaries, blood depends on the (g) _____ effect of active (h) _____ to be
moved along and (i) _____ to prevent its backward flow.

   6. Blood is a liquid tissue. It contains (a) _____ blood
cells, which are manufactured in the (b) _____ marrow, and which
are about 90% hemoglobin. (c) _____ blood cells are less numerous
than the (d) _____ ones, their chief function being to protect
the body from (e) _____. Among these blood cells are the neutrophils and monocytes, which combat foreign agents by the process of (f)
_____. The lymphocytes give rise to (g) _____ and so combat
disease.

   7. The fluid in which the blood cells are suspended is called
the (a) _____. Mostly (b) _____, it carries food molecules
to our cells, removes metabolic wastes, carries vitamins and hormones,
and transports heat through the body. It also contains (c) _____,
an essential component of the clotting process. So the three major
functions of blood are to (d) _____ materials to and from our body
tissues, to defend the body against (e) _____ and _____,
and to help maintain a constant body (f) _____.

   8. (a) _____ actively transports oxygen from the lungs to
cells throughout the body. It also transports the metabolic waste (b)
_____ _____ from the cells to the (c) _____.

   9. A substantial amount of the (a) _____ entering a capillary passes into the tissue space, most but not all of it entering the
(b) _____ end of the capillary. That remaining fluid is picked
up by tiny vessels of the (c) _____ system. When lymph begins
to accumulate in the tissues and distend them, the condition is called

(d) _____.

10. The stimulus that maintains the heart's (a) _____ beat is self-contained in the wall of the right (b) _____ and is called the (c) _____.

11. Two auxiliary nervous control centers of heartbeat are located in the (a) _____. During stress, the (b) _____ nerves cause an (c) _____ in the rate and strength of the heartbeat. During unnecessary or excessive heart activity, the (d) _____ nerves of the medulla transmit impulses that cause a (e) _____ of the heart-beat.

12. The flow of blood through arterioles is also regulated by nerves and by the hormone (a) _____, which increase the flow of blood to skeletal muscles during periods of stress. During increased activity, individual tissues themselves can (b) _____ (increase/decrease) their own blood supply. In mammals, another organ that monitors blood pressure is the (c) _____, which responds to (d) _____ (high/low) pressure by releasing an enzyme called (e) _____.

13. Warm-blooded animals do not always maintain a body temperature higher than their surroundings and are more properly referred to as (a) _____ animals. Cold-blooded animals are (b) _____. Birds and mammals are (c) _____ and generate their heat (d) _____ _____ (externally/internally). Thus, they are also (e) _____.

14. Many mammals can generate more heat by increasing their (a) _____ or their (b) _____ _____. Mechanisms that can conserve heat include reduced blood flow to (c) _____ and (d) _____ _____.

15. When blood vessels are cut, blood loss must be prevented. This is accomplished by clotting factors that give rise to a substance called (a) _____, which gradually forms a (b) _____ in which the blood cells become embedded, and hence bleeding gradually stops.

C. TESTING TERMS AND CONCEPTS

1. The pump forcing an earthworm's blood to the capillaries consists of
   a) a five-chambered heart.
   b) cardiac muscle lining the blood vessels.
   c) cytoplasmic streaming in the individual cells.
   d) five pairs of aortic loops.

2. An animal is described to you as having no capillaries at all. Which of the following represents a reasonable conclusion?
   a) The demand on the circulatory system for the transport of oxygen must be very low or even zero.
   b) If this animal has a circulatory system, it must be "closed."
   c) This animal could be either an earthworm or a squid.
   d) The animal probably does not have a hemocoel either.

3. In a fish, blood is received under full pressure by
    a) the brain.
    b) muscles and internal organs.
    c) all tissues of the body.
    d) the gills.

4. Veins always carry blood
    a) that is oxygenated.
    b) that is deoxygenated.
    c) to the heart.
    d) from the heart.

5. Blood is forced through veins primarily by
    a) ventricular contraction.
    b) skeletal muscle contraction.
    c) movement of valve flaps.
    d) systolic pressure.

6. Oxygen-deficient blood from the body is first received by the
    a) right atrium.
    b) left atrium.
    c) right ventricle.
    d) left ventricle.

7. You carefully feel a blood vessel visible on the inner surface of
    your forearm and find it has no pulse. That vessel is
    a) a capillary.
    b) an artery.
    c) carrying blood to the left atrium.
    d) carrying oxygen-poor blood.

8. The tissues of the heart receive nutrients from blood in
    a) capillaries of the coronary system.
    b) capillaries of the systemic system.
    c) the aorta.
    d) the chambers of the heart itself.

9. Lymphocytes combat disease by
    a) endocytosis.
    b) forming antibodies.
    c) forming antigens.
    d) forming fibrinogen.

10. The chief function of albumin is to
    a) maintain normal blood pressure.
    b) fight bacterial disease.
    c) contribute to the formation of a blood clot.
    d) nourish body tissues.

11. Hemoglobin is a(n)
    a) lipid with heme.
    b) fatty acid.
    c) protein.
    d) antibody.

12. Interstitial fluid reenters the venous end of the capillaries
    a) enriched with protein.
    b) enriched with oxygen.
    c) under osmotic pressure.
    d) under pressure of the heartbeat.

13. A rabbit is quietly feeding in your garden when you appear suddenly and fire a shotgun over its head. Which of the following will not occur?
    a) nervous increase in the heart rate
    b) hormonal increase in blood pressure
    c) increased production of interstitial fluid
    d) increased action of the vagus nerve

14. Excess fluid in the tissue spaces is picked up and returned to the bloodstream by
    a) veins.
    b) lymph nodes.
    c) lymphatic vessels.
    d) glands.

15. A lizard is an ectotherm. How does its circulatory system help most to achieve a high body temperature each morning?
    a) by carrying heat from its skin to deeper body structures
    b) by carrying heat from deep structures to the skin
    c) by carrying heat from active skeletal muscles to other parts of the body
    d) by reducing heat loss through countercurrent exchange

D. UNDERSTANDING AND APPLYING TERMS AND CONCEPTS

Decide whether the following statements are true (T) or false (F).

_____ 1. The grasshopper lacks an oxygen-carrying pigment in its blood.

_____ 2. In the frog, as in the cat, oxygen-rich and oxygen-poor blood are kept separated.

_____ 3. Red blood cells are effective in destroying harmful agents in the blood.

_____ 4. An antibody usually combines with a variety of antigens.

_____ 5. Glucose, amino acids, short-chain fatty acids, vitamins, hormones, nitrogenous wastes (e.g. urea), and many ions are all transported dissolved in the plasma of the blood.

_____ 6. The reaction between hemoglobin and oxygen is reversible.

_____ 7. About one-third of the plasma entering a capillary passes into the tissue space and is stored there.

_____ 8. In times of danger or other stress, the arterioles supplying

218

_____ 9. Blood pressure in times of stress may be partially controlled by muscles in the capillary walls.

10. The squid and octopus have closed circulatory systems, while other mollusks (e.g. clams, snails) have open systems. What feature of the squid's activity demands the more elaborate system? _____

11. A drop of blood will pass through the following major blood vessels, heart chambers, and capillary bed on its way from a jugular vein to the aorta. Number these structures in the correct order of blood flow.

_____ jugular vein   _____ left atrium

_____ aorta   _____ right atrium

_____ pulmonary vein   _____ left ventricle

_____ pulmonary artery   _____ right ventricle

_____ superior vena cava   _____ pulmonary capillaries

12. Choose the best description for each term. Descriptions may be used more than once or not at all.

_____ a) erythrocyte

_____ b) thrombocyte

_____ c) lymphocyte

_____ d) neutrophil

_____ e) histamine

_____ f) plasma

_____ g) hemoglobin

_____ h) adrenalin

_____ i) fibrin

A. increases blood flow in specific areas

B. decreases rate and strength of heartbeat

C. engulfs bacteria in times of health and illness

D. attacks bacteria without engulfing and digesting them

E. involved in one of the first steps of clot formation

F. long, insoluble, protein threads

G. a transport protein

H. a cell that transports both oxygen and carbon dioxide

I. transports hormones and nutrients

13. Trace the electrical pathway within the heart: from the (a) _____, over the (b) _____, to the (c) _____, down the (d) _____ _____ of the septum, to the (e) _____.

14. In general, capillary beds are used irregularly, flowing full some of the time and not flowing at all at other times. Why is this the case? a)_____

    Name a part of the body that is especially irregularly supplied with blood ((b) _____) and one that must have a more uniform supply ((c) _____). What structure is used to adjust the flow of blood through a capillary bed? (d) _____ What would happen if all capillaries were allowed to fill at once? (e) _____

15. In what three ways does the circulatory system contribute to the maintenance of a high body temperature in homeotherms?
    a)_____
    b)_____
    c)_____

ANSWERS TO CHAPTER EXERCISES

Reviewing Terms and Concepts

1. a) active b) cytoplasmic c) circulatory

2. a) vessels b) pump c) exchanges d) closed e) open f) three

3. a) single b) ventricle c) gills d) atrium e) atria f) ventricle
   g) atrium h) body tissues i) veins j) atrium k) lungs

4. a) flour b) atrium c) deficient d) ventricle e) lungs f) carbon
   dioxide g) oxygen h) atrium i) ventricle

5. a) capillaries b) blood c) cells d) capillaries e) heart
   f) low g) squeezing h) muscles i) valves

6. a) red b) bone c) White d) red e) infection f) endocytosis
   g) antibodies

7. a) plasma b) water c) fibrinogen d) transport e) infection,
   injury f) temperature

8. a) Hemoglobin b) carbon dioxide c) lungs

9. a) plasma b) venule c) lymphatic d) edema

10. a) rhythmic b) atrium c) pacemaker

11. a) medulla b) accelerator c) increase d) vagus (or inhibitory)
    e) slowing

12. a) adrenalin b) increase c) kidney d) low e) renin

13. a) homeothermic b) poikilothermic c) homeothermic d) internally
    e) endothermic

14. a) metabolism b) physical activity c) extremities d) counter-
    current exchange

15. a) fibrin b) meshwork

Testing Terms and Concepts

1. d	6. a	11. c
2. a	7. d	12. c
3. d	8. a	13. d
4. c	9. b	14. c
5. b	10. a	15. a

Understanding and Applying Terms and Concepts

1. T 2. F 3. F 4. F 5. T 6. T 7. F 8. F 9. F

10. It is an active predator, requiring a larger oxygen supply.

11. 1, 10, 7, 5, 2, 8, 3, 9, 4, 6

12. a) H b) E c) D d) C e) A f) I g) G h) A i) F

13. a) S-A node b) atria c) A-V node d) conducting fibers
    e) ventricles

14. a) active tissues require greater blood flow than inactive

    b) skeletal muscles or digestive tract c) heart or brain

    d) precapillary sphincters e) There would be a rapid decrease in

    blood pressure and insufficient flow to brain, causing fainting

    (shock).

15. a) distribute heat from metabolically or physically active tissues
    to other parts of the body

    b) decrease loss of heat from body surface by decreasing peripheral
    blood flow

    c) utilize countercurrent exchange to decrease peripheral loss of
    heat without decreasing peripheral flow

24

THE IMMUNE RESPONSE

I. REVIEWING THE CHAPTER

A. CHAPTER HIGHLIGHTS

24.1 Introduction

It is common knowledge that certain diseases never strike the same individual twice. After contracting the disease initially, a change occurs within the body making that individual immune to the disease. Immunity results from either of two mechanisms, cell-mediated immunity carried out by certain lymphocytes and active against viruses and perhaps cancer, and antibody immunity carried out by protein molecules in blood serum (the "antibodies") and active against bacterial diseases.

24.2 The Structure of the Immune System

The central organs of the immune system are the bone marrow and the thymus. The bone marrow produces a variety of white blood cells, but the two most important categories are the T lymphocytes, responsible for the cell-mediated response, and the B lymphocytes, which produce antibodies. Immature T lymphocytes leave the bone marrow, travel in the bloodstream to the thymus, and mature there. B lymphocytes mature in the bone marrow itself. Both types then disperse to clusters of lymphoid tissue throughout the body, including the spleen, tonsils, appendix, and a multitude of lymph nodes. These are the structures where foreign materials, called antigens, are most likely to be encountered, and here macrophages engulf the invaders and the immune responses are initiated.

24.3 The Structure of Antibodies

Antibodies are serum globulins (globular proteins) with a basic structure that is uniform among many different antibodies. The molecule consists of two identical long polypeptided chains (the heavy chains), to which are attached two small carbohydrates and two identical short polypeptide chains (the light chains). About two-thirds of this molecule has the same structure from one antibody to another, within a given type, and is called the constant (C) region. The remaining one-third differs in its amino acid sequence from antibody to antibody and is called the variable (V) region, and contains small sections that are especially variable called hypervariable regions. These last are crucial to antibody function.

24.4 The Interaction of Antibodies and Antigens

Antibodies act by combining, in a "lock and key" manner, with a small section of antigen molecule, called the antigenic determinant. This behavior is similar to that of an enzyme and its substrate. The specific combining site that unites with the antigenic determinant comprises hypervariable regions of both the light and heavy chains of the antibody. Once antibodies combine with an antigen, the antigen can be engulfed by phagocytic white blood cells or otherwise destroyed by the body.

24.5 Antigens

The immune system manufactures antibodies against noninfectious materials, as well as against infectious organisms, such as ragweed pollen, insect venoms, and the Rh antigen found in the blood of "Rh positive" individuals. T lymphocytes are also manufactured against noninfectious materials, for example the active ingredient in poison ivy. In general, any macromolecule not a normal component of the body will act as an antigen. This includes polysaccharides, proteins, and nucleic acids. The immune system distinguishes between "self" and "non-self."

24.6 How are Antibodies Elicited?

The steps leading to antibody production begin when the antigen is engulfed by phagocytes. There is evidence to show that the body has lymphocytes capable of recognizing the antigen and combining with it after the phagocyte action. The reaction between lymphocytes and antigen occurs at receptors on the cell surface that are identical in the combining site to the antibodies that will ultimately appear in the blood. Once the appropriate lymphocyte binds to an antigen, it begins to divide and build up a clone of cells capable of producing large amounts of that one antibody. For four or five days, these B lymphocytes are concerned mainly with building up the clone; then they begin to produce significant amounts of antibodies and are called "plasma cells."

24.7 The Secondary Response

A second exposure to an antigen, even years after the first, often elicits a much faster "secondary response" to that antigen. Such immunity develops because, during the primary reaction many lymphocytes neither continue to divide, contributing to the clone, nor produce antibodies, but revert to small lymphocytes called "memory cells." Memory cells recognize antigens as the original B lymphocytes do, they have lifespans of 20 years or more, and their large number insures that the secondary response to a disease will be rapid and massive. Vaccines are preparations of antigenic material that will produce this population of memory cells without producing the symptoms of the disease.

24.8 The Genetic Basis of Antibody Diversity

The ability of an organism to manufacture specific antibodies for a million or more different antigens places a heavy burden on the genome that must encode for these millions of proteins. A human genome contains only thirty thousand to forty thousand genes, not millions. In fact, each antibody is not encoded by a single, stable gene. There are separate genes for the constant and variable regions of the heavy and light chains, and the variable genes exist as two (light chain) or three (heavy chain) separate segments. As B lymphocytes differentiate, a few different constant genes and a few hundred different variable gene segments link randomly to yield an enormous diversity of finished heavy and light chains. Secondly, the variable segments are unstable, and point mutations produce new gene segments and further increase the diversity of the final antibody chains. Thirdly, the various heavy and light chains combine randomly to again increase the diversity of the finished antibody molecules. In this way, hundreds of antibody genes can direct the production of millions of different antibody molecules, an antibody for literally any antigen you could possibly encounter.

24.9 Transplantation and the Uniqueness of the Individual

Landsteiner found antigenic substances now called A and B antigens on red blood cells. Individuals with A antigens have group A blood, those with B antigens, group B blood. Those with both A and B antigens have group AB blood, while those with neither have group O blood. To facilitate transfusions, blood introduced must not contain red blood cells that the patient's antibodies can agglutinate.

Successful organ transplants depend on careful tissue typing and the use of drugs, called immunosuppressives, which act by inhibiting mitosis. Otherwise, the foreign tissue is destroyed by a cell-mediated immune response. The rejection is slow in the first exposure but rapid after a second transplant from the same donor. This "second-set" phenomenon appears to be equivalent to the secondary "memory" response found in antibody reactions.

24.10 Cancer and Immune Surveillance

Another function of the cell-mediated immune mechanism is to destroy cancerous cells. Cancer cells are recognized as "non-self," even though they have arisen within the organism itself. This watch-dog activity is known as "immune surveillance."

24.11 Allergies

Allergies are cell- or antibody-mediated immune reactions to otherwise harmless environmental antigens (foods, pollen, dust). Hay-fever involves the vigorous production of type IgE antibodies. These antibodies are bound to mast cells which, upon antigen contact, explosively expel histamine. To combat the symptoms caused by histamine, the patient is given antihistamines.

B. KEY TERMS

antigen   plasma cell

clonal selection   allergy

B lymphocyte   secondary response

cell-mediated   vaccines

clone   T lymphocyte

immune surveillance   antihistamine

antibodies   desensitize

antigenic determinant

II. MASTERING THE CHAPTER

A. LEARNING OBJECTIVES

When you have mastered the material in this chapter, you should be able to:

1. Define all key terms.

2. Discuss the difference between cell-mediated and humoral or antibody-mediated immunity.

3. List the two central organs and four secondary organs of the immune system.

4. Describe the origins and actions of the T and B lymphocytes.

5. Describe and explain the structure of an antibody molecule.

6. Explain how antibodies combine with antigens.

7. List three categories of macromolecules that may act as antigens, and one category that does not.

8. Cite the results of experiments done by Ada and Byrt and by Edelman to determine how antibodies are elicited.

9. Explain the clonal selection theory.

10. Describe the secondary response and explain how it occurs so rapidly.

11. Explain how relatively few genes can direct the production of a great diversity of antibodies.

12. Explain how a vaccine can confer immunity without the necessity of contracting the disease.

13. Distinguish among the four Landsteiner blood groups and list the

antigens and antibodies a person with each blood group would have.

14. Discuss the reasons for rejection of organ transplants and the steps one can take to prevent it.

15. State why cancer cells are usually destroyed by the immune system and why this protection sometimes fails.

16. Explain the immune mechanism that causes a person to suffer the symptoms of hayfever.

B. REVIEWING TERMS AND CONCEPTS

1. It is common knowledge that certain diseases never strike the same individual twice. After contracting the disease initially, a change occurs within the body making that individual (a) _____ to the disease. Immunity results from either of two mechanisms, (b) _____ immunity carried out by certain lymphocytes and active against viruses and perhaps cancer, and (c) _____ immunity carried out by protein molecules in blood serum and active against bacterial diseases.

2. The central organs of the immune system are the (a) _____ and the (b) _____. The (c) _____ produces two, important categories of white blood cell: the (d) _____, responsible for the cell-mediated response, and the (e) _____, which produce antibodies.

3. Antibodies are globular (a) _____ with a basic structure that is uniform among many different antibodies. The molecule consists of two identical heavy chains, to which are attached two small (b) _____ and two identical (c) _____ _____. About two-thirds of this molecule has the same structure from one anti-body to another, within a given type, and is called the (d) _____ region. The remaining one-third differs in its amino acid sequence from antibody to antibody and is called the (e) _____ region, and contains small, functionally important sections called (f) _____ regions.

4. Antibodies act by combining with the material that elicited them, the (a) _____. This behavior is similar to that of an (b) _____ and its substrate. In fact, the antigen combines with only a small part of the antigen, called the (c) _____ _____.

5. In general, any (a) _____ not a normal component of the body will act as an antigen. Antigens may be: (1) (b) _____, as in Type III pneumococcus; (2) (c) _____, as in the diptheria toxin or the outer coat of a polio virus; or (3) (d) _____. The immune system may make antibodies against (e) _____ materials, such as ragweed pollen or the Rh antigen, as well as against (f) _____ organisms.

6. Antibody production begins when the (a) _____ is en-gulfed by (b) _____. There is evidence to show that the body

226

has (c) _____ capable of recognizing the antigen and combining with it after the phagocyte action. Once the appropriate lymphocyte binds to an antigen, it begins to divide and build up a (d) _____ of cells capable of producing large amounts of that one antibody. When these cells do begin to produce significant amounts of antibodies, they are called (e) _____ _____.

7. A second exposure to an antigen, even years after the first, often elicits a much faster (a) _____ _____ to that antigen. Such immunity develops because, during the primary reaction many lymphocytes neither continue to (b) _____, contributing to the clone, nor produce (c) _____, but revert to small lymphocytes called (d) _____ _____. (e) _____ _____ recognize antigens as the original B lymphocytes do, and their large number insures that the secondary response to a disease will be (f) _____ and _____.

8. Millions of different antibodies can be produced by only a few (a) _____ genes because the genes code for not whole antibodies but parts of antibodies, and these parts come together in (b) _____ combinations. Furthermore, the variable genes are prone to (c) _____, resulting in new genes and therefore an increased variety of antibodies.

9. Landsteiner found antigenic substances on red blood cells, now called (a) _____ and _____ antigens. Individuals with A antigens have group (b) _____ blood, those with B antigens, group (c) _____ blood. Those with both have group (d) _____ blood and those with neither have group (e) _____ blood.

10. Successful organ transplants depend on careful tissue (a) _____ and the use of drugs, called (b) _____, which act by (c) _____ mitosis. Otherwise, the foreign tissue is destroyed by a (d) _____ immune response. The rejection is slow in the first exposure but rapid after a second transplant from the same donor. This (e) _____ phenomenon appears to be equivalent to the secondary (f) _____ response found in antibody reactions.

11. Another function of the cell-mediated immune mechanism is to destroy (a) _____ cells, recognized as non-self, even though they have arisen within the organism itself. This watchdog activity is known as (b) _____ _____.

12. Allergies are cell- or (a) _____-mediated immune reactions to otherwise harmless environmental antigens. Hayfever involves the vigorous production of type (b) _____ antibodies. These antibodies are bound to (c) _____ cells which, upon antigen contact, explosively expel (d) _____. To combat the hayfever symptoms, the patient is given (e) _____.

C. TESTING TERMS AND CONCEPTS

1. Cell-mediated immunity

a) is a mechanism that resides in the serum of the blood.
b) involves the release of antibodies by a kind of white blood cell.
c) acts against many viral diseases and cancers.
d) combats bacterial disease and is responsible for the symptoms of hayfever.

2. An important difference between T and B lymphocytes is that T lymphocytes
   a) take up ultimate residence in the tonsils.
   b) undergo a period of differentiation in the thymus.
   c) are inactive until after they leave the bone marrow.
   d) must develop into "plasma cells" before they can be effective.

3. Which statement about the role of phagocytosis in immunity is false?
   a) Phagocytosis of foreign material somehow triggers the immune response.
   b) In cell-mediated immunity, invaders are phagocytized by lymphocytes.
   c) T lymphocytes recruit macrophages, and both cells act against foreign materials together.
   d) Antibodies are often effective because they make phagocytosis of the invader easier.

4. Edelman's model of a myeloma protein consists of
   a) a long chain and two short chains connected by S-S bridges.
   b) two long chains and two short chains connected by S-S bridges.
   c) two long chains and a short chain intertwined around each other.
   d) two long chains, a short chain, and a carbohydrate.

5. Which statement is true about the variability of antibody molecules?
   a) Antibodies are quite constant in structure, one to another.
   b) Variable regions exist in long chains but short chains are constant.
   c) Both short and long chains have constant and variable regions; in addition, long chains have hypervariable regions.
   d) Both short and long chains have all three types of regions.

6. The antibody is secreted from the cell by
   a) active transport.
   b) diffusion.
   c) endocytosis.
   d) exocytosis.

7. An antibody attaches to an antigen
   a) along its whole length.
   b) in a white blood cell.
   c) by a lock and key arrangement.
   d) at one of the constant regions of a polypeptide chain.

8. Antibody formation is not elicited by
   a) glucose.
   b) nucleic acids.
   c) proteins.
   d) noninfectious materials.

9. Passive protection is a form of immunity
   a) often found in older people.
   b) often elicited by vaccination.
   c) that is less powerful but longer lasting than active protection.
   d) in which the antibodies are not made by one's own immune system.

10. The secondary response against an antigen is more rapid than the primary response, because
   a) there are many more lymphocytes active against that antigen.
   b) there are more lymph nodes able to trap that antigen.
   c) the appropriate lymphocytes have more accurate information about that antigen.
   d) the antibodies become more destructive against the antigen.

11. Antibodies are
   a) proteins, and synthesized like other proteins.
   b) proteins, but do not contain amino acids.
   c) polysaccharides assembled in Golgi bodies.
   d) nucleic acids assembled on rough E.R.

12. Preparations of infectious agents altered so as not to cause disease are
   a) toxiods.
   b) antibodies.
   c) vaccines.
   d) immunosuppressives.

13. The number of genes necessary to code for millions of different antibodies is
   a) millions of different genes.
   b) hundreds of different genes.
   c) variable—some species require millions; others need very few.
   d) zero—genes are not involved.

14. If a skin graft fails,
   a) the same donor will not be used for fear of a more powerful "second-set" reaction.
   b) the same donor might be used again to "desensitize" the recipient.
   c) a different donor will not be used because this patient has already been "sensitized."
   d) a different donor might be used in an effort to avoid eliciting the same antibodies.

15. The "universal donor" is blood group
   a) A.
   b) B.
   c) AB.
   d) O.

D. UNDERSTANDING AND APPLYING TERMS AND CONCEPTS

1. List four components of the lymphatic system involved in the

immune response.

a)_____

b)_____

c)_____

d)_____

Why is the immune system more closely associated with the lymphatic system than with the blood circulatory system? e)_____

2. What category of organic molecule (carbohydrate, lipid, protein, nucleic acid) do antibodies fall into? a) _____
What is it about that category that ideally suits it for the task of the antibody? b)_____

3. What is the significance of the hypervariable regions of an antibody? _____

4. Do we posess the ability to manufacture almost any conceivable antibody, or do we acquire the ability as we are exposed to various antigens? _____

Decide whether the following statements are true (T) or false (F).

_____ 5. A lymph node is more than 99% effective in removing and destroying bacteria.

_____ 6. Only macromolecules can stimulate the immune response.

_____ 7. The immune system is organized in such a way that antigens "instruct" the lymphocytes how to build antibodies against them.

_____ 8. Organisms often have genes for antibodies against totally unnatural, man-made antigens.

_____ 9. Antibody genes are copied by mitosis much less accurately than other genes.

_____ 10. Human myeloma proteins have constant regions and variable regions.

_____ 11. Antibodies combine with a small part of the antigen, called the antigenic determinant.

_____ 12. The interaction between antibody and antigen is precise and specific.

_____ 13. The clonal selection theory says that antibody-producing clones arise as a result of the selective stimulation of pre-existing cells.

_____ 14. A given antigen gives rise to a single antibody molecule.

_____ 15. Blood group is an inherited characteristic.

_____ 16. In order to reject a graft, a mouse must not have had a
functioning thymus gland at the time its immune machinery was
set up.

_____ 17. A mouse will accept as "self" a skin graft or tumor from
another mouse of the same inbred strain.

ANSWERS TO CHAPTER EXERCISES

Reviewing Terms and Concepts

1. a) immune b) cell-mediated c) antibody

2. a) bone marrow b) thymus c) bone marrow d) T lymphocytes
    e) B lymphocytes

3. a) proteins b) carbohydrates c) light chains d) constant
    e) variable f) hypervariable

4. a) antigen b) enzyme c) antigenic determinant

5. a) macromolecule b) polysaccharides c) proteins d) nucleic
    acids e) noninfectious f) infectious

6. a) antigen b) phagocytes c) lymphocytes d) clone e) plasma cells

7. a) secondary response b) divide c) antibodies d) memory cells
    e) memory cells f) rapid, massive

8. a) hundred b) random c) mutation

9. a) A, B b) A c) B d) AB e) O

10. a) typing b) immunosuppressives c) inhibiting d) cell-mediated
    e) second-set f) memory

11. a) cancerous b) immune surveillance

12. a) antibody b) IgE c) mast d) histamine e) antihistamines

Testing Terms and Concepts

1. c	6. d	11. a
2. b	7. c	12. c
3. b	8. a	13. b
4. b	9. d	14. a
5. d	10. a	15. d

Understanding and Applying Terms and Concepts

1. a) thymus b) tonsil c) appendix d) lymphnode e) The blood is
    more isolated from the environment and from the possibility of

invasion. Bacteria and other foreign materials tend to enter the lymphatic system.

2. a) protein b) More than any other type of molecule, proteins exist in a great variety of three-dimensional shapes.

3. It is here that each type of antibody differs from every other type and so attaches to a different antigen.

4. We have the ability before exposure.

5. T 6. T 7. F 8. T 9. T 10. T 11. T 12. T 13. T 14. F 15. T 16. F 17. F

25

EXCRETION AND HOMEOSTASIS

I. REVIEWING THE CHAPTER

A. CHAPTER HIGHLIGHTS

25.1 Excretion in Plants

To complete the story of metabolism, we must examine the ways in which living things return their waste products of metabolism to the environment, in a process called excretion. The most abundant of these products are carbon dioxide, water, and ammonia. Although "wastes" in excess amounts, these products do play important regulatory roles in the body.

Excretion in green plants does not pose serious problems since green plants use much of their waste products of respiration in photosynthesis. Also, metabolism in plants is based on carbohydrates rather than proteins. Terrestrial plants for the most part store their wastes in heartwood and/or leaves, while aquatic plants diffuse them into the external environment.

25.2 Excretion in the Amoeba

Many single-celled freshwater organisms eliminate their metabolic wastes by diffusion, as do aquatic plants. Excess water, however, is collected in a contractile vacuole and expelled by energy expenditure.

25.3 Excretion in the Invertebrates

Excretion in many invertebrates is achieved by means of structures called nephridia. They pick up fluid containing both wastes and useful materials. The useful materials are reclaimed by the cells of the tubule while the wastes are expelled through openings called nephridiopores. Insects have slightly different structures called Malpighian tubules, which expel wastes not directly into the environment but into the gut. More water is conserved by this arrangement than by the nephridium. Excretion of nitrogen in the form of uric acid, rather than ammonia, also conserves water in insects.

25.4 Structure of the Human Kidney/25.5 The Formation of Urine

The main excretory organ of the vertebrates is the kidney. Each kidney is composed of about one million microscopic tubules, called nephrons. The first section of the nephron is called Bowman's capsule, which, with the glomerulus, filters the blood of small molecules. Of these, needed nutrients are reabsorbed by active transport in the proximal tubule and water is reabsorbed by osmosis. Additional adjustment of the contents of the nephric filtrate occurs in the loop of Henle and distal tubule, producing the finished product, urine.

25.6 Control of the Kidney

A substance called ADH is secreted by the pituitary gland and regulates the water concentration of the blood. In the event of excess water, ADH production is inhibited and less water is reabsorbed by the distal tubule and collecting ducts, thus giving rise to large amounts of dilute urine. In time of water shortage in the blood—due to excessive perspiring, for example—ADH production is increased and stimulates greater reabsorption of water by the distal and collecting tubules. When people lose the ability to secrete ADH, they become victims of the disease diabetes insipidus, one symptom being an enormous production of watery urine accompanied by terrible thirst.

25.7 Mechanics of Elimination

Several nephrons drain into a single collecting duct. The collecting ducts drain into the renal pelvis, which empties into the ureter and bladder. The bladder empties to the outside through the urethra.

25.8 Kidney Disease

Some of the conditions that can cause temporary or permanent kidney failure are shock, blood loss, certain poisons, and certain infectious diseases. Treatment involves the use of an external, artificial kidney or dialysis machine, or the transplantation of a new kidney.

25.9 The Nitrogenous Wastes of Humans

Most of the nitrogenous wastes in the body arise from the breakdown of amino acids in a process called deamination. This yields ammonia, a very poisonous substance. The liver quickly converts this ammonia into urea, a less poisonous material, and the kidney removes it from the body.

25.10 Aquatic Vertebrates

Freshwater fish face the problem of coping with a continual inflow of water from their hypotonic environment. To avoid serious dilution of its body fluids, the freshwater fish must excrete excess water. It does so by means of its kidney. Most of the nitrogenous wastes leave the body of freshwater fish in the form of ammonia by diffusion out of the gills.

Saltwater bony fishes face the opposite problem in their hypertonic environment. They tend to lose water to the environment. They replace it by drinking seawater and desalting it. The salt is returned to the sea by active transport at the gills.

25.11 Terrestrial Vertebrates

The amphibian kidney, like that of the freshwater fish, functions chiefly as a mechanism for excreting excess water. When in water, frogs take in water by osmosis through the skin. When on land, frogs conserve water by adjusting the rate of filtration at the glomerulus and by reabsorbing water from its bladder.

Reptiles, such as the desert rattlesnake, survive in their dry environment by being able to convert waste nitrogen compounds into uric acid. Since this substance is quite insoluble, it can be excreted without the use of too much water. Birds also excrete uric acid.

B. KEY TERMS

pelvis	Malpighian tubules
uric acid	nephron
medulla	proximal tubule
glomerulus	collecting duct
distal tubule	urine
tubular reabsorption	ADH
tubular secretion	bladder
ureter	cloaca
deamination	nephridia
contractile vacuole	cortex
Bowman's capsule	diabetes insipidus
loop of Henle	urethra
nephric filtrate	urea

II. MASTERING THE CHAPTER

A. LEARNING OBJECTIVES

When you have mastered the material in this chapter, you should be able to:

1. Define all key terms.

2. Cite three reasons why excretion in a green plant is less of a problem than is excretion in a cat.

3. State the fundamental problem and the principal organ of excretion in the amoeba, earthworm, grasshopper, and dog.

4. Diagram the principal structures composing the human kidney.

5. Distinguish between tubular secretion and filtration-reabsorption, and cite the principal function of each.

6. Relate the chains of events preventing excess water loss when you begin to perspire profusely on a hot summer day.

7. Tell what happens to excess amino acids from the time of deamination to the time their toxic product is eliminated from the body.

8. Cite the two ways in which marine fishes have "solved" the problem of losing vital body water to their hypertonic environment.

9. Distinguish between the excretory systems—and their end products—of a rattlesnake and a human.

10. Explain how a dialysis machine cleanses blood of wastes.

B. REVIEWING TERMS AND CONCEPTS

1. Living things return their waste products of metabolism to the environment in a process called (a) _____. These products are chiefly carbon dioxide, (b) _____, and _____. Although "wastes," these products play important (c) _____ roles in the body.

2. Excretion in green plants does not pose serious problems since green plants use much of their waste products of respiration in (a) _____. Also, metabolism in plants is based on (b) _____ rather than proteins.

3. The amoeba and many single-celled freshwater organisms eliminate their metabolic wastes by (a) _____; so do aquatic plants. Excess water, however, is collected in a contractile (b) _____ and expelled by expenditure of (c) _____.

4. Excretion in many invertebrates is achieved by means of structures called (a) _____. They pick up fluid containing both (b) _____ and useful materials. The useful materials are reclaimed while the (c) _____ are expelled through openings called (d) _____. Insects have excretory structures called (e) _____ _____, which release their wastes into the (f) _____. This arrangement is particularly effective in conserving (g) _____. A second (h) _____-saving feature in insects is the excretion of nitrogen as (i) _____ _____.

5. The main excretory organs of vertebrates are the (a) _____. A structure called the (b) _____ filters waste out of the (c) _____, and manufactures a liquid called (d) _____. The chief nitrogenous waste in all mammals is (e) _____, which is

manufactured in the (f) _____ . Most of the nitrogenous wastes
in the body arise from the breakdown of (g) _____ _____
in a process called (h) _____ .

6. Water balance in mammals is regulated by the hormone (a)
_____ , secreted by the (b) _____ gland. This hormone
specifically alters the walls of the (c) _____
and _____ _____ _____ so that (d) _____ _____ (more/
less) water is reabsorbed and a (e) _____ (hypotonic/hypertonic)
urine is produced.

7. After urine is manufactured it flows through collecting ducts
into the renal (a) _____ , into a tube called the (b) _____ ,
to the (c) _____ , and out of the body through the (d) _____ .

8. Freshwater fish face the problem of coping with a continual
inflow of (a) _____ from their hypotonic environment. To
avoid serious dilution of its body fluids, the freshwater fish must
excrete excess water. It does so by means of its (b) _____ .
Most of the nitrogenous wastes leave the body of freshwater fish by
(c) _____ out of the (d) _____ .

9. Saltwater bony fishes tend to lose (a) _____ to the environment. they replace it by drinking seawater and (b) _____
it. The (c) _____ is returned to the sea by (d) _____
transport at the (e) _____ .

10. The amphibian kidney, like that of freshwater fish, functions
chiefly as a mechanism for excreting excess (a) _____ . When in
water, frogs take in water by (b) _____ through the (c) _____ .
When on land, frogs conserve water by retaining in their bladder
copious amounts of (d) _____ _____ (placed there
while in the water), which is reabsorbed into the blood and so replaces
the water lost by evaporation through the skin.

11. Reptiles, such as the desert rattlesnake, survive in their
dry environment by being able to convert waste nitrogen compounds into
(a) _____ _____ . Since this substance is quite
(b) _____ , it can be excreted without the use of too much
water.

C. TESTING TERMS AND CONCEPTS

1. Mammals use considerable protein as a fuel, as well as a building
    material. What problem does this cause for the body?
    a) The catabolism of protein yields uric acid, a poison to living
       cells.
    b) The catabolism of protein yields ammonia, a potent poison.
    c) The breakdown of protein consumes unusually large amounts of
       water.
    d) The use of protein as a fuel liberates large amounts of carbon
       dioxide.

2. The metabolism in plants is based principally on

237

a) proteins.
b) carbohydrate.
c) hydrocarbons.
d) lipids.

3. In herbaceous, terrestrial plants, wastes tend to be
   a) eliminated through root epidermis.
   b) stored in root cortex.
   c) stored in heartwood.
   d) stored until the tops die.

4. Why is excretion less of a problem for aquatic animals than for terrestrial animals?
   a) Aquatic animals tend to be smaller.
   b) Terrestrial animals have a problem with water conservation.
   c) Aquatic animals have a great tolerance for urea buildup.
   d) Terrestrial animals generally feed on coarser foods.

5. A human nephron is most similar, in how it manufactures urine, to a(n)
   a) protozoan's contractile vacuole.
   b) insect's Malpighian tubule.
   c) marine fish's nephron.
   d) earthworm's nephridium.

6. Which of the animal pairs below are mismatched in their dominant nitrogenous waste?
   a) amoeba—freshwater fish
   b) grasshopper—bird
   c) snake—frog
   d) human—kangaroo rat

7. Urine is produced by the
   a) medulla.
   b) pelvis.
   c) loop of Henle.
   d) nephron.

8. The reabsorption of water in the human kidney is due to
   a) active transport.
   b) endocytosis.
   c) osmosis.
   d) filtration.

9. The conversion of ammonia to urea occurs in the
   a) liver.
   b) pancreas.
   c) glomerulus.
   d) distal tubule.

10. Maintaining a constant blood pH is aided by the kidney action called
    a) secretion.
    b) filtration.

c) reabsorption.
d) dialysis.

11. The hormone ADH is secreted by the
    a) medulla.
    b) cortex.
    c) blood.
    d) pituitary.

12. Of the filtrate removed from the blood by the nephrons, what per-
    cent is reabsorbed back into the bloodstream?
    a) 2
    b) 42
    c) 87
    d) 99

13. The reabsorption of glucose and other useful materials from the
    nephric filtrate takes place in the
    a) Bowman's capsule.
    b) proximal tubule.
    c) glomerulus.
    d) collecting duct.

14. Through variable active transport and varying amounts of ADH, we
    excrete very different amounts of materials from the kidneys,
    from one day to another. Why do our excretory processes vary?
    a) The amount of nitrogenous wastes we produce varies considerably
       day to day.
    b) We have varying needs for these "wastes."
    c) Our tolerance for poisons varies.
    d) The amount of wastes excreted by other means varies.

15. Which answer is false? Shark's blood
    a) contains large amounts of urea.
    b) contains small amounts of ammonia.
    c) is processed by filtration-reabsorption.
    d) is hypotonic to seawater.

D. UNDERSTANDING AND APPLYING TERMS AND CONCEPTS

1. List four reasons why plants have no serious problems in excretion.
   a)_____
   b)_____
   c)_____
   d)_____

2. Describe the similar problem in excretion faced by an amoeba and
   a freshwater fish. (a)_____
   _____
   Describe the different solutions to this problem found in the
   amoeba (b) and fish (c).
   b)_____
   c)_____

3. Describe the <u>similar</u> problem in excretion faced by a grasshopper and a kangaroo rat. (a)_____

_____
Describe the <u>different</u> solutions to this problem found in the grasshopper (b) and rat (c).
b)_____
c)_____

4. In what order would a molecule of urea pass through the following structures on its way out of the body? Number them, placing a zero before the structures <u>not</u> on a direct route out of the body.

_____a) renal pelvis   _____i) aorta

_____b) glomerulus   _____j) loop of Henle

_____c) proximal tubule   _____k) urethra

_____d) distal tubule   _____l) renal artery

_____e) collecting duct   _____m) renal vein

_____f) afferent arteriole   _____n) bladder

_____g) efferent arteriole   _____o) ureter

_____h) Bowman's capsule

5. Match the correct activities or functions to the excretory structures described on the left.

_____a) distal tubule

_____b) proximal tubule

_____c) glomerulus

_____d) Bowman's capsule

_____e) capillary network

_____f) collecting duct

_____g) renal pelvis

_____h) loop of Henle

A. Responsible for secretion of wastes

B. Filtration of blood

C. First receives nephric filtrate

D. Obligate reabsorption of most needed nutrients and water

E. Produces steep salt gradient in kidney medulla

F. No secretion but does reabsorb varying amounts of water, depending on body's need

G. Receives urine

H. Receives reabsorbed nutrients

6. Match the terms in Column B with the descriptions in Column A.

240

Column A	Column B
a) The nitrogenous waste of insects is _____	A. ADH
	B. ammonia
b) The chief nitrogenous waste in humans is _____	C. urea
c) In reptiles these are very small or entirely lacking _____	D. glomeruli
	E. nephron
d) The relative permeability of the walls of the distal tubules at a given moment depends on the presence or absence of _____	F. liver
	G. nephridia
	H. osmosis
e) Urea is manufactured in the _____	I. tubular secretion
	J. amino acids
f) That structure of the human kidney that manufactures urine is the _____	K. uric acid
	L. gills
g) Frogs take in water by _____	
h) Excretory structures in the earthworm are known as _____	
i) Nitrogenous wastes in the human body result from the breakdown of _____	
j) The principal organ for nitrogenous waste exretion in freshwater fishes is the _____	
k) Deamination results in the accumulation of _____	
l) This process is an ion exchange process _____	

ANSWERS TO CHAPTER EXERCISES

Reviewing Terms and Concepts

1. a) excretion b) water, ammonia c) regulatory

2. a) photosynthesis b) carbohydrates

3. a) diffusion b) vacuole c) energy

4. a) nephridia b) waste c) wastes d) nephridiopores e) Malpighian
    tubules f) gut g) water h) water i) uric acid

5. a) kidneys b) nephron c) blood d) urine e) urea f) liver
    g) amino acids h) deamination

6. a) ADH b) pituitary c) distal tubules, collecting ducts d) more
    e) hypertonic

7. a) pelvis b) ureter c) bladder d) urethra

8. a) water b) kidney c) diffusion d) gills

9. a) water b) desalting c) salt d) active e) gills

10. a) water b) osmosis c) skin d) watery urine

11. a) uric acid b) insoluble

Testing Terms and Concepts

1.	b	6.	c	11.	d
2.	b	7.	d	12.	d
3.	d	8.	c	13.	b
4.	b	9.	a	14.	b
5.	d	10.	a	15.	d

Understanding and Applying Terms and Concepts

1. a) Plants have a low catabolic rate and produce few catabolic
    wastes. b) Plants use many wastes (water, carbon dioxide,
    nitrogen compounds) to a degree animals can't. c) Plants rely
    more on carbohydrates, which yield fairly benign wastes, than
    on proteins, whose wastes are more poisonous. d) Cell walls
    make excess water less of a threat.

2. a) Both are hypertonic to their environments and must excrete
    large quantities of water. b) contractile vacuole that expels
    water c) nephron with glomerulus and Bowman's capsule that
    produces copious hypotonic urine

3. a) Both are terrestrial and must excrete wastes with as small a
    water loss as possible. b) The grasshopper has Malpighian
    tubules that collect uric acid and deposit it into the gut,
    where almost all water is reclaimed. c) The rat has kidneys
    with nephrons with unusually long loops of Henle. These allow
    the rat to produce a solution of urea 17 times more concentrated
    than its blood.

4. a) 10 b) 4 c) 6 d) 8 e) 9 f) 3 g) 0 h) 5 i) 1 j) 7 k) 13
    l) 2 m) 0 n) 12 o) 11

5. a) A b) D c) B d) C e) H f) F g) G h) E

242

6. a) K b) C c) D d) A e) F f) E g) H h) G i) J j) L
    k) B l) I

RESPONSIVENESS AND COORDINATION
IN PLANTS

I. REVIEWING THE CHAPTER

A. CHAPTER HIGHLIGHTS

26.1 Importance of Internal Communication

One of the chief distinguishing features of living things, as
opposed to nonliving things, is that living things are capable of
actively responding to certain changes in their environment. Environmental changes serve as stimuli that trigger a definite response of
the part of the organism.

In animals, two systems of internal communication are found. The
nervous system is a very fast-acting system. Specialized cells called
neurons conduct electrochemical impulses from one part of the body to
another. The endocrine system is somewhat slower in its action.
Specialized glands, the endocrine glands, release hormones to the circulatory fluid. Plants do not have a nervous system and therefore
rarely exhibit quick, localized responses. Plants achieve their
responsiveness and coordination through a system of plant hormones.

26.2 Growth Movements

Two kinds of growth movements in response to outside stimuli are
recognized in plants: (1) nastic movements, a response independent of
the direction from which an external stimulus strikes the organism;
and (2) tropisms, growth movements whose directions are determined by
the direction from which the stimulus strikes the plant. The response
of plants to a light stimulus is called phototropism; the response of
plants to a gravity stimulus is called geotropism.

26.3 The Mechanism of Phototropism/26.4 The Discovery and Role of Auxin

The first experiments on plant phototropism were conducted by
Charles Darwin and his son, Francis. Ultimately F. W. Went experimented with Avena sativa to show that a chemical growth stimulator,
known as auxin, is responsible for the phototropic response in plants.
Auxin is produced in the shoot tip and translocated to the shady side
and down the shoot, where it stimulates cell elongation and causes the
bending of the shoot toward the light.

26.5 Other Auxin Activities

Auxins participate in the coordination of a variety of other plant activities: (1) apical dominance, or the inhibition of lateral growth; (2) fruit development; (3) inhibition of abscission, the shedding of leaves and fruit; and (4) root initiation, particularly that of adventitious roots, arising from stems or leaves rather than from the regular root system of the plant.

26.6 How do Auxins Work?

Auxin has been shown to affect plant genes themselves, stimulating RNA production and protein synthesis. Other, more rapidly acting effects in the cell have not yet been discovered.

26.7 The Gibberellins

Gibberellins are a family of hormones that stimulate vertical stem growth (not curved, as with auxin), particularly in "dwarf" varieties. They also stimulate root growth, flower development, the sprouting of buds, and the germination of seeds.

26.8 The Cytokinins

Cytokinins, acting together with auxin, strongly stimulate mitosis in meristematic tissues. Cytokinins also promote differentiation of cells in meristems and slow down the aging of plant parts.

26.9 Abscisic Acid (ABA)/26.10 Ethylene

Abscisic acid is an inhibitor that speeds up abscission and causes dormancy in meristems, as a means of protecting them from cold weather. Ethylene stimulates the ripening of fruit.

26.11 Factors that Initiate Flowering

In some plants, such as the tomato, the stimulus that initiates flowering is internal, a matter of adequate growth and food reserves. In most plants, however, external stimuli are important. A period of cold temperatures is necessary for flowering in biennials. A particular length of daylight and night, or photoperiod, is required in many other species.

26.12 The Mechanism of Photoperiodism/26.13 The Discovery of Phyto-chrome/26.14 Other Phytochrome Activities

Plants that are controlled by their photoperiod fall into two classes: short-day or long-night plants, which flower only when exposed to a period of darkness longer than some critical period typical of the species, and long-day or short-night plants, which flower only when exposed to a period of darkness shorter than its critical period.

The photoperiod is detected by a pigment called phytochrome, found in the leaves, where the hormone florigen is produced. Florigen moves through the phloem to the flower bud and causes it to open.

245

Phytochrome triggers other responses to light, including the germination of some seeds and the rapid elongation of stems in the dark, known as etiolation.

B. KEY TERMS

stimuli   geotropism

endocrine system   apical dominance

tropisms   abscission

auxin   abscisic acid

nervous system   photoperiodism

hormones   adventitious roots

phototropism   florigen

bioassay   gibberellins

neurons   cytokinins

nastic movements   phytochrome

actinomycin D

II. MASTERING THE CHAPTER

A. LEARNING OBJECTIVES

When you have mastered the material in this chapter, you should be able to:

1. Define all key terms.

2. Describe the two related systems of communication in animals, and state briefly why plants have only one.

3. Describe the mechanisms by which plants can respond to their environment with movement.

4. Cite the two kinds of growth movements in response to outside stimuli in plants, and give one example of each.

5. Distinguish between phototropism and geotropism, and give an example of each.

6. List four ways in which auxins participate in the coordination of plant activities.

7. Describe specifically how auxin causes differential growth in stems.

8. Distinguish between auxins, gibberellins, and cytokinins by citing one example of the action of each.

9. Describe how the inhibitory or "aging" hormones abscisic acid and ethylene contribute to the coordination of a plant's activities.

10. Cite three factors that may be required to induce flowering.

11. Distinguish between the roles of light and dark in the inducement of flowering.

12. Describe the roles of phytochrome and florigen in the control of flowering.

B. REVIEWING TERMS AND CONCEPTS

1. One of the chief distinguishing features of living things, as opposed to nonliving things, is that living things are capable of actively (a) _____ to certain changes in their (b) _____. Environmental changes serve as (c) _____ that trigger a definite (d) _____ on the part of the organism.

2. In animals, two systems of internal communication are found. The (a) _____ system is a very fast-acting system. Specialized cells called (b) _____ conduct electrochemical impulses from one part of the body to another. The (c) _____ system is some-what slower in its action. Specialized glands, the (d) _____ _____, release (e) _____ to the (f) _____ fluid.

3. Plants achieve their responsiveness and coordination through a system of plant (a) _____. Two kinds of growth movements in response to outside stimuli are recognized in plants: (b) _____ movements, responses independent of the direction from which an external stimulus strikes the organism, and (c) _____, growth movements whose directions are determined by the direction from which the stimulus strikes the plant.

4. The growth response of plants to a light stimulus is called (a) _____; the response of plants to a gravity stimulus is called (b) _____. F. W. Went experimented with Avena sativa to show that a chemical growth stimulator, known as (c) _____, in varying concentrations is responsible for the phototropic response in plants. This hormone is produced in the (d) _____ _____ and translocated to the (e) _____ (shady/sunny) side and down the shoot, where it causes differential growth.

5. Auxins participate in the coordination of a variety of plant activities other than phototropism: (a) _____ _____, or the inhibition of lateral growth; (b) _____ development; (c) _____, involved in the shedding of leaves and fruit; and (d) _____ initiation, particularly those arising from (e) _____ or _____ rather than from the regular (f) _____ system.

6. (a) _____ are plant hormones that stimulate stem growth, root growth, bud sprouting, and germination of seeds. (b) _____ stimulate mitosis and promote (c) _____ of cells in meristems. (d) _____ _____ is an inhibitor that causes dormancy in meristems. Ethylene causes the (e) _____ of fruit.

7. In some plants, the stimulus that initiates flowering is internal, a matter of adequate (a) _____ or (b) _____. In most plants, however, external stimuli are important. A period of cold is necessary for flowering in (c) _____. A particular length of daylight and night, or (d) _____, is required in other species.

8. Photoperiodism is controlled by a pigment called (a) _____, a (b) _____ (what color?) pigment that absorbs (c) _____ light. Phytochrome is also involved in the (d) _____ of some seeds and the regulation of stem growth.

C. TESTING TERMS AND CONCEPTS

1. Which statement about coordination of plant activity is true?
   a) A plant's endocrine system is more important than its nervous system.
   b) Plants are very passive, compared to animals, and when mature, rarely respond to environmental stimuli.
   c) Plant hormones usually travel through the plant in the xylem.
   d) Plant hormones regulate the degeneration and death of plant parts, as well as the growth and healthy functioning of those parts.

2. The opening and closing of some flowers and leaves in the plant kingdom is an example of
   a) a positive phototropism.
   b) a negative phototropism.
   c) a geotropism.
   d) a nasty.

3. A plant organ that exhibits positive geotropism is the
   a) root.
   b) stem.
   c) leaf.
   d) flower.

4. If the region below the tip of an oat coleoptile is both shaded from the light and chemically blocked in its metabolic activity, light directed from the side will cause
   a) a bending toward the light.
   b) a bending away from the light.
   c) increased vertical growth.
   d) no change.

5. If a mica water is inserted halfway into the right side of an oat coleoptile, just below the tip, and light is shone vertically downward onto the plant, the plant will

a) continue to grow vertically, at a normal rate.
b) continue to grow vertically, at a reduced rate.
c) begin to bend to the right.
d) begin to bend to the left.

6. Auxins inhibit the development of
   a) coleoptiles.
   b) abscission layers.
   c) the ovary wall of the pistil.
   d) adventitious roots.

7. Auxin production assures that
   a) fruits will be shed.
   b) leaves will be retained.
   c) seeds will break dormancy.
   d) lateral buds will break dormancy.

8. Gibberellins do not stimulate the
   a) sprouting of buds in spring.
   b) germination of seeds.
   c) growth of dwarf plants.
   d) turning of houseplants toward the window.

9. A plant organ that exhibits negative phototropism is the
   a) leaf.
   b) stem.
   c) flower.
   d) root.

10. The development of a uniform yellow color in lemons is hastened
    by
    a) abscisic acid.
    b) cytokinin.
    c) ethylene.
    d) florigen.

11. The differentiation of meristematic cells is stimulated by
    a) gibberellins.
    b) cytokinins.
    c) auxins.
    d) abscisic acid.

12. During the germination of a seed
    a) auxin stimulates cell division and the emergence of the young
       embryo.
    b) auxin weakens the seed coat and aids the emergence of the young
       embryo.
    c) gibberellin stimulates enzyme production and digestion of food
       for the embryo.
    d) cytokinins are not active.

13. In Maryland Mammoth tobacco, flowering is induced by
    a) a certain temperature.
    b) a certain growth achieved.

c) a certain photoperiod.
d) development of sufficient food reserves.

14. The receptor of the photoperiodic stimulus for the development of flower primordia is the
a) flower bud.
b) meristem.
c) lateral bud.
d) leaf.

15. At nightfall, all the phytochrome in a plant is in the
a) red-absorbing form.
b) far-red-absorbing form.
c) florigen form.
d) a form that depends on the species of plant.

D. UNDERSTANDING AND APPLYING TERMS AND CONCEPTS

1. There are two mechanisms by which a plant can move, in response to changes in its environment. The more rapid one involves (a)_____.
The slower one involves (b)_____.

2. What is the basic evidence for the existence of communication between one part of a plant and another in the case of positive phototropism? (a)_____
and photoperiodism? (b)_____

3. Describe a role played by each of these hormones in the germination of a seed:
a) auxin_____
b) gibberellin_____
c) abscisic acid_____
d) cytokinin_____

4. A long-day plant will flower when the days reach 14 hours and the nights 8 hours. If such a plant is exposed to this 14-8 photoperiod but a box is placed over the plant for a half hour in the middle of the day, will it still flower? (a)_____ Why or why not? (b)_____

Decide whether the following statements are true (T) or false (F).

____ 5. Quick, localized responses are not common in the plant kingdom.

____ 6. An equal distribution of auxin in plant stems is responsible for the bending in phototropism.

____ 7. It is possible to stimulate fruit development in nonpollinated flowers.

_____ 8. Apical dominance seems to result from the upward transport of auxin produced in the root meristem.

_____ 9. Treatment of plant cells with auxin produces an increase in RNA synthesis and in protein synthesis.

_____ 10. Gibberellin overcomes the hereditary limitations in many "dwarf" plant types.

_____ 11. Many newly formed seeds must undergo a period of enforced dormancy before they can germinate.

_____ 12. A positive geotropism is a growth upward, against the pull of gravity.

_____ 13. A bioassay is a test using a living organism to determine the identity of an experimental substance.

_____ 14. Grafting a long-day plant to short-day plant yields a combination that flowers at no season of the year.

_____ 15. Phytochrome has been implicated in the germination of seeds and ripening of fruits, as well as in flowering.

ANSWERS TO CHAPTER EXERCISES

Reviewing Terms and Concepts

1. a) responding b) environment c) stimuli d) response

2. a) nervous b) neurons c) endocrine d) endocrine glands
   e) hormones f) circulatory

3. a) hormones b) nastic c) tropisms

4. a) phototropism b) geotropism c) auxin d) shoot tip e) shady

5. a) apical dominance b) fruit c) abscission d) root e) stems,
   leaves f) root

6. a) Gibberellins b) Cytokinins c) differentiation d) Abscisic
   acid e) ripening

7. a) growth b) food reserves c) biennials d) photoperiod

8. a) phytochrome b) blue c) orange-red d) germination

Testing Terms and Concepts

1.	d	6.	b	11.	b
2.	d	7.	b	12.	c
3.	a	8.	d	13.	c
4.	d	9.	d	14.	d
5.	c	10.	c	15.	b

Understanding and Applying Terms and Concepts

1. a) changes in turgor pressure in cells supporting leaves or other
    plant parts b) cell division and cell elongation

2. a) Shading the tip of a shoot shows this meristem is the part
    sensitive to the direction of the light, but the region of
    elongation is the part that ultimately responds. A signal must
    move from the former to the latter. b) Again, shading shows
    leaves are sensitive to the photoperiod, but it is the flower
    bud that responds.

3. a) stimulates cell elongation and tropic responses of stem and
    root. b) causes enzymes to be released that digest endosperm.
    c) maintains dormancy. d) stimulates first cell divisions and
    differentiation.

4. a) yes b) It is the period of darkness that is critical, and our
    experiment has not changed that.

5. T	9. T	13. F
6. F	10. T	14. F
7. T	11. T	15. T
8. F	12. F	

27

ANIMAL ENDOCRINOLOGY

I. REVIEWING THE CHAPTER

A. CHAPTER HIGHLIGHTS

27.1 Introduction

     Animals need some mechanism by which the various cells, tissues,
and organs can communicate. Two systems exist in most animals. The
nervous system consists of neurons that transmit electrical impulses
throughout the body. The endocrine system secretes chemical substances
called hormones that are transported throughout the body by the blood.
In some cases, all cells of the body are affected by a hormone; more
often, only certain "target" organs respond to it.

27.2 Insect Hormones

     In insects, the endocrine control of molting and metamorphosis
has been extensively studied. The brain hormone prothoracicotropic
hormone stimulates the prothoracic glands to produce ecdysone, which
initiates molting. In a young larva, the corpora allata produce
juvenile hormone (JH), which causes the molt to be postponed to another
larval stage. As the insect ages, less and less JH is produced, resulting in a pupal molt and finally, when no JH is present, metamorphosis
into the adult.

     Knowledge of how certain hormones regulate metamorphosis has pro-
mise for pest control. For example, when solutions of organic compounds structurally similar to a hormone called juvenile hormone are
sprayed on mature caterpillars, or on the foliage on which they are
feeding, normal pupation is prevented.

27.3 Research Techniques in Endocrinology

     The research techniques used in the study of hormones have become
quite standardized: the suspected endocrine tissue is first removed,
consequent changes are observed, the tissue is replaced and the
symptoms are expected to disappear, an extract of the tissue (presumably containing the unknown hormone) is prepared that should successfully replace the tissue itself, and finally the extract is purified and the hormone identified.

27.4 The Thyroid Gland

     The thyroid gland, located in the neck, produces the hormone
thyroxin, which controls the rate of body metabolism. Hypothyroidism

(insufficient production) before maturity results in cretinism, abnormal physical and mental development. In adults it leads to myxedema, an overweight condition and coarsening of the features. Inadequate amounts of iodine lead to enlargement and extra activity of the thyroid gland, a condition called goiter. The rate of activity of the thyroid gland is governed by the pituitary hormone TSH.

27.5 The Parathyroid Glands

The parathyroid glands, embedded in the rear of the thyroid gland, moniter blood calcium levels and increase blood Ca^{++} when levels fall too low. Its targets are the bones of the body, the small intestine, and the kidneys. Loss of function of the parathyroid glands can be compensated for by careful addition of Ca^{++} or vitamin D to the diet. Hyperparathyroidism causes extremely brittle bones and kidney stones.

The thyroid also monitors blood calcium levels, and when levels rise too high, release the hormone calcitonin, which shifts Ca^{++} from blood to bones. Two hormones working together give much finer control over blood Ca^{++} than either could do alone.

27.6 The Skin

When ultraviolet radiation strikes the skin, it triggers the production of claciferol, also called vitamin D, a hormone that enhances the absorption of Ca^{++} from the intestine. Inadequate amounts of this hormone prevent the normal deposition of calcium in bone tissue and cause rickets in children and osteomalacia in adults.

27.7 The Stomach and Intestine

The stomach secretes the hormone gastrin, which stimulates the production of hydrochloric acid by the parietal cells of the stomach and so aids in digestion. The intestine secretes cholecystokinin and secretin, hormones that stimulate the secretion of pancreatic digestive fluid and bile.

27.8 The Islets of Langerhans

The islets of Langerhans, scattered throughout the pancreas, produce the hormone insulin, which regulates the level of glucose in the blood. It does this by stimulating the liver to convert glucose to glycogen and fats by making the cells of the body more permeable to the entrance of glucose. Insufficient production of this hormone results in diabetes mellitus, a disease resulting in the inability to convert glucose to glycogen or fats.

The pancreas also produces glucagon, which stimulates the breakdown of glycogen and increases blood sugar levels. Finally, somatostatin is released as blood sugar and fat levels rise. This inhibits the action of glucagon. Thus, three hormones interact to hold blood glucose levels constant.

27.9 The Pituitary Gland

The pituitary gland is often called the "master" gland, because it controls the activity of several other endocrine glands. It is located at the base of the brain and consists of two major lobes. The anterior lobe produces eight different hormones. Human growth hormone indirectly promotes growth of the body, during childhood and adolescence, and a variety of anabolic reactions, during stress, in the adult. Prolactin is best known for stimulating milk production following birth; it causes other kinds of maternal activity in other vertebrates. Thyroid-stimulating-hormone causes the production of thyroxin and is in turn depressed by thyroxin. Similarly, adrenocorticotropic hormone stimulates the adrenal cortex, and follicle-stimulating hormone (FSH) acts upon the gonads. In females, FSH stimulates the ovarian follicles; in males, it stimulates the seminiferous tubules. Luteinizing hormone also acts on the gonads. In females it triggers ovulation; in males it stimulates the interstitial cells. Beta-lipotropin enhances the metabolism of fats in lab animals. Melanocyte-stimulating hormone has little effect in humans, but in other vertebrates it causes a dramatic darkening of the skin.

The posterior lobe of the pituitary gland is not really an endocrine gland; it apparently stores and releases two hormones produced in the hypothalamus of the brain. Oxytocin stimulates smooth muscle contractions, especially in the uterus at birth and in the mammary glands during nursing. Antidiuretic hormone increases the reabsorption of water by the kidneys and causes vasoconstriction in arterioles, thus increasing blood pressure.

27.10 The Hypothalamus

The hypothalamus is a component of both the nervous and endocrine systems and acts as a communicating link between them. This makes it possible for endocrine responses to be partially based on information gathered by the nervous system, even from the organism's surrounding environment.

In addition to the hormones released by the posterior pituitary, the hypothalamus produces three releasing hormones and one inhibitory hormone, which control the activity of the anterior pituitary.

27.11 The Adrenal Glands

The adrenal glands, located one atop each kidney, have two parts. The exterior part of the gland is called the adrenal cortex. It releases glucocorticoids that help in the conversion of fats and protein into intermediary metabolites that are ultimately converted into glucose. Glucocorticoids also supress inflammation and are crucial in the long-term resistance to stress. The adrenal cortex also produces many mineralocorticoids that help in the reabsorption of Na^+ and Cl^- in the tubules of the kidney and maintain adequate blood pressure. These cortical hormones are absolutely essential to life. The interior part of the adrenal glands, called the adrenal medulla, produces the hormones adrenaline and noradrenaline. The former stimulates activities which enable the body to cope with stress, and the latter causes an increase in blood pressure; thus, both help prepare the animal for action.

27.12 The Gonads

The gonads release hormones in both males and females. In males, the testes release androgens (for example, testosterone) into the bloodstream. This hormone triggers the development of secondary male sexual characteristics. In females, the ovaries release hormones called estrogens. The two major functions of the estrogens are to promote secondary female sexual characteristics and to prepare the body for possible pregnancy. A second ovarian hormone is progesterone, which continues to prepare the uterus for pregnancy and inhibits the development of new eggs.

27.13 The Placenta

In addition to providing for the nourishment of the embryo, the placenta produces a variety of hormones: first, human chorionic gonadotropin, which stimulates the corpus luteum and so maintains high levels of female sex hormones, and, later, the sex hormones themselves. In this way, the uterus is kept in a pregnant condition.

27.14 The Pineal Gland/27.15 The Kidney

The pineal produces melatonin, which inhibits the gonads in some mammals and may be involved in the regulation of annual reproductive cycles. The kidney produces renin, which maintains blood pressure, and erythropoietin, which stimulates red blood cell production.

27.16 Hormones and Homeostasis/27.17 The Mechanism of Action of Hormones

The endocrine system plays a major role in the maintenance of a constant internal environment. For instance, the concentration of sugar, water, and various salt ions in the ECF is maintained within narrow limits by hormonal action. The competence of cells to respond to a given hormone is important. All hormones are carried by the blood from the endocrine gland producing them to every cell of the body. Once admitted to a target cell, a given hormone simply releases the cell's potentialities to respond to a given situation. Steroid hormones enter the target cell and directly induce gene activity. Protein hormones bind to a receptor site on the cell surface and indirectly affect the cell's activity.

27.18 The Pheromones

Pheromones, unlike hormones, are substances released into the external environment by exocrine glands. They provide a means of communication with other members of the same species. Some pheromones, when released by some members of a given species, initiate physiological changes in the others. Others trigger immediate action when detected.

B. KEY TERMS

nervous system   gastrin

molting	insulin
goiter	hormones
glucagon	pupa
endocrine system	secretin
metamorphosis	diabetes mellitus
thyroxin	PTTH
PTH	ecdysone
cholecystokinin	corpora allata
somatostatin	juvenile hormone
TSH	diapause
LH	mineralocorticoids
glucocorticoids	testosterone
follicle	estrogens
relaxin	pheromones
melatonin	prolactin
erythropoietin	FSH
HGH	oxytocin
ACTH	GnRH
MSH	adrenal cortex
TRH	progesterone
adrenal medulla	beta-lipotropin
cortisol	ADH
aldosterone	somatostatin
HCG	CRF
renin	

II. MASTERING THE CHAPTER

A. LEARNING OBJECTIVES

When you have mastered the material in this chapter, you should be able to:

1. Define all key terms.

2. List the three activities involved in chemical coordination in animals.

3. Describe the endocrine control of molt and metamorphosis in insects.

4. Describe some uses of insect hormones in pest control.

5. List five steps involved in the study of endocrinology.

6. Distinguish between the functions of the thyroid and parathyroid glands.

7. Distinguish between hypoparathyroidism and hyperparathyroidism, and cite the "cure" for each.

8. Name one hormone each produced by the skin, stomach, and intestine, and cite its function in each case.

9. Describe the actions of insulin, glucagon, and somatostatin, and explain how they complement one another.

10. State which lobe of the pituitary is the "master" gland and discuss the roles played by its tropic hormones.

11. List the nontropic hormones released by the pituitary and describe their roles.

12. List the four polypeptides synthesized in the hypothalamus and state the function of each.

13. Name the two major regions of the adrenal glands, and state the function of each.

14. Distinguish between testosterone and estrogens, and give at least two functions of each.

15. Discuss the endocrine function(s) of the placenta, pineal gland, and kidney.

16. State three ways in which hormones maintain homeostasis in the body.

17. Discuss the mechanisms of action of steroid hormones and protein or polypeptide hormones at the level of the target cell.

B. REVIEWING TERMS AND CONCEPTS

1. Animals need some mechanism by which the various cells, tissues,

and organs can communicate. Two systems exist in most animals. The
(a) _____ system consists of (b) _____ that transmit
electrical impulses throughout the body. The (c) _____ system
transmits chemical substances called)d) _____ that are transported throughout the body by the (e) _____. In some cases, all
cells of the body are affected by a hormone; more often, only certain
(f) _____ organs respond to it.

    2. In insects, the brain hormone prothoracicotropic hormone
stimulates the prothoracic glands to produce (a) _____, which
initiates (b) _____. In a young larva, the corpora allata produce (c) _____ _____, which causes the molt to be
postponed to another larval stage. As the insect ages, less and less
(d) _____ is produced, resulting in a pupal molt and finally
metamorphosis into the (e) _____.

    3. The research techniques used in the study of hormones have
become quite standardized: the suspected endocrine tissue is first
(a) _____, consequent changes are observed, the tissue is (b)
_____ and the symptoms are expected to (c) _____, an
extract of the tissue (presumably containing the (d) _____)
is prepared, which should successfully replace the tissue itself, and
finally the extract is purified and the (e) _____ identified.

    4. The thyroid gland, located in the neck, produces the hormone
(a) _____, which controls the rate of body metabolism. Insufficient production of this hormone—a condition called (b)_____—
leads to abnormal physical and (c) _____ development in the
young. Inadequate amounts of iodine lead to enlargement and extra
activity of the thyroid gland, a condition called (d) _____.
The rate of activity of the thyroid gland is governed by the hormone
(e) _____.

    5. The parathyroid glands, embedded in the rear of the (a)
_____ gland, increase blood (b) _____ when levels fall
too low. Its targets are the (c) _____ of the body, the (d)
_____ _____, and the (e) _____.

    6. The thyroid also monitors blood calcium levels, and when
levels rise too high, release the hormone (a) _____, which
shifts Ca^{++} from (b) _____ to bones.

    7. When ultraviolet radiation strikes the skin, it triggers the
production of (a) _____, also called (b) _____, a hormone
that enhances the absorption of Ca^{++} from the (c) _____.

    8. The stomach secretes the hormone (a) _____, which sti-
mulates the production of hydrochloric acid and so aids in digestion.
The intestine secretes cholecystokinin and secretin, hormones that
stimulate the secretion of (b) _____ _____ and (c)
_____.

    9. The islets of Langerhans, scattered throughout the (a)
_____, produce the hormone (b) _____, which regulates

the level of (c) _____ in the blood. It does this by stimulating the liver to convert (d) _____ to (e) _____ and fats and by making the cells of the body more permeable to the entrance of glucose. Insufficient production of this hormone results in (f)_____
_____ _____.

10. The pancreas also produces (a) _____, which stimulates the breakdown of glycogen and increases blood (b) _____ levels. Finally, (c) _____ is released as blood sugar and fat levels rise, and it inhibits the action of (d) _____.

11. The (a) _____ gland is often called the "master" gland, because it controls the activity of several other endocrine glands. It is located at the base of the (b) _____ and consists of two major lobes. The (c) _____ lobe produces eight different hormones. (d) _____ _____ _____ production. Thyroid-stimulating hormone causes the production of (f) _____ and is in turn depressed by (g) _____. Similarly, adrenocorticotropic hormone stimulates the (h) _____ _____, and follicle-stimulating hormone acts upon the (i) _____. In females, FSH stimulates the ovarian follicles; in males, it stimulates the (j) _____ _____. Luteininzing hormone triggers (k) _____ in females; in males it stimulates the (l) _____ _____.

12. The posterior lobe of the pituitary gland is not really an (a) _____ _____; it apparently stores and releases two hormones produced in the (b) _____. (c) _____ stimulates smooth muscle contractions, especially in the uterus at birth. (d) _____ _____ increases the reabsorption of water by the kidneys and regulates blood pressure.

13. The hypothalamus is a component of both the (a) _____ and _____ systems and acts as a communicating link between them. In addition to the hormones released by the (b) _____ _____, the hypothalamus produces three releasing hormones and one inhibitory hormone, which control the activity of the (d) _____ _____ _____.

14. The exterior part of the adrenal gland is called the adrenal (a) _____. It releases (b) _____ that help in the conversion of fate and protein into glucose and are crucial in the long-term resistance to (c) _____. The adrenal (d) _____ also produces many (e) _____ that help in the reabsorption of Na$^+$ and Cl$^-$ in the kidney and maintain adequate (f) _____ _____. The interior part of the adrenal glands, called the adrenal (g) _____, produces the hormones (h) _____ and _____, which enable the body to cope with stress.

15. The gonads release hormones in both males and females. In males, the (a) _____ release androgens (for example, (b) _____) into the bloodstream. This hormone triggers the development of secondary male sexual characteristics. In females, the (c) _____ release hormones called (d) _____, which promote secondary

female sexual characteristics and prepare the body for possible (e)
_____. A second ovarian hormone is (f) _____, which con-
tinues to prepare the uterus for (g) _____ and inhibits the
development of new (h) _____.

16. The placenta produces human chorionic gonadotropin, which
stimulates the (a) _____ _____ and so maintains high
levels of female (b) _____ _____, and, later, the (c)
_____ _____ themselves. In this way, the uterus
is kept in a pregnant condition.

17. The endocrine system plays a major role in the maintenance of
internal (a) _____. For instance, the concentration of sugar,
water, and various salt ions in the ECF is maintained within narrow
limits by (b) _____ action. The competence of cells to respond
to a given hormone is important. All hormones are carried by the (c)
_____ from the endocrine gland producing them to (d) _____
cell of the body. Once admitted to a target cell, a given hormone
simply releases the cell's potentialities to respond to a given situation. Steroid hormones (e) _____ the target cell and (c)
_____ induce gene activity. Protein hormones bind to a receptor
site on the (g) _____ _____ and indirectly affect the
cell's activity.

18. Pheromones, unlike hormones, are substances released into the
(a) _____ by (b) _____ glands. They provide a means of
communication with other members of the same species.

C. TESTING TERMS AND CONCEPTS

1. Chemical coordination in animals involves
   a) the use of chemical byproducts of normal bodily processes to
      influence other bodily processes.
   b) activation of whatever tissues a hormone contacts.
   c) the movement of a hormone from a producing tissue through a
      tube to the bloodstream, whereby it is carried throughout the
      entire body.
   d) the movement of a hormone from producing cells directly to
      target cells, through the process of diffusion.

2. In the Cecropia moth, formation of the pupa is directly initiated
   by
   a) juvenile hormone.
   b) the brain.
   c) the corpora allata.
   d) ecdysone.

3. If a juvenile hormone antagonist was sprayed onto a caterpillar
   crop pest, the pests
   a) would continue to molt but would not pupate and would not be
      able to reproduce.
   b) would molt and pupate normally but would neither metamorphose
      nor reproduce.
   c) would quickly pupate but be unable to reproduce as adults.

261

d) would be unable to molt and would soon die.

4. In amphibians, metamorphosis from the larval to the adult stage is triggered by
    a) ecdysone.
    b) JH.
    c) axolotl.
    d) thyroxin.

5. Vitamin D, and important regulator of Ca^{++}, is produced in the
    a) liver.
    b) kidney.
    c) skin.
    d) small intestine.

6. Removal of the parathyroid glands is associated with a resulting low level in the blood of
    a) calcium.
    b) iodine.
    c) thyroxin.
    d) oxygen.

7. Hypothyroidism in adults causes myxedema, whose symptoms include
    a) low calcium and iodine levels in blood.
    b) low activity levels and overweight.
    c) nervous irritability and insomnia.
    d) weakened bones and kidney stone.

8. TSH is produced in the
    a) thyroid gland.
    b) pituitary gland.
    c) hypothalamus.
    d) blood.

9. Insulin acts to
    a) stimulate production of glycogen and fats.
    b) inhibits synthesis of proteins.
    c) make cells less permeable to the entrance of glucose.
    d) increase the absorption of glucose from the intestine.

10. Gigantism results from a
    a) hyposecretion of TSH.
    b) hyposecretion of MSH.
    c) hypersecretion of TRH.
    d) hypersecretion of HGH.

11. Which two hormones have effects similar to an effect of adrenalin?
    a) glucagon and cortisol
    b) insulin and aldosterone
    c) noradrenalin and oxytocin
    d) HGH and thyroxin

12. An animal does not respond to a stressful experience by producing
    a) adrenalin.

b) noradrenalin.
c) glucocorticoids.
d) growth hormone.

13. The stimulation of red blood cell production is triggered by the
a) skin.
b) pineal gland.
c) kidney.
d) islets of Langerhans.

14. A hormone associated with secondary sexual characteristics in
males is
a) FSH.
b) LH.
c) mineralocorticoids.
d) estrogens.

15. The hormone that works in conjunction with LH in stimulating the
secretion of estrogens is
a) FSH.
b) MSH.
c) CSH.
d) TSH.

D. UNDERSTANDING AND APPLYING TERMS AND CONCEPTS

1. If anatomical study reveals a tissue you suspect has an endocrine
function, how should you proceed to prove it? List four steps.
a)_____
b)_____
c)_____
d)_____

2. Once a steroid has entered a target cell, what are two mechanisms
by which it could activate that cell?
a)_____
b)_____

3. Several endocrine control mechanisms involve the production of a
hypothalamic hormone, which affects the production of an anterior
pituitary hormone, which affects the production of yet a third
hormone. Of the following diagrams, which correctly indicate this
kind of influence between or among hormones?

_____	a) CRF	ACTH	glucocorticoids
_____	b) TRH	TSH	thyroxin
_____	c) MSH	FSH	estrogen
_____	d) LH	progesterone	
_____	e) ACTH	ADH	
_____	f) LH	oxytocin	
_____	g) somatostatin	HGH	
_____	h) MSH	TSH	FSH
_____	i) GnRH	FSH	estrogen

4. Unlike insulin, human growth hormone cannot be replaced by that of

263

other mammals. How can large amounts of HGH be produced for
clinical use? _____
_____
_____

5. Many hormones can be obtained from a crude homogenate of the tis-
    sue that produces it, and this has long been a standard procedure.
    Why can't insulin be obtained this way? _____
    _____

6. Both glucagon and the glucocorticoids raise blood glucose levels.
    Describe two important <u>differences</u> between the actions of these
    hormones.
    a)_____
    b)_____

7. Match the terms in Column B with the descriptions in Column A.

Column <u>A</u>   Column <u>B</u>

a) The hormone that directly   A. cholecystokinin
    initiates molting in the
    Cecropia Moth _____   B. ecdysone

b) A hormone produced by the   C. estrogen
    skin _____
                                       D. endocrine system
c) A hormone associated with
    the level of glucose in   E. parathyroid
    the blood _____
                                       F. testosterone
d) Noradrenaline is produced
    in the _____   G. metamorphosis

e) Two communication systems   H. pituitary
    exist in most animals _____
    and _____   I. vitamin D

f) A gland that controls the   J. adrenal medulla
    rate of body metabolism
                                       K. nervous system
    _____
                                       L. insulin
g) Two hormones produced in
    the duodenum _____ and   M. thyroid

    _____   N. calcitonin

h) A hormone especially im-   O. hormones
    portant after a pregnancy
                                       P. prolactin
    _____
                                       Q. goiter
i) A female hormone that pre-
    pares the organism for
    possible pregnancy

j) Chemical messengers that
control body functions

    _____

R. juvenile hormone

S. secretin

k) A hormone that acts as a
brake on metamorphosis

    _____

l) These glands maintain a
sufficient level of cal-
cium ions in the blood

    _____

m) A hormone produced by the
islets of Langerhans _____

n) A male hormone associated
with the onset of adolescence

    _____

o) The marked transformation in
body structure occurring in
insects _____

p) A condition brought on by
iodine deficiency _____

q) The gland that secretes a
hormone that regulates human
growth _____

r) A hormone that acts antagonistically
to PTH _____

8. Name the correct pituitary or hypothalamic hormones described
below.

a) During childhood and adolescence, this hormone indirectly stimulates growth of bone, muscle, cartilage, and other connective
tissues. _____

b) This polypeptide increases the permeability to water of the distal
tubules and collecting ducts of the kidney and causes arteriole
constriction throughout the body. _____

c) A polypeptide causes a darkening of the skin in many vertebrates.
_____

d) These hormones provide a controlling link between the nervous
system and the general metabolic rate of the body. _____
_____

e) More studied in females, this protein stimulates milk production
and causes salt and water retention by the kidneys. _____

f) This polypeptide stimulates the outer layer of the adrenal glands.

g) It _inhibits_ the production of growth hormone and thyroid-stimulating hormone. _____

h) This polypeptide stimulates labor contractions at the time of birth. _____

i) This provides a mechanism by which the brain can stimulate the reproductive system. _____

j) This polypeptide enhances the metabolism of fats in some mammals and may do so in humans. _____

k) This stimulates egg development in females and sperm production in males. _____

l) It stimulates progesterone production in females and testosterone production in males. _____

m) This increases the production of ACTH, during times of mental or physical stress. _____

ANSWERS TO CHAPTER EXERCISES

Reviewing Terms and Concepts

1. a) nervous b) neurons c) endocrine d) hormones e) blood
    f) target

2. a) ecdysone b) molting c) juvenile hormone d) JH e) adult

3. a) removed b) replaced c) disappear d) hormone e) hormone

4. a) thyroxin b) hypothyroidism c) mental d) goiter e) TSH

5. a) thyroid b) calcium c) bones d) small intestine e) kidneys

6. a) calcitonin b) blood

7. a) calciferol b) vitamin D c) intestine

8. a) gastrin b) pancreatic fluid c) bile

9. a) pancreas b) insulin c) glucose d) glucose e) glycogen
    f) diabetes mellitus

10. a) glucagon b) sugar c) somatostatin d) glucagon

11. a) pituitary b) brain c) anterior d) Human growth hormone
     e) milk f) thyroxin g) thyroxin h) adrenal cortex i) gonads
     j) seminiferous tubules k) ovulation l) interstitial cells

12. a) endocrine gland b) hypothalamus c) Oxytocin d) Antidiuretic hormone

13. a) nervous, endocrine b) posterior pituitary c) anterior
    pituitary

14. a) cortex b) glucocorticoids c) stress d) cortex e) mineralo-
    corticoids f) blood pressure g) medulla h) adrenaline, nor-
    adrenaline

15. a) testes b) testosterone c) ovaries d) estrogens e) pregnancy
    f) progesterone g) pregnancy h) eggs

16. a) corpus luteum b) sex hormones c) sex hormones

17. a) homeostasis b) hormonal c) blood d) every e) enter
    f) directly g) cell surface

18. a) environment b) exocrine

Testing Terms and Concepts

1. a	6. a	11. a
2. d	7. b	12. d
3. c	8. b	13. c
4. d	9. a	14. b
5. c	10. d	15. a

Understanding and Applying Terms and Concepts

1. a) Remove it and look for symptoms. b) Replace it and look for
    recovery. c) Remove the tissue, replace an extract of that
    tissue, and expect recovery. d) Purify the extract and identify
    the specific compound responsible for the recovery (the hormone).

2. a) A hormone can activate inactive enzymes already present in the
    cell. b) Other hormones activate transcription at specific
    genes.

3. a, b, d, g, i

4. The gene for HGH is inserted into the genome of E. coli. During
    the rapid reproduction and metabolism of this bacterium, large
    amounts of the hormone will be produced.

5. The pancreas produces digestive enzymes, as well as insulin. In
    a homogenate, these enzymes digest the insulin before it can be
    extracted.

6. a) The glucocorticoids have longer lasting effects. b) The
    glucocorticoids make glucose available from noncarbohydrate
    sources; sometimes this is referred to as "new" sugar.

7. a) B b) I c) L d) J e) D, K f) M g) A, S h) P i) C j) O
    k) R l) E m) L n) F o) G p) Q q) H r) N

8. a) HGH b) ADH c) MSH d) TRM, TSH e) prolactin f) ACTH

THE ELEMENTS OF NERVOUS
COORDINATION

I. REVIEWING THE CHAPTER

A. CHAPTER HIGHLIGHTS

28.1 The Three Components of Nervous Coordination

Nervous coordination differs from endocrine coordination in being faster and generally more localized in its action. Although both plants and animals carry on chemical coordination by means of transported hormones, nervous coordination is characteristic of animals only.

For an animal to respond to changes in the environment, it must have stimulus receptors, impulse conductors, and effectors. Our eyes and other sense organs are stimulus receptors. Nerves make up the impulse conductors and are composed of two kinds of neurons—sensory neurons, which transmit impulses from the stimulus receptor to the central nervous system and the brain and spinal cord; and motor neurons, which transmit impulses from the central nervous system to the part of the body that will take action. The most important effectors in humans are the muscles and glands.

28.2 The Reflex Arc

In humans, the simplest unit of nervous response is the reflex arc. A stimulus detected by receptors in the skin initiates nerve impulses in the sensory neurons. These impulses enter the spinal cord and initiate impulses in interneurons, which in turn initiate impulses in motor neurons. A muscle is then stimulated to contract. A reflex arc bypasses the brain completely and so saves valuable time. When you touch a hot object, the reflex arc prevents you from being seriously burned.

28.3 The Neuron

A neuron is a cell specialized to conduct electrochemical impulses. In many neurons, the nerve impulses are generated in the dendrites or in the cell body, then conducted along the axon.

Sensory neurons link the stimulus receptors with the central nervous system. Interneurons, also called association neurons, provide an unimaginably large number of cross-connections to produce a virtually limitless number of circuits within the central nervous system. Motor neurons transmit impulses from the central nervous system to the muscles and glands.

28.4 The Nerve Impulse

The interior of a neuron is negatively charged with respect to
the exterior. The charge is called the resting potential and is due
principally to a concentration of sodium ions in the ECF, maintained
by active transport, that is ten times that of the cytoplasm. Certain
types of stimuli increase the permeability of the membrane to sodium
ions. When the threshold is reached, the membrane becomes completely
permeable to sodium ions and they rush in, temporarily reversing the
charge on the membrane and making adjacent areas of the membrane
permeable to sodium ions. Thus a wave of depolarization, the action
potential, sweeps down the neuron. If an axon reacts at all, it reacts to its fullest extens: an action potential is an "all-or none"
response. However, the number of action potentials does vary with the
strength of the stimulus.

28.5 The Synapse

The points at which the axon terminals of one neuron are in contact with other neurons are called synapses. When an action potential
reaches the end of a presynaptic neuron, a "transmitter" substance,
such as acetylcholine, is released, which can cause a new action
potential to form in the postsynaptic neuron. Inhibitory synapses
also exist, where the transmitter substance does not depolarize the
postsynaptic neuron, but hyperpolarizes it (makes the interior more
negative). The ability to stimulate and inhibit effectors allows a
much finer degree of coordination than stimulation alone could do.

28.6 Touch and Pressure

Touch is a category of light mechanical stimulation, where an
outside force physically deforms a part of the body. Touch receptors,
such as nerve endings associated with a hair follicle, are located
close to the surface of the skin.

The Pacinian corpuscle is a pressure receptor that responds to
physical deformation, but it is located deeper in the skin.

Proprioceptors are sense receptors in muscle and tendon tissue
that initiate nerve impulses when a muscle or tendon is stretched or
contracts. These impulses enable the brain to determine the state of
the contraction of the muscle. For example, if you start to lose your
balance, proprioceptors come to your rescue.

Pain arises from massive mechanical stimulation, and is sensed by
naked nerve endings in the skin.

28.7 Hearing

Sound waves in the air pass down the auditory canal of each ear
and strike the tympanic membrane, causing it to vibrate. These
vibrations are transmitted across the middle ear by three tiny, linked

bones, the ossicles. The innermost bone, the stirrup, transmits
mechanical vibration to the cochlea of the inner ear, where sensitive
hair cells serve as the vibration receptors. Electrical impulses
arising in these cells then initiate nerve impulses that travel along
the auditory nerve to the brain.

28.8 Equilibrium

The movement of fluid in other parts of the inner ear, including
the semicircular canals, stimulates hair cells and detects the position
of the body with respect to gravity and the motion of the body.

28.9 The Compound Eye

The arthropod eye functions quite differently from that of vertebrates and is called a compound eye because it is made up of repeating
units, the ommatidia, each of which functions as a separate visual
receptor. Each ommatidium responds to a very small portion of the
total view; all ommatidia together create a mosaic of light and dark
dots, a rather coarse, grainy view of the world.

Some insects, such as honeybees, have more than one pigment in
each ommatidium and therefore can distinguish colors, though not
necessarily the same colors humans recognize.

28.10 The Structure of the Human Eye/28.11 The Detection of Light

The eyes of both mollusks and vertebrates operate on the same
basic principle as a camera. Light is admitted into the interior of
the eye and brought into focus by the cornea and lens. In the front
of the eye, the middle or choroid coat forms the iris, which may be
pigmented, and the opening in the center is called the pupil, its
variable size depending on the amount of light reaching it. The inner
coat of the eye, the retina, contains the light receptors—the rods
and cones. The rods are used chiefly for vision in dim light. A pigment called rhodopsin is the agent that absorbs light in the rods.
The cones operate only in bright light and enable us to see colors.
All impulses generated by the rods and cones travel back to the brain
by way of neurons in the optic nerve.

28.12 Heat Receptors

Distributed through the skin are receptors that, when warmed, give
rise to the sensation of warmth. Other receptors are responsible for
the sensation of cold. The human body also has receptors that detect
internal temperature changes, located in the hypothalamus of the brain.

28.13 Taste

In order for a substance to be tasted, it must be soluble in the
moisture of the mouth. Four types of taste buds can be distinguished
morphologically on the tongue's surface and on the soft palate of the
mouth, and although they are not correlated with the bud types, there
appear to be only four primary taste sensations: sweet, sour, salty,

and bitter. Other tastes result from combinations of the primary tastes and from additional smell, temperature, and touch effects.

28.14 Smell

Humans detect odors by means of receptor cells located in the two olfactory epithelia high in the nasal cavity. Only two kinds of receptor cells can be distinguished morphologically, but as many as seven may be distinguished according to their function. In most cases the shape of a molecule determines which receptors will respond to it.

28.15 Internal Chemical Receptors

Humans also have receptors that detect chemical changes in the internal environment. Our brain and circulatory system contain cells sensitive to carbon dioxide, oxygen, and glucose, and our sensation of thirst arises as a result of stimulation of special cells in the hypothalamus of the brain, sensitive to changes in the osmotic pressure of the blood.

28.16 Magnetoreceptors and Electroreceptors

Many animals can home in response to the earth's magnetic field. In pigeons, microscopic grains of magnetite near the surface of the brain apparently respond to magnetic stimulation. A number of electric and nonelectric fishes have receptors sensitive to weak electric fields. Electric fishes navigate and communicate, using these signals.

B. KEY TERMS

myelin sheath	ganglia
cell body	dendrites
subthreshold	nerve impulse
threshold	synapses
sensory neurons	acetylcholine (ACh)
motor neurons	action potential (AP)
axon	presynaptic neuron
interneurons	EPSP
resting potential	receptor
generator potential	touch
refractory period	thirst
postsynaptic neuron	photoreceptors

IPSP	ommatidia
mechanoreceptors	choroid coat
sclerotic coat	astigmatism
iris	vitreous humor
cataracts	retinal
rods	conductors
cones	proprioceptors
effectors	Eustachian tube
chemoreceptors	auditory nerve
cornea	olfactory epithelia
retina	tympanic membrane
aqueous humor	cochlea
rhodopsin	statocysts
optic nerve	pressure receptor
Pacinian corpuscle	fatigue
ossicles	pain
organ of Corti	lens
taste buds	fovea

II. MASTERING THE CHAPTER

A. LEARNING OBJECTIVES

When you have mastered the material in this chapter, you should be able to:

1. Define all key terms.

2. Cite three ways in which nervous coordination differs from endocrine coordination, and give an example of each.

3. Name the three components required for an organism to respond to changes in its environment.

4. List the three distinct groups into which neurons can be placed, and cite the function of each group.

5. List the five structures and their actions involved in a withdrawal reflex.

6. List the parts of a neuron and describe their functions.

7. Distinguish between the action potential, resting potential, and generator potential of a neuron, and cite three stimuli capable of changing the generator potential.

8. Distinguish between an excitatory synapse and an inhibitory synapse.

9. Differentiate between touch and pressure, and describe a receptor sensitive to each.

10. Discuss the value of proprioreceptors and pain receptors.

11. State the function of the ossicles, tympanic membrane, and organ of Corti.

12. Describe three broad functions of the ear.

13. Draw a diagram of the human eye, locating and labeling ten components.

14. Differentiate between a compound and a camera-type eye.

15. Discuss the differences between smell and taste.

16. Distinguish between rods and cones, and state the function of each.

17. Discuss the value of having receptors deep in the interior of the body.

18. Discuss the variety of receptor abilities found in nonhuman animals.

B. REVIEWING TERMS AND CONCEPTS

1. Nervous coordination differs from endocrine coordination in being (a) _____ and generally more (b) _____ in its action. Although both plants and animals carry on chemical coordination by means of transported (c) _____, nervous coordination is characteristic of (d) _____ only.

2. For an animal to respond to changes in the environment, it must have (a) _____ receptors, (b) _____ conductors, and (c) _____. Our eyes and other sense organs are (d) _____ receptors. (e) _____ make up the impulse conductors and are composed of two kinds of neurons—(f) _____ neurons, which transmit impulses from the stimulus receptor to the (g) _____ nervous system, and (h) _____ neurons, which transmit impulses from the (i) _____ nervous system to the

274

part of the body that will take action. The most important effectors in humans are the (j) _____ and _____ .

3. In humans, the simplest unit of nervous response is the (a) _____ _____ . A stimulus detected by (b) _____ in the skin initiates nerve impulses in the (c) _____ neurons. These impulses enter the spinal cord and initiate impulses in (d) _____ , which in turn initiate impulses in (e) _____ neurons. A muscle is then stimulated to contract. A reflex arc bypasses the (f) _____ completely and so saves valuable time.

4. A (a) _____ is a cell specialized to conduct electro-chemical impulses. In many (b) _____ , the nerve impulses are generated in the (c) _____ or in the cell body, then conducted along the (d) _____ .

5. (a) _____ neurons link the stimulus receptors with the central nervous system. (b) _____ , also called (c) _____ neurons, provide an unimaginably large number of cross-connections to produce a virtually limitless number of circuits within the central nervous system. (d) _____ neurons transmit impulses from the central nervous system to the (e) _____ and _____ .

6. The interior of a neuron is (a) _____ charged with respect to the exterior. The charge is called the (b) _____ and is due principally to a concentration of (c) _____ ions in the ECF, maintained by active transport, that is ten times that of the cytoplasm. Certain types of stimuli increase the (d) _____ of the membrane to (e) _____ ions. When the threshold is reached, the membrane becomes completely perme-able to these ions and they rush in, temporarily (f) _____ the charge on the membrane. A wave of depolarization, the (g) _____ , sweeps down the neuron.

7. The points at which the axon terminals of one neuron are in contact with other neurons are called (a) _____ . When an action potential reaches the end of a (b) _____ neuron, a "transmitter" substance, such as acetylcholine, is released, which can cause a new action potential to form in the (c) _____ neuron. Inhibitory synapses also exist, where the transmitter substance does not depolarize the postsynaptic neuron, but (d) _____ it (makes the interior more negative).

8. (a) _____ receptors, such as nerve endings associated with a hair follicle, are located close to the surface of the skin. The Pacinian corpuscle is a (b) _____ receptor located deeper in the skin. (c) _____ are sense receptors in muscle and tendon tissue that initiate nerve impulses when a muscle or tendon is stretched or contracts. (d) _____ arises from massive mechanical stimulation, and is sensed by naked nerve endings in the skin.

9. Sound waves in the air pass down the (a) _____ canal of each ear and strike the (b) _____ membrane, causing it to vibrate. These vibrations are transmitted across the (c) _____

ear by three tiny, linked bones, the (d) _____. The innermost bone transmits the vibration to the (e) _____ of the (f)
_____ ear, where sensitive (g) _____ _____
serve as the vibration receptors. Electrical impulses then travel
along the (h) _____ _____ to the brain. The movement of fluid in other parts of the inner ear, including the (i)
_____ _____, stimulates hair cells and detects the
position of the body with respect to (j) _____, and the motion
of the body.

10. The arthropod eye functions quite differently from that of
vertebrates and is called a (a) _____ eye because it is made up
of repeating units, the (b) _____, each of which functions as a
separate visual receptor. Some insects, such as honeybees, have more
than on pigment in each (c) _____ and therefore can
distinguish (d) _____.

11. The eyes of both mollusks and vertebrates operate on the same
basic principle as a (a) _____. Light is admitted into
the interior of the eye and brought to focus by the (b) _____
and _____. In the front of the eye, the choroid coat forms
the (c) _____, which may be pigmented, and the opening in the
center is called the (d) _____, its variable size depending on
the amount of (e) _____ reaching it. The inner coat of the
eye, the (f) _____, contains the light receptors—the (g)
_____ and _____. The rods are used chiefly
for vision in (h) _____ light. The cones operate only in
(i) _____ light and enable us to see (j) _____.

12. In order for a substance to be tasted, it must be (a) _____
in the moisture of the mouth. There appear to be only four primary
taste sensations: (b) _____, _____, _____,
and _____. Other tastes result from combinations of the
primary tastes and from additional (c) _____, _____,
and _____ effects.

13. Humans detect odors by means of receptor cells located in the
two (a) _____ epithelia high in the nasal cavity. Only (b)
_____ kinds of receptor cells can be distinguished morphologically, but as many as (c) _____ may be distinguished according
to their function. In most cases the (d) _____ of a molecule
determines which receptors will respond to it. Humans also have
receptors that detect chemical changes in the internal environment.
Our (e) _____ and _____ system contain cells sensitive
to carbon dioxide, oxygen, and glucose.

C. TESTING TERMS AND CONCEPTS

1. As you stand on a small boat, the noise of the surf, the sting
   of the spray, and the rocking back and forth of the boat all
   stimulate your
   a) chemoreceptors.
   b) mechanoreceptors.
   c) effectors.
   d) conductors.

2. Not a part of a reflex arc is
   a) the brain.
   b) the spinal cord.
   c) a sensory neuron.
   d) a synapse.

3. A wave of depolarization sweeping down a neuron is called the
   a) generator potential.
   b) resting potential.
   c) action potential.
   d) graded potential.

4. ATP is used by neurons to actively expel
   a) action potentials.
   b) electrons.
   c) sodium.
   d) potassium.

5. A chemical transmitter of nerve impulses is
   a) AP.
   b) ACh.
   c) EPSP.
   d) IPSP.

6. Impulses travel most rapidly
   a) across a peripheral synapse.
   b) over the cell body.
   c) along a small-diameter dendrite.
   d) along a myelinated axon.

7. Impulses are transmitted from the central nervous system to the muscles and glands by
   a) motor neurons.
   b) interneurons.
   c) sensory neurons.
   d) association neurons.

8. Which answer is false? Neural signals can move in either direction
   a) over a cell body.
   b) along a dendrite.
   c) along an axon.
   d) across a synapse.

9. Exocytosis is an important process in
   a) cell bodies.
   b) Schwann cells.
   c) synaptic knobs.
   d) postsynaptic membranes.

10. If neurons are firing repeatedly and rapidly, causing muscular convulsions, some abnormal condition must have caused an
    a) overproduction of ATP.
    b) under accumulation of chloride ions.
    c) underproduction of acetylcholine.
    d) underproduction of acetylcholinesterase.

11. When you start to lose your balance, the brain is informed by
    a) proprioceptors.
    b) pressure receptors.
    c) the tympanic membrane.
    d) Pacinian corpuscles.

12. If humans had compound eyes, they
    a) would have a greater ability to detect fine details.
    b) could detect a greater variety of colors.
    c) would have better night vision.
    d) would be more sensitive to movement.

13. The ability to distinguish between the sound of a high C and mid-
    dle C is a function of the
    a) ossicles.
    b) organ of Corti.
    c) Eustachian tube.
    d) semicircular canals.

14. In order for us to taste a substance, the substance must be
    a) sweet, sour, salty, or bitter.
    b) organic or nutritive.
    c) soluble in water.
    d) at least somewhat odorous.

15. In humans, chemoreceptors are not located in
    a) the skin.
    b) the hypothalamus.
    c) blood vessels.
    d) the nasal passages.

D. UNDERSTANDING AND APPLYING TERMS AND CONCEPTS

1. When you walk barefoot through a field and step on a thorn, a
   reflex is initiated. What kind of neurons carry impulses toward
   the spinal cord? (a)_____
   Once there, interneurons carry signals toward which three
   destinations?
   (b)_____
   (c)_____
   (d)_____

2. Briefly describe the functions of these parts of the nervous
   system:
   (a) dendrite_____
   (b) axon_____
   (c) axon hillock_____
   (d) cell body_____
   (e) Schwann cell_____
   (f) synaptic vessicle_____

3. In an axon, what is responsible for causing the polarization that
   exists before an action potential forms?
   (a)_____

278

What events cause the rapid reversal of polarity during the first half of the action potential? (b)_____

What events cause the equally rapid recovery of the original, negative condition? (c)_____

4. Number the following structures to show the order in which sound waves pass through the human ear. Place a zero before any structure not involved in the transmission of sound information.

_____ a) external ear
_____ b) ossicles
_____ c) tympanic membrane
_____ d) round window
_____ e) oval window
_____ f) Eustachian tube
_____ g) semicircular canals
_____ h) organ of Corti
_____ i) lymph of cochlea
_____ j) auditory canal

5. Number the following structures to show the order in which light passes through the human eye.

_____ a) aqueous humor
_____ b) vitreous humor
_____ c) lens
_____ d) rods and cones
_____ e) ganglion and bipolar cells
_____ f) choroid coat
_____ g) cornea
_____ h) pupil

6. It is commonly said that humans have five primary senses. Is this true either on the basis of receptor structure or sensation produced? (a)_____
Explain. (b)_____

Decide whether the following statements are true (T) or false (F).

_____ 7. All organisms have cells that are specialized for conducting electrochemical impulses.

_____ 8. Sensory neurons run from the central nervous system to the various types of effectors.

_____ 9. The membrane of the resting neuron is virtually impermeable to the passage of sodium ions.

_____ 10. If enough acetylcholine contacts a postsynaptic neuron, an action potential forms on the dendrite and travels toward the cell body.

_____ 11. Most nerves contain only motor or sensory axons.

_____ 12. The most important effectors in humans are the muscles and the glands (both exocrine and endocrine).

_____ 13. With a single pigment it becomes possible to distinguish colors.

_____ 14. It takes up to five minutes in the dark for our eyes to become fully adapted to the dark.

_____ 15. Heat is electromagnetic radiation of wavelengths shorter than those of light.

_____ 16. The Pacinian corpuscle is a heat receptor.

_____ 17. At the back of the eye, the choroid coat reflects light and is responsible for "eyeshine" at night.

_____ 18. Heat and cold are detected with the same kind of receptor.

_____ 19. Many animals, including humans, can sense and respond to the earth's magnetic field.

ANSWERS TO CHAPTER EXERCISES

Reviewing Terms and Concepts

1. a) faster b) localized c) hormones d) animals

2. a) stimulus b) impulse c) effectors d) stimulus e) Nerves
    f) sensory g) central h) motor i) central j) muscles, glands

3. a) reflex arc b) receptors c) sensory d) interneurons
    e) motor f) brain

4. a) neuron b) neurons c) dendrites d) axon

5. a) Sensory b) Interneurons c) association d) Motor e) muscles,
    glands

6. a) negatively b) resting potential c) sodium d) permeability
    e) sodium f) reversing g) action potential

7. a) synapses b) presynaptic c) postsynaptic d) hyperpolarizes

8. a) Touch b) pressure c) Proprioceptors d) Pain

9. a) auditory b) tympanic c) middle d) ossicles e) cochlea
    f) inner g) hair cells h) auditory nerve i) semicircular
    canals j) gravity

10. a) compound b) ommatidia c) ommatidium d) colors

11. a) camera b) cornea, lens c) iris d) pupil e) light
    f) retina g) rods, cones h) dim i) bright j) colors

12. a) soluble b) sweet, sour, salty, bitter c) smell, temperature,
    touch

13. a) olfactory b) two c) seven d) shape e) brain, circulatory

Testing Terms and Concepts

1.	b	6.	d	11.	a
2.	a	7.	a	12.	d
3.	c	8.	d	13.	b
4.	c	9.	c	14.	c
5.	b	10.	d	15.	a

Understanding and Applying Terms and Concepts

1. a) sensory b) to the brain, where you experience pain c) to
   flexor muscles of leg, which lift the leg d) to extensor
   muscles of leg, which are inhibited

2. a) receives signals from receptors or other neurons
   b) carries signals to other cells
   c) site where action potential first forms
   d) maintains dendrites and axon
   e) nourishes and insulates axon
   f) stores transmitter chemical

3. a) The resting potential is caused by a net expulsion of sodium
   by active transport. b) The axon membrane becomes permeable
   to sodium, and it rushes in, down its concentration gradient.
   c) The membrane becomes impermeable to sodium but permeable to
   potassium, which rushes out, down its concentration gradient.

4. a) 1 b) 4 c) 3 d) 8 e) f f) 0 g) 0 h) 7 i) 6 j) 2

5. a) 2 b) 5 c) 4 d) 7 e) 6 f) 8 g) 1 h) 3

6. a) no b) There are many more than five kinds of both receptors
   and sensations: sense of pain, warmth, equilibrium. . . (sense
   of red, green, sweet, sour? Both receptors and sensations are
   distinctly different.)

7.	F	11.	F	15.	F	19.	T
8.	F	12.	T	16.	F		
9.	T	13.	F	17.	F		
10.	F	14.	F	18.	F		

29

THE NERVOUS SYSTEM

I. REVIEWING THE CHAPTER

A. CHAPTER HIGHLIGHTS

29.1 The Spinal Cord

The central nervous system consists of the spinal cord and the brain. The peripheral nervous system informs the central nervous system of stimuli that have been detected and causes the muscles and glands to respond. The central nervous system serves as a coordinating center for the actions to be carried out.

The spinal cord carries out two main functions in nervous coordination. First, it connects the peripheral nervous system to the brain. Second, it acts as a minor coordinating center itself. Simple reflex responses, such as the knee-jerk reflex, can take place through the sole action of the spinal cord.

29.2 The Brain

The brain receives neural input from all over the body, organizes this input, and generates conscious throught and coordinated body action. The brain is divided into the hindbrain, midbrain, and forebrain.

29.3 The Hindbrain/29.4 The Midbrain/29.5 The Forebrain

Two portions of the hindbrain are the cerebellum and the medulla oblongata. The latter regulates heartbeat and breathing and other vital bodily functions. The former seems to coordinate muscular activity in the body. The midbrain helps us maintain our balance and relays nerve impulses between the forebrain and hindbrain and between the forebrain and eyes. The most prominent part of the human forebrain is the cerebrum. The forebrain also includes the thalamus, the hypothalamus, part of the pituitary, and the pineal gland. The exterior of the cerebrum is the cortex composed of masses of cell bodies (gray matter), large areas of which are associated with sensations, learning, memory, logical analysis, foresight, and creativity. Damage to the frontal lobes of the cerebrum may produce changes in human personality and behavior. The thalamus passes sensory information on to the cerebrum. The hypothalamus is the center for many emotions and other feelings such as hunger or thirst. The hypothalamus also produces a variety of hormones.

29.6 The Processing of Visual Information

The processing of visual information begins in the retina. Here, not only is light perceived, but neurons are so arranged that some particularly respond to contrast between light and dark. In the first cells of the visual cortex, lines of light of various orientations are distinguished, and these cells converge on cells that then recognize corners and other, more complex shapes. In general, information processing seems to involve selective filtering out of some information received by the receptors and the consequent heightening and therefore noticing of other aspects of that input.

29.7 The Sensory-Somatic System

The sensory-somatic system consists of 12 pairs of cranial nerves and 31 pairs of spinal nerves, which carry perceived sensory information to the central nervous system and conscious commands out to the skeletal muscles.

29.8 The Autonomic Nervous System

The autonomic nervous system consists of sensory and motor neurons linking the central nervous system with our various internal organs. The actions of this nervous system are largely involuntary.

29.9 The Sympathetic Nervous System

The sympathetic nervous system is a part of the autonomic nervous system. Its fibers release adrenaline or noradrenaline onto various organs, including the adrenal gland, and its effects are those of the "fight or flight" response.

29.10 The Parasympathetic Nervous System

The parasympathetic system is also a part of the autonomic nervous system. Acetylcholine is the transmitter released onto the effectors, and the results are exactly opposite to those of the sympathetic nervous system. The heart is slowed; digestion increases. The parasympathetic system returns the body to normal after the sympathetic system has guided it through some emergency.

29.11 Drugs and the Nervous System

Nervous system activity can be altered by chemicals that mimic, block, or otherwise alter its biochemical activities. Perception, muscular coordination, and emotions are among the functions of the brain that chemicals can alter. There are three major classes of psychoactive drugs. Caffeine, cocaine, nicotine, and amphetamines are examples of stimulants, which act on the sympathetic nervous system and accelerate heart rate, dilate the pupils, and increase blood sugar. Ethanol, barbiturates, tranquilizers, opiates, and anesthetics are examples of depressants, which reduce nervous system activity. Mescaline, psilocybin, LSD, and STP are examples of hallucinogens, which have a powerful distorting effect on visual and auditory perceptions and produce marked enhancement of emotional responses.

29.12 Opiate-like Peptides in the Brain

Some neurons in the central nervous system manufacture and release peptides that bind to the same cells morphine and other opiates do and have the same effects, e.g. pain relief. Two of these peptides, the enkephalins, inhibit pain transmitting neurons and so reduce the strength of the pain signals reaching the cerebral cortex. This system of intrinsic pain killers protects us from extreme or long-term pain.

B. KEY TERMS

rods	peripheral nervous system
ganglion cells	mixed nerves
stimulants	nerve roots
cones	tracts
lateral geniculate body	meninges
depressants	medulla oblongata
bipolar cells	cerebellum
vagus	reticular formation
hallucinogens	cerebrum
central nervous system	

II. MASTERING THE CHAPTER

A. LEARNING OBJECTIVES

When you have mastered the material in this chapter, you should be able to:

1. Define all key terms.

2. Distinguish between the central nervous system and the peripheral nervous system.

3. Cite the two main functions carried out by the spinal cord in nervous coordination.

4. Distinguish between the hindbrain, midbrain, and forebrain by describing and citing functions of each.

5. Discuss the processing of information in the cerebral cortex.

6. Distinguish between the sensory-somatic system and the autonomic system, and cite functions of each.

7. Distinguish between the sympathetic nervous system and the parasympathetic system, and cite functions of each.

8. Discuss the effects of extrinsic and intrinsic drugs on the central nervous system.

B. REVIEWING TERMS AND CONCEPTS

1. The (a) _____ nervous system consists of the spinal cord and brain. The (b) _____ nervous system informs the central nervous system of stimuli that have been detected and causes the (c) _____ and _____ to respond. The (d) _____ nervous system serves as a coordinating center for the actions to be carried out.

2. The spinal cord carries out two main functions in nervous coordination. First, it connects the (a) _____ nervous system to the (b) _____. Second, it acts as a minor coordinating center itself. Simple (c) _____ responses can take place through the sole action of the (d) _____ _____.

3. The brain is composed of the (a) _____-brain, (b) _____-brain, and (c) _____-brain. Two portions of the hindbrain are the (d) _____ and the (e) _____ _____. The latter regulates (f) _____ and _____. The former seems to coordinate (g) _____ activity in the body.

4. The (a) _____-brain helps us maintain our (b) _____ and relays nerve impulses between the (c) _____ and _____. The most prominent part of the human forebrain is the (d) _____. The forebrain also includes the (e) _____, the _____, part of the (f) _____, and the (g) _____ _____. The exterior of the cerebrum is the (h) _____, which is associated with sensations, learning, memory, logical analysis, foresight, and creativity.

5. The processing of visual information begins in the (a) _____. Here, not only is light perceived, but (b) _____ are particularly noticed. In successive levels within the CNS, certain information is (c) _____ _____ and other information is therefore (d) _____.

6. The sensory-somatic system consists of (a) _____ pairs of cranial nerves and (b) _____ pairs of spinal nerves, which carry sensory information to (c) _____ (conscious/subconscious) brain centers and (d) _____ commands out to (e) _____ _____.

7. The (a) _____ nervous system consists of (b) _____ and _____ neurons linking the central nervous system with out various internal organs. The actions of this nervous system are largely (c) _____ (voluntary/involuntary).

8. The (a) _____ nervous system is that part of the autonomic system that prepares the body for emergencies. The (b) _____

285

system stops these "fight-or-flight" responses and returns the body to normal.

9. Nervous system activity can be altered by chemicals that mimic, block, or otherwise alter its biochemical activities. (a) _____, (b) _____ _____, and (c) _____ are among the functions of the brain that chemicals can alter. There are three major classes of psychoactive drugs. Caffeine, cocaine, nicotine, and amphetamines are examples of (d) _____, which act on the (e) _____ nervous system and accelerate (f) _____ _____, dilate the (g) _____, and increase blood (h) _____. Ethanol, barbiturates, tranquilizers, opiates, and anesthetics are examples of (i) _____, which reduce nervous system activity. Mescaline, psilocybin, LSD, STP are examples of (j) _____, which have a powerful distorting effect on (k) _____ and (l) _____ perceptions and produce marked enhancement of emotional responses.

10. Some neurons in the CNS release peptides that bind to the same cells that (a) _____ do and have the same pain-killing effect. The enkephalins inhibit (b) _____ neurons. These mechanisms protect us from (c) _____ or _____ pain.

C. TESTING TERMS AND CONCEPTS

1. Located exclusively within the spinal cord and brain are the
   a) sensory neurons.
   b) interneurons.
   c) sympathetic neurons.
   d) parasympathetic neurons.

2. "Gray matter" is commonly associated with intelligence because it
   a) consists of long fibers capable of rapid transmission of information.
   b) contains many well-insulated axons and dendrites.
   c) contains many cell bodies, responsible for much of the metabolic activity of the brain.
   d) contains many synapses and therefore many different potential circuits.

3. The sensory pathway into the brain does not include
   a) a spinal tract.
   b) a ventral root.
   c) the outer layer of the spinal cord.
   d) the thalamus.

4. Humans are most set apart from other vertebrates by virtue of their
   a) cerebrum.
   b) cerebellum.
   c) reticular formation.
   d) pineal gland.

5. Changes in human personality and behavior may be produced by damage to the

a) midbrain.
b) frontal lobes.
c) cerebellum.
d) hypothalamus.

6. Breathing and heartbeat control are regulated by the
    a) cerebellum.
    b) midbrain.
    c) medulla oblongata.
    d) hypothalamus.

7. A drug that is a noradrenaline mimic would have what effect in
    the autonomic nervous system?
    a) stimulate the postganglionic fibers
    b) stimulate the preganglionic fibers
    c) have a parasympathetic effect on internal organs
    d) have a sympathetic effect on internal organs

8. The peripheral nervous system is connected to the brain by
    a) motor neurons.
    b) sensory neurons.
    c) the spinal cord.
    d) the pons.

9. Nerves that transmit impulses from our receptors (chiefly of ex-
    ternal stimuli) to the central nervous system belong to the
    a) sensory-somatic system.
    b) autonomic system.
    c) sympathetic system.
    d) parasympathetic system.

10. The actions of the autonomic nervous system are
    a) largely voluntary.
    b) largely involuntary.
    c) both; it depends on the action.
    d) neither; the ANS is sensory only.

11. The most dramatically addictive drugs are
    a) barbiturates.
    b) tranquilizers.
    c) stimulants.
    d) opiates.

12. Nervous system activity is reduced by
    a) drinking coffee.
    b) drinking alcohol.
    c) smoking a cigarette.
    d) taking amphetamines.

13. Nicotine and caffeine act on the
    a) sensory-somatic system.
    b) autonomic system.
    c) sympathetic system.
    d) cerebrum.

14. The opiates, such as morphine,
    a) enter neuron cell bodies to have their effects.
    b) enter synaptic knobs.
    c) bind to neuron cell membranes.
    d) bind to nonneuronal "accessory" cells in the brain.

15. Experimental stimulation of enkephalin neurons
    a) enhances sensory activity.
    b) depresses sensory activity.
    c) alters the electrical activity of certain interneurons.
    d) alters the synaptic activity of certain interneurons.

D. UNDERSTANDING AND APPLYING TERMS AND CONCEPTS

1. What specific region of the CNS is responsible for
    _____ a) memory
    _____ b) subconscious adjustment of fine-tuning of on-going
            locomotion
    _____ c) conscious control of locomotion
    _____ d) reflex limb withdrawal
    _____ e) direct regulation of blood pressure
    _____ f) alerting the conscious centers to significant sensory
            input
    _____ g) conscious sensation
    _____ h) feelings of hunger and thirst

2. A person is in an automobile accident. What symptoms would you
    observe if X-rays revealed damage to the
    a) right posterior cerebrum? _____
    b) temporal lobes of the cerebrum? _____
    c) very top of the cerebrum, around the fissure of Rolando?
    _____
    d) cerebellum? _____
    e) medulla oblongata? _____

3. What if the accident somehow tore loose all the dorsal spinal roots
    along the left side of the lower half of the spinal cord? What
    part of the body would be affected?
    (a)_____
    Describe the effect. (b)_____
    _____

4. Where in the central nervous system do these drugs exert their
    effects?
    a) ethyl alcohol_____
    b) opiates_____
    c) stimulants_____
    d) mescaline and LSD_____

5. The processing of visual input and maybe of all sensory information
    seems to involve the phenomenon of convergence: many neurons at
    the next level. Describe two ways in which meaning can be
    extracted from sensory input with this arrangement of neurons.
    a)_____

b) _____

Decide whether the following statements are true (T) or false (F).

_____ 6. There are 43 pairs of peripheral nerves.
_____ 7. The peripheral nerves are "mixed" nerves, with both sensory and motor neurons in each.
_____ 8. The sympathetic nervous system releases noradrenaline, which stimulates the internal organs (heart, intestines, etc.) in times of emergency.
_____ 9. Parasympathetic stimulation <u>constricts</u> the pupil, <u>increases</u> flow of saliva, and <u>causes</u> urination.
_____ 10. Sensory-somatic sensations tend to go to the cerebrum; autonomic sensations tend to go to the hypothalamus.
_____ 11. Important sympathetic nerves originate at the medulla oblongata and branch to many different thoracic and abdominal organs.
_____ 12. Many stimulants act on the sympathetic nerves of the autonomic nervous system, while the depressants act on the parasympathetic nerves.

ANSWERS TO CHAPTER EXERCISES

Reviewing Terms and Concepts

1. a) central b) peripheral c) muscles, glands d) central

2. a) peripheral b) brain c) reflex d) spinal cord

3. a) hind b) mid c) fore d) cerebellum e) medulla oblongata
   f) heartbeat, breathing g) muscular

4. a) mid b) balance c) forebrain, hindbrain d) cerebrum e) thalamus,
   hypothalamus f) pituitary g) pineal gland h) cortex

5. a) retina b) contrasts c) filtered out d) emphasized

6. a) 12 b) 31 c) conscious d) conscious e) skeletal muscles

7. a) autonomic b) sensory, motor d) involuntary

8. a) sympathetic b) parasympathetic

9. a) Perception b) muscular coordination c) emotions d) stimulants
   e) sympathetic f) heart rate g) pupils h) sugar i) depressants
   j) hallucinogens k) visual l) auditory

10. a) opiates b) pain-transmitting c) extreme, long-term

Testing Terms and Concepts

1. b   6. c   11. d

2. d	7. d	12. b
3. b	8. c	13. c
4. a	9. a	14. c
5. b	10. b	15. d

Understanding and Applying Terms and Concepts

1. a) cerebrum b) cerebellum c) cerebrum d) spinal cord
    e) medulla f) reticular formation g) cerebrum h) hypothalamus

2. a) left eye blindness b) deafness c) disrupted sensation and
    motor control of legs and torso d) clumsy locomotion demanding
    much more conscious attention e) cessation of breathing and
    death

3. a) left leg and left side of abdomen b) loss of all sensation but
    not of motor control

4. a) Depending on the dose, alcohol can depress the entire brain.
    b) They depress pain pathways in spinal cord and brain, and the
    medulla oblongata. c) They stimulate the hypothalamus. d) The
    hallucinogens apparently stimulate the cerebrum.

5. a) Some input (presumably useless) will be filtered out, for
    instance, diffuse light causes little cortical activity. b)
    Specific patterns trigger corresponding pathways and are therefore
    especially noticed, for instance, edges and corners result in
    activity in specific cortical cells.

6. T	9. T	12. F
7. F	10. T	
8. F	11. F	

30

MUSCLES AND OTHER EFFECTORS

I. REVIEWING THE CHAPTER

A. CHAPTER HIGHLIGHTS

30.1 Kinds of Muscles

The structures with which animals carry out actions are called effectors. The most important of these are the ones that secrete substances (the glands) and those that bring about motion. In vertebrates, the most important effectors for creating motion are the muscles.

Vertebrates have three distinctly different kinds of muscles. The wall of the heart is made up of cardiac muscle. The walls of all hollow organs of the body—except the heart—are made up of smooth muscle. All muscles attached to our bones are called skeletal muscles. Cardiac and smooth muscle tissue generally are not under voluntary control, but skeletal muscle is.

30.2 The Structure and Organization of Skeletal Muscle

A single skeletal muscle consists of a narrow, relatively fixed "origin," a thickened belly of parallel muscle fibers (individual cells), and a narrow, movable "insertion." Muscles are arranged in pairs, across joints. This allows one to bend or flex the joint and the other to straighten or extend it.

30.3 The Activation of Skeletal Muscle

Both cardiac and smooth muscle can contract without being stimulated by the nervous system. Nerves reaching those muscles simply change the rate or strength of the contractions. In contrast, skeletal muscle is totally dependent on nervous stimulation. When an action potential reaches the neuromuscular junction, ACh is released, just as at a neural synapse. The muscle fiber membrane is depolarized, an action potential sweeps over its surface, and the cell contracts. Like a neuron, a muscle fiber responds in an all-or-none manner.

30.4 The Physiology of the Entire Muscle

The functional unit of a skeletal muscle is the "motor unit," a few to one or two thousand muscle fibers and the one, branched, axon that innervates them. In most muscles, even at rest, a few motor units are contracting at all times, producing a weak contraction called muscle tonus. Above this, a graded response can be elicited by stimulating more and more motor units.

30.5 The Muscle Fiber/30.6 The Chemical Composition of Skeletal Muscle/30.7 The Sliding Filament Hypothesis

A skeletal muscle is made up of muscle fibers, which in turn are composed of myofibrils. These latter structures are made up of arrays of parallel thick filaments, composed of the protein myosin, and of thin filaments composed of the three proteins actin, tropomyosin, and troponin. Skeletal muscles contract and relax as these thick and thin filaments slide back and forth past each other in a ratchet-like fashion.

30.8 Coupling Excitation to Contraction

A muscle cell action potential not only spreads over the entire surface of the cell but into its interior as well, along tubes of the T-system. This insures that the entire volume of these large cells is stimulated at once. This stimulation causes large quantities of calcium ions to be released from the sarcoplasmic reticulum, which triggers the reaction between actin and myosin.

30.9 The Chemistry of Muscular Contraction

The immediate source of energy for muscular contraction is ATP. The store of ATP in a muscle fiber is quickly exhausted during any activity, whereupon a similar, energy-rich molecule, creatine phosphate, supplies the needed energy. When these sources are gone, more ATP is produced by cellular respiration or by anaerobic glycolysis.

30.10 Cardiac Muscle

Cardiac muscle also is composed of interlocking, striated fibers, but the myofibrils in cardiac muscle are often branched. Unlike skeletal muscle, cardiac muscle's ability to contract is self-generated, and oxygen deprivation leads to death of the muscle tissue.

30.11 Smooth Muscle

Smooth muscle lacks striations, although it is composed of thick and thin filaments of actin and myosin. Smooth muscle can contract spontaneously, but it is controlled by the sympathetic and parasympathetic nervous systems. The action of smooth muscle is much slower than that of skeletal muscle.

30.12 Cilia and Flagella

Cilia and flagella are whiplike appendages on the outside of various eukaryotic cells. Their function is to move the extracellular fluid surrounding the cell. This action may either move the fluid medium in relation to the cell, or move the cell in relation to the medium. Flagella tend to beat in a whiplike action while cilia tend to move in a rowing action. Cilia and flagella are composed of microtubules that operate in a sliding manner similar to that in skeletal muscle, the motion being similarly powered by ATP.

30.13 Electric Organs

Certain fishes have electrical organs used as both offensive and defensive weapons. Others use their electrical organs to serve a variety of signaling functions, including navigation and locating and courting mates.

30.14 Chromatophores/30.15 Luminescent Organs

Certain other organisms have the ability to change their color patterns by changing the distribution of pigments in special cells called chromatophores. Many marine animals give off visible light through the use of luminescent organs, or through luminescent bacteria adhering to their bodies. Fireflies are another example of light-emitting animals.

B. KEY TERMS

cardiac muscle	sarcomere
tendon	myosin
antagonistic pair	flagella
isometric	luciferin
myofibrils	skeletal muscle
tetanus	flexor
creatine phosphate	motor unit
cilia	thick filament
chromatophores	thin filament
smooth muscle	tropomyosin
extensor	troponin
neuromuscular junction	oxygen debt
isotonic	tonus
EPP	actin
action potential	basal body
latent period	luciferase

II. MASTERING THE CHAPTER

LEARNING OBJECTIVES

When you have mastered the material in this chapter, you should be able to:

1. Define all key terms.

2. Cite the three distinctly different kinds of muscles found in vertebrates, and give an example of each.

3. Distinguish between the three basic parts of any skeletal muscle.

4. Explain how a motor neuron triggers muscle cell contraction.

5. Define "motor unit" and explain its significance to skeletal muscle action.

6. Explain the sliding-filament hypothesis of muscle contraction.

7. Cite three sources of ATP for muscle contraction.

8. Cite two ways in which the makeup of cardiac muscle differs from that of skeletal muscle.

9. Distinguish between tonus in smooth muscle and in skeletal muscle.

10. List four ways in which the functioning of the filaments in skeletal muscle and that of the microtubules in cilia are similar.

11. List four ways in which electric fish use their ability to generate electricity.

B. REVIEWING TERMS AND CONCEPTS

1. The structures with which animals carry out actions are called (a) _____. The most important of these are the ones that secrete substances, the (b) _____, and those that bring about (c) _____. In vertebrates, the most important effectors for creating motion are the (d) _____.

2. Vertebrates have (a) _____ distinctly different kinds of muscles. The wall of the heart is made up of (b) _____ muscle. The walls of all hollow organs except the heart are made up of (c) _____ muscle. All muscles attached to our bones are called (d) _____ muscles. (e) _____ and (f) _____ muscle tissue generally are not under voluntary control, but (f) _____ muscle is.

3. A single skeletal muscle consists of a narrow, relatively fixed (a) _____, a thickened (b) _____ of parallel muscle fibers, and a narrow, movable (c) _____. Muscles are arranged in (d) _____ across joints.

4. Both (a) _____ and _____ muscle can contract without being stimulated by the nervous system. Nerves reaching those muscles simply change the (b) _____ or _____ of the contractions. In contrast, (c) _____ muscle is totally dependent on nervous stimulation. When an action potential reaches the (d) _____ _____, ACh is released, just as

at a neural synapse. The muscle fiber membrane is (e) _____,
and (f) _____ _____ sweeps over its surface, and
the cell contracts. Like a neuron, a muscle fiber responds in an (g)
_____ manner.

5. The functional unit of a skeletal muscle is the (a) _____
_____ _____, a few to one or two thousand (b)
_____ _____ and the one, branched, (c) _____
that innervates them. In most muscles, even at rest, a few motor units
are contracting at all times, producing a weak contraction called (d)
_____ _____. Above this, a graded response can be
elicited by stimulating more and more (e) _____ _____.

6. A skeletal muscle is made up of muscle fibers, which in turn
are composed of (a) _____. These latter structures are made
up of arrays of parallel thick filaments, composed of the protein (b)
_____, and of thin filaments composed of the three proteins,
(c) _____, tropomyosin, and troponin. Skeletal muscles contract as these thick and thin filaments (d) _____ _____
each other.

7. A muscle cell action potential not only spreads over the
entire surface of the cell but into its interior as well, along tubes
of the (a) _____. This insures that the entire volume of
these large cells is stimulated at once. This stimulation causes
large quantities of (b) _____ ions to be released from the
(c) _____ _____, which triggers the reaction between
(d) _____ and _____.

8. The immediate source of energy for muscular contraction is (a)
_____. The store of (b) _____ in a muscle fiber is
quickly exhausted during any activity, whereupon a similar, energy-
rich molecule, (c) _____ _____, supplies the
needed energy. When these sources are gone, more is produced by (d)
_____ _____ or by _____ _____.

9. Cardiac muscle also is composed of interlocking, striated
fibers, but the myofibrils in cardiac muscle are often (a) _____.
Unlike skeletal muscle, cardiac muscle's ability to contract is (b)
_____, and oxygen deprivation leads to (c) _____ of
the muscle tissue.

10. (a) _____ muscle lacks striations, although it is
composed of thick and thin filaments of (b) _____ and _____.
Smooth muscle can contract spontaneously, but it is controlled by the
(d) _____ and _____ nervous systems. The
action of smooth muscle is much (d) _____ than that of
skeletal muscle.

11. (a) _____ and _____ are whiplike appen-
dages on the outside of various eukaryotic cells. Their function is
to (b) _____ the extracellular fluid surrounding the cell.
They are composed of (c) _____ that operate in a (d)
_____ manner similar to that in skeletal muscle, the motion

being similarly powered by (e) _____.

12. Certain fishes have (a) _____ organs used as weapons and in navigation and courtship. Certain other organisms have the ability to change their color patterns by changing the distribution of pigments in special cells called (b) _____. Many marine animals give off (c) _____ through the use of luminescent organs, or through luminescent bacteria adhering to their bodies.

C. TESTING TERMS AND CONCEPTS

1. Which listing below shows muscle parts in the correct order, from the relatively immovable end near the body axis, to the more movable end further away?
    a) belly tissue, tendon, origin
    b) origin, belly tissue, insertion
    c) origin, smooth tissue, belly tissue
    d) insertion, smooth tissue, tendon

2. Which statement about neural control of muscle tissue is false?
    a) The autonomic nervous system usually triggers heart contraction.
    b) The sensory-somatic system usually triggers contractions in skeletal muscles.
    c) The stomach and intestines can contract without input from the peripheral nervous system.
    d) The stomach and intestines can remain contracted without neural input.

3. The biceps muscle of the arm
    a) bends the elbow but cannot straighten it back out again.
    b) is one of the many "extensors" of the body.
    c) is composed of cells that do not go into "tetanus."
    d) does not exhibit "tonus."

4. A neuromuscular junction of a skeletal muscle differs from a sensory-somatic synapse in
    a) having ACh in its terminal vesicles.
    b) that the interior of the muscle cell is positively charged.
    c) that calcium ions flow into the muscle cell to trigger the action potential, rather than sodium.
    d) that only one neural action potential is enough to trigger a muscle cell action potential.

5. A phenomenon that does not help explain how we develop maximum tensions in skeletal muscles is
    a) clonus.
    b) tonus.
    c) tetanus.
    d) short refractory period.

6. A muscle contraction unable to budge a heavy load is said to be
    a) fatigued.
    b) isotonic.
    c) tonic
    d) isometric.

7. Upon stimulation, contraction of all the sarcomeres throughout a muscle fiber
   a) occurs in a rapid-fire sequence.
   b) is simultaneous.
   c) is delayed until Ca^{++} levels rise high enough.
   d) is delayed until Na^+ levels rise high enough.

8. The immediate source of energy for muscle contraction is
   a) respiration.
   b) glycolysis.
   c) ATP.
   d) creatine phosphate.

9. Skeletal muscle contracts as a direct result of the action of the
   a) Z-lines.
   b) H-zone
   c) sarcomeres.
   d) sarcoplasmic reticulum.

10. A muscle fiber can be made to contract by injecting it with
    a) lactic acid.
    b) pyruvic acid.
    c) ADP.
    d) Ca^{++}.

11. If muscle fibers were treated so that the actin filaments were removed,
    a) the cells could respond only weakly because the myosin would be acting alone.
    b) the cells could no longer contract because the actin filaments normally bind to the myosin and actively pull on it.
    c) the cells could no longer contract because calcium normally forms bridges between actin and myosin and so permits the pulling of one on the other.
    d) the cells could no longer contract because the actin filaments provide the link to the Z-lines, so it is their pull that ultimately shortens the sarcomere.

12. The action of skeletal muscle and that of cilia both require all of these except
    a) action potentials involving an influx of Na^+.
    b) cross bridges between sliding filaments.
    c) ATP.
    d) Ca^{++}.

13. If an individual is exercising at a rate that will not require long heavy breathing afterward, to repay an oxygen debt, he is apparently
    a) not drawing on his creatine phosphate reserves.
    b) not drawing on his glycogen reserves.
    c) relying on cellular respiration for ATP synthesis.
    d) transporting all lactic acid to the liver for processing.

14. The T-tubules contribute to a muscle contraction by transporting
    a) action potentials.
    b) acetylcholine.
    c) calcium ions.
    d) glucose.

15. The effectors <u>dominantly</u> involved in camouflage are
    a) chromatophores.
    b) luminescent organs.
    c) exocrine glands.
    d) electric organs.

D. UNDERSTANDING AND APPLYING TERMS AND CONCEPTS

1. Distinguish each of the three types of muscle tissue listed below
   by characterizing its <u>location</u> in the body, its <u>cell</u> <u>structure</u>,
   and the <u>nature</u> of its action.

	Skeletal	Cardiac	Smooth
Location	(a)_____	(b)_____	(c)_____

   Cell

   Action

2. At a neural synapse, the electrochemical event at the postsynaptic
   membrane is called an (a) _____ . At a neuromuscular
   junction, the similar event is called an (b) _____ .
   Describe an important functional difference between the two: (c)
   _____

3. Consider the events associated with a single twitch of a muscle
   fiber. What two are graded?
   a)_____
   b)_____
   What two are all-or-none?
   c)_____
   d)_____

4. The fibers of skeletal muscles are organized into motor units, each
   innervated by a different motor neuron. What would the effect be
   if muscles were not divided into motor units but were each innervated
   by a single motor neuron?
   a)_____
   Such a muscle could theoretically respond normally if, in addition
   to lacking motor unit organization, its cells responded differently
   than real muscle fibers do. Can you propose such a difference in
   cellular response? (b)_____
   _____
   _____

5. After a period of jogging, two people were sweating equally but
   one continued to breathe heavily long after the other. Why?

_____
_____
_____

6. List three similarities in the functioning of a muscle fiber and the electroplate of an electric organ.
   a)_____
   b)_____
   c)_____

ANSWERS TO CHAPTER EXERCISES

Reviewing Terms and Concepts

1. a) effectors b) glands c) motion d) muscles

2. a) three b) cardiac c) smooth d) skeletal e) Cardiac
   f) smooth g) skeletal

3. a) origin b) belly c) insertion d) pairs

4. a) cardiac, smooth b) rate, strength c) skeletal d) neuromuscular
   junction e) depolarized f) action potential g) all-or-none

5. a) motor unit b) muscle fibers c) axon d) muscle tonus
   e) motor units

6. a) myofibrils b) myosin c) actin d) slide past

7. a) T-system b) calcium c) sarcoplasmic reticulum d) actin, myosin

8. a) ATP b) ATP c) creatine phosphate d) cellular respiration,
   anaerobic glycolysis

9. a) branched b) self-generated c) death

10. a) Smooth b) actin, myosin c) sympathetic, parasympathetic
    d) slower

11. a) Cilia, flagella b) move c) microtubules d) sliding e) ATP

12. a) electrical b) chromatophores c) light

Testing Terms and Concepts

1. b	6. d	11. d
2. a	7. b	12. a
3. a	8. c	13. c
4. d	9. c	14. a
5. b	10. d	15. a

Understanding and Applying Terms and Concepts

1. a) attached to bones b) in heart wall c) in other hollow organs

(stomach) d) long, multinucleate, striated e) branched, uninucleate, striated f) spindle-shaped, uninucleate, not striated
g) voluntary, can maintain active tonus h) involuntary, no tonus,
cannot tolerate oxygen deprivation i) involuntary, slow, maintains
passive tonus

2. a) EPSP b) EPP c) The EPP normally always triggers an action
potential.

3. a) the amount of ACh released b) the EPP c) the AP d) the
contraction itself

4. a) The entire muscle would respond in an all-or-none manner; we
could not vary our actions to suit the task at hand. b) If muscle
fibers varied substantially in their thresholds, one AP would trigger only some of the cells in our hypothetical muscle, even though
all cells were innervated by that one neuron. More action potentials
would reach the threshold of more cells, and a graded response by
the muscle could be achieved. Another mechanism that would yield
graded responses in this muscle would be muscle fibers that each
responded in a graded manner, but such cells are harder to imagine.

5. They both ran the same distance, produced and used the same amount
of ATP, and therefore produced the same amount of excess heat.
However, the conditioned runner has a more efficient cardiovascular
system that delivers oxygen to the muscles faster and so allows
the production of ATP through aerobic respiration. The out-of-
shape runner had to rely on anaerobic glycolysis and so built up
a lactic acid oxygen debt.

6. a) Both maintain a negative resting potential. b) Both are innervated by a cholinergic motor neuron. c) In both, it is an in-
flux of Na^+ that initiates the action.

THE ELEMENTS OF BEHAVIOR

I. REVIEWING THE CHAPTER

A. CHAPTER HIGHLIGHTS

31.1 What is Behavior?

Behavior is action that alters the relationship between the organism and its environment. It may occur as a result of an external stimulus or as the result of an internal stimulus. There are two main types of behavior—that which is learned, and that which is innate. Innate behavior tends to be quite inflexible, a given stimulus giving rise to a given response. Learned behavior tends to be more or less permanently altered as a result of the experience of the individual organism.

31.2 Behavior in Plants

Behavior in plants seems to be innate, all members of a given species responding in the same way to a given stimulus. Plant behavior is restricted to growth movements and turgor movements. Movements whose direction is determined by the direction from which a stimulus strikes the plant are called tropisms. Those that are undirected are called nastic movements.

31.3 Taxes

Some animals respond to a stimulus by automatically moving toward or away from it. Such responses are called taxes. A moth's motion toward light, for example, is a phototaxis. E. coli's migration to a sweetened area of its liquid medium is a chemotaxis.

31.4 Reflexes

Reflexes are the simplest innate responses in animals having a nervous system. These are automatic responses of part of the body to a stimulus. For example, the stretch reflex in the thigh automatically stiffens the leg is the knee begins to buckle, and it monitors conscious leg movements and makes adjustments if they are greater or less than the brain intended.

31.5 Instincts

Instincts are complex behavior patterns which, like reflexes, are inborn, rather inflexible, and of value in adapting the animal to its environment. They differ from reflexes in their degree of complexity,

the entire body participating in instinctive behavior. A spider
building a web is an example of this type of complex behavior. Instincts are inherited just as the structure of tissues and organs is
inherited.

31.6 Releasers of Instinctive Behavior

Once the body is prepared internally for certain types of instinctive behavior, an external stimulus is needed to initiate the
behavior. This is the "releaser" of the instinctive behavior. Con-
stellations of stars release migratory behavior in certain birds, and
pheromones are important releasers for social insects.

31.7 Rhythmic Behavior and Biological "Clocks"

The migration of birds in the fall is an example of innate behavior that is repeated at definite intervals. Such behavior is
described as rhythmic or periodic. Cycles of such behavior may be as
short as two hours or as long as a year. These cycles are partially
triggered by external, environmental cycles, like the daily cycle of
day and night or the annual cycle of the seasons. There is also some
kind of an internal timer, a "biological clock," that maintains these
behavioral cycles in artificially constant conditions.

31.8 The Life History of the Honeybee/31.9 The Work of the Hive/31.10 Tools of the Honeybee

Probably no group of animals has developed such a varied
repertory of instinctive behavior as the colonial insects—the ants,
termites, and honeybees. An active, normal beehive contains one queen,
a few hundred drones, and many thousand workers. The appearance of
several distinct body types in a species is called polymorphism. In
the case of the honeybees, the different forms are each adapted to
carry out specific functions. The queen's function is to lay eggs.
Among other things, the workers clean the hive, gather food, and care
for the grubs. These tasks are accomplished through the use of
sensitive sense receptors and appendages containing a variety of
specialized tools. The drones' function is to mate with the queen.

31.11 Communications among Honeybees

Foragers returning to the hive with food inform bees within the
hive about food source by performing a tail-wagging dance, which tells
the others, first, that food is available; second, the direction to
the food; third, the distance to the food; and fourth, the odor of the
food.

31.12 Habituation/31.13 Imprinting

Almost all animals are able to learn not to respond to repeated
stimuli that have proved to be harmless, a learning phenomenon known
as habituation. One of the most narrowly specialized, clear-cut
examples of learning is imprinting, in which a newly hatched gosling
will, upon hatching, adopt and follow as its mother the first object

it sees moving and emitting appropriate sounds.

31.14 The Conditioned Response

Perhaps the simplest form of learned behavior is the conditioned response. This is a response which, as a result of experience, comes to be caused by a stimulus different from that which originally triggered it. A well-known example is the ability to train a dog to salivate in anticipation of food at the sound of a bell.

31.15 Instrumental Conditioning/31.16 Motivation/31.17 Concepts

Most animals solve mazes and other problems by trial and error. As long as sufficient motivation is present they try each alternative and gradually, through repeated successes and failures, restrict their behavior to the "correct" response. Insight involves putting familiar things together in new ways, and so is truly creative. Insight also depends on the development of concepts, such as the idea of "middleness" or "oddness."

31.18 Language

Language involves two levels of abstraction. The idea of a chair is an abstraction that can be represented by a symbol, the second level of abstraction. The use of language among chimps has been investigated, and it has been found that chimps can minipulate object-symbols (instead of spoken words). One chimp amassed a vocabulary of about 130 words including nouns, verbs, and adjectives.

31.19 Memory

All learning depends on memory. If an animal is to change its behavior on the basis of experience, it must be able to remember its experiences. Once something is learned, it must be remembered. Memory may be a dynamic process in which sensations give rise to nerve impulses that circulate indefinitely through the network of neurons in the central nervous system. Another theory says that our memories may be stored in a chemical code within the brain. Some look to RNA, others to proteins, as the substances in which memories are coded.

31.20 The Adaptive Significance of Behavior

Most behaviors are of value because they permit the animal to obtain food, avoid enemies, survive in the physical conditions of their environment, or pass their genes on to the next generation. Solo or cooperative foraging behaviors tend to change in a way that maximizes the cost/benefit ratio of the effort. Defensive behaviors vary from simple fleeing to the use of weapons, camouflage, mimicry, or co-operative mobbing. Survival in the physical environment involves avoidance of unsuitable conditions and, in some cases, active alteration of one's environment, as in bees cooling their hive. Reproduction usually involves the females choosing their mates and the males competing for the opportunity through the use of display, fighting, or territoriality.

B. KEY TERMS

tropisms	scout bee
phototaxis	spindle fibers
instinct	polymorphism
releasers	stretch reflex
imprinting	phonemes
concept	chemotaxis
taxes	muscle spindle
magnetotaxis	habituation
ethologist	operant conditioning
queen	instrumental conditioning
worker	insight
drone	kinesis
forager	

II. MASTERING THE CHAPTER

A. LEARNING OBJECTIVES

When you have mastered the material in this chapter, you should be able to:

1. Define all key terms.

2. Distinguish between innate and learned behavior, citing two examples of each.

3. Distinguish between tropisms and taxes, citing two examples of each.

4. Discuss the mechanism of the stretch reflex and list three of its functions.

5. Explain the concepts of the "releaser" and biological "clock."

6. List four "responsibilities" of a worker honeybee, all of which activities are examples of innate behavior.

7. List the four kinds of information about a food source a dancing bee is able to communicate to other bees, and explain how that information is communicated.

8. Distinguish between habituation, imprinting, and conditioning, and cite one example of each.

9. Explain the relationship between motivation and learning.

10. State how insight learning differs from all other types.

11. Cite the two basic theories of memory.

12. List the four basic needs that behaviors meet and give two examples of each.

B. REVIEWING TERMS AND CONCEPTS

1. There are two main types of behavior—that which is (a) _____, and that which is _____. (b) _____ behavior tends to be quite inflexible, a given stimulus giving rise to a given response. (c) _____ behavior tends to be more or less permanently altered as a result of the experience of the individual organism.

2. Behavior in plants seem to be (a) _____, all members of a given species responding in the same way to a given stimulus. Plant behavior is restricted to (b) _____ movements and _____ movements. Movements whose direction is determined by the direction from which a stimulus strikes the plant are called (c) _____.

3. Some animals respond to a stimulus by automatically moving toward or away from it. Such responses are called (a) _____. A moth's motion toward light, for example, is termed (b) _____. E. coli's migration to a sweetened area of its liquid medium is termed (c) _____.

4. (a) _____ are the simplest innate responses in animals having a nervous system. These are automatic responses of part of the body to a (b) _____.

5. (a) _____ are complex behaviors which, like (b) _____ are inborn and inflexible. They involve the entire body and are (c) _____ just as the structure of tissues and organs are.

6. Once the body is prepared internally for certain behaviors, an external stimulus is required to (a) _____ the behavior. (b) _____ are important chemical releasers of insect social behavior.

7. The migration of birds in the fall is an example of (a) _____ behavior that is repeated at definite intervals. Such behavior is described as (b) _____ or _____. Cycles of such behavior may be as short as two hours or as long as a year.

8. Foraging bees return to the hive and perform a (a) _____

dance to communicate that a food source is nearby. The (b) _____
dance tells that food is more distant, and gives the distance and
direction, relative to the (c) _____ position.

9. Almost all animals are able to learn not to respond to repeated stimuli that have proved to be harmless, a phenomenon known as
(a) _____, which is a form of (b) _____. One of the
most narrowly specialized, clear-cut examples of learning is (c)
_____, in which a newly hatched gosling will, upon hatching,
adopt and follow as its mother the first object it sees moving and
making appropriate noises.

10. Perhaps the simplest form of learned behavior is the (a)
_____ response.

11. Most animals solve mazes and other problems by (a) _____
and _____. As long as sufficient motivation is present,
they try each alternative and gradually, through repeated successes
and failures, learn to solve the problem. (b) _____ involves
putting familiar things together in new ways, and so is truly creative.

12. Language involves two levels of abstraction. The idea of a
chair is an abstraction that can be represented by a (a) _____,
the second level of abstraction. All learning depends on (b)
_____. If an animal is to change its behavior on the basis of
experience, it must be able to remember its experiences. Memory may
be a (c) _____ process in which sensations give rise to
nerve impulses that circulate indefinitely through the network of
neurons in the (d) _____ nervous system. Another theory says
that our memories may be stored in a (e) _____ code within the
brain. Some look to (f) _____, others to _____, as
the substances in which memories are coded.

13. Most behaviors are of value because they permit the animal to
obtain (a) _____, avoid (b) _____, avoid harsh
(c) _____ _____, or attract a (d) _____.

C. TESTING TERMS AND CONCEPTS

1. Behavior can best be defined as
    a) externally directed activities.
    b) externally directed activities triggered by external stimuli.
    c) both externally directed activities and internal changes.
    d) externally directed activities and internal changes triggered
       by external or internal stimuli.

2. A moth flying toward a light source at night is exhibiting a
    a) phototaxis.
    b) phototropism.
    c) positive photokinesis.
    d) negative photokinesis.

3. Instinctive behavior tends not to be
    a) complex.

b) simple.
c) invariable in form.
d) unaffected by practice.

4. If the gamma motor fibers to the stretch receptors of your leg
   extensor muscle are cut,
   a) a tap at your knee tendon will no longer elicit a leg kick.
   b) a tap at the knee tendon will elicit a stronger than normal kick.
   c) inadequate control over the leg flexor will cause awkward
      voluntary leg movements.
   d) poorer control over the extensor will cause awkward voluntary
      movements.

5. Which of these is not a releaser of instinctive behavior?
   a) hormone
   b) pheromone
   c) song or call
   d) conspicuous crest of feathers or ruff of fur

6. The stretch reflex is an example of a(n)
   a) learned behavior.
   b) innate behavior.
   c) instinct.
   d) taxis.

7. A worker bee does not feed the colony's grubs
   a) a salivary secretion.
   b) bee bread.
   c) royal jelly.
   d) flower nectar.

8. Drone honeybees are produced from
   a) fertilized eggs.
   b) unfertilized eggs.
   c) eggs laid in the fall.
   d) grubs fed a special diet.

9. The greater the speed with which a tail-wagging dance is performed,
   a) the greater the quality of the food source.
   b) the greater the distance to the food source.
   c) the lesser the quality of the food source.
   d) the lesser the distance to the food source.

10. If a honeybee discovers an excellent food source directly outside
    its hive and in the direction of the sun at that moment, and if it
    is then trapped and anesthetized for six hours, what dance will it
    perform on being returned in good condition to the hive?
    a) no dance at all
    b) a round dance
    c) a tail-wagging dance oriented horizontally
    d) a tail-wagging dance oriented vertically

11. Language ability is most highly developed in
    a) honeybees.
    b) dogs.

307

c) songbirds.
d) chimpanzees.

12. Learning to ignore stimuli that prove to be harmless is called
a) imprinting.
b) habituation.
c) instrumental conditioning.
d) operant conditioning.

13. A white-crowned sparrow learns the details of its courtship song as a nestling, but it can learn this only during a few specific weeks of its life. Songs heard before this period or later in its life teach the sparrow nothing. This might best be called an example of
a) habituation.
b) imprinting.
c) simple conditioning.
d) insight learning.

14. It is unlikely that memory is coded in
a) active neural circuits.
b) protein molecules.
c) RNA.
d) ATP.

15. Which human behavior is least likely to be an instinctive behavior?
a) infant's sucking response
b) infant's recognition of its mother
c) adult tendency to "jet lag"
d) a smile

D. UNDERSTANDING AND APPLYING TERMS AND CONCEPTS

1. When pursued by a bird, a grayling butterfly turns toward the sun and flies toward it, thus blinding its pursuer. Describe the characteristic of this behavior that determines it is not a
a) reflex_____
b) positive phototropism_____
c) positive photokinesis_____
d) negative geotaxis_____
e) learned escape response_____
What type of behavior is it? (f)_____

2. Suppose you hear a songbird sing a complex song. Describe four different types of study that would help determine if the song was learned or instinctive.
a)_____
b)_____
c)_____
d)_____

3. An adult herring gull has a yellow bill with a red spot near the tip. A hungry chick will direct more begging pecks at a pencil with three red stripes around the end than at its parent's own bill.

Describe another example of a strong response to an inappropriate releaser. (a)_____

_____

4. Many people can read or study with a TV set on nearby and not be distracted. Is this an example of adaptation of sensory receptors, or learning? (a) _____ Explain. (b)_____

_____

_____

5. Does learning only occur if some motivation is present or some reward given? (a) _____ What causes a rat to explore a maze and learn to run it successfully when no food or other "reward" is placed at the other end? What causes a monkey to explore a puzzle and learn to solve it when the effort earns no "reward"? (b)_____

_____

_____

ANSWERS TO CHAPTER EXERCISES

Reviewing Terms and Concepts

1. a) learned, innate b) Innate c) Learned

2. a) innate b) growth, turgor c) tropisms

3. a) taxes b) phototaxis c) chemotaxis

4. a) Reflexes b) stimulus

5. a) Instincts b) reflexes c) inherited

6. a) release b) Pheromones

7. a) innate b) rhythmic, periodic

8. a) round b) tail-wagging c) sun's

9. a) habituation b) learning c) imprinting

10. a) conditioned

11. a) trial, error c) Insight

12. a) symbol b) memory c) dynamic d) central e) chemical f) RNA, proteins

13. a) food b) enemies c) physical conditions d) mate

Testing Terms and Concepts

1. a	6. b	11. d
2. a	7. d	12. b

3.	b	8.	b	13.	b
4.	d	9.	d	14.	d
5.	a	10.	b	15.	b

Understanding and Applying Terms and Concepts

1. a) It is too complex, involving action of the whole body.
   b) It is an <u>animal</u> behavior. c) It is <u>directed</u> toward the sun;
   kineses are <u>not</u> goal-directed. d) It is <u>unlikely</u> that an insect
   would have the opportunity to learn this strategy—its nervous
   system is too simple and life too short. f) positive phototaxis

2. a) Study variability within the species. Instincts tend to be
   relatively invariable. b) Compare complexity of behavior with that
   of nervous system and with length of life span. c) Look for an
   opportunity to learn during the normal life history of the bird,
   or artificailly deprive the bird of that opportunity. d) Interbreed different subspecies or species and expect a "hybrid" behavior if instinctive.

3. a) A male robin will attack a tuft of red feathers as if the tuft
   were a trespassing male. b) The CNS often filters out and ignors
   much of its stimulus input, focusing on some key part of that input. In the gull example, the chick doesn't normally peck at "a
   parent gull's head" but at "a thin object with contrasting red at
   tip." That is all the chick really "sees."

4. a) learning b) Adaptation implies an inability to perceive and
   therefore to respond to the TV. At any time, in this situation,
   one could attend to the show and become distracted. What we <u>do</u>
   have is an acquired ability to ignore uninteresting stimuli, a
   CNS phenomenon. This is learning.

5. a) yes b) Knowledge of the rat's or monkey's environment <u>is</u> a
   reward. Evolution has produced central nervous systems that are
   "uncomfortable" with ignorance. In the past, those rats who
   didn't explore and learn didn't know the best escape routes and
   were caught by predators. Thus curiosity evolved. Curiosity is
   as strong a motivation as hunger.

EVOLUTION: THE EVIDENCE

I. REVIEWING THE CHAPTER

A. CHAPTER HIGHLIGHTS

32.1 The Evidence from Paleontology

The evidence in support of the theory of evolution is considerable and comes from a wide variety of sources. Although not the first to present evidence in support of evolution, Charles Darwin was the first to build an overwhelmingly impressive case for its occurrence. In his influential book The Origin of Species, he said that all living things have, with modification, descended from a common ancestor. Further, he offered a large number of facts in evidence that were best explained by evolution. Of special significance is the fact that Darwin proposed a theory to explain how evolution works, namely the theory of natural selection.

The science of paleontology and the fossils (footprints, bones, shells, whole organisms that are trapped in resin or frozen, or those formed by petrification) that it studies have provided direct evidence for evolution. The fossils, formed in a variety of ways, show a decreasing complexity and variety as one travels backward in time. The fossil record is not a perfect one. Geological processes, the low likelihood of fossilization occurring, and the fact that many transient life forms existed in small numbers all contribute to the problem of "missing links," i.e., gaps in the fossil record.

32.2 The Evidence from Comparative Anatomy

Studies of comparative anatomy reveal basic patterns that provide evidence of inheritance from a common ancestor. Such patterns are described as homologous. These studies reveal that the more recently two species have shared a common ancestor, the more homologous organs they have in common. Of special significance are those homologous organs that are vestigial; these are interpreted as being remnants of a species's evolutionary heritage.

32.3 The Evidence from Embryology

The embryonic development of all vertebrates, particularly during the early stages, displays striking similarities. During development, embryos may go through stages that produce structures that are only temporary yet are found as permanent features in other species. The more closely related two species are, the longer their embryonic development proceeds in a parallel fashion. This repeating of ancestral forms during embryonic development is referred to as recapitulation and occurs biochemically as well as anatomically.

32.4   The Evidence from Comparative Biochemistry

Particularly strong evidence for evolution is provided by studying the biochemical homologies so abundant in living organisms. The amino-acid sequence of cytochrome c is an example. The more closely related two species are, the greater is the similarity displayed by the amino-acid sequences of the enzyme. Furthermore, since the amino acid sequences are so similar, it indicates the similarity of nucleotide sequences—hence similar genes. Comparative serology also provides similar evidence which is difficult to interpret in any other way than an evolutionary one.

32.5   The Evidence from Chromosome Structure

Homologies in chromosome structure reveal that the more closely related two species are on the basis of other criteria, the more similar are their karyotypes. For example, chromosomes from various species can be compared on the basis of visual markers such as, bands, loops, inversions, and duplications.

32.6   The Evidence from Protective Resemblance

A classic example of evolution in action comes from the studies of "industrial melanism." This case involves the darkening of moths as their habitat, trees, became darker due to soot. The explanation was that the lighter moths were selected against (preyed upon by birds).

32.7   The Evidence from Geographical Distribution

Further support for evolutionary theory comes from the distribution of plants and animals around the world, particularly the oceanic islands.

32.8   The Evidence from Domestication

The domestication of plants and animals, with their subsequent genetic manipulation by humans, adds even greater support to the theory of evolution and particularly to the cornerstone of Darwin's work, the theory of natural selection. Namely, just as we can drastically alter a domestic species by selective breeding, so too can nature select varieties that will survive and reproduce.

32.9   Summary

Darwin's The Origin of Species laid the foundation for the processes by which species develop—evolution by natural selection. Evidence from many disciplines has been obtained subsequently to support the ideas.

B.   KEY TERMS

evolution	homologous
natural selection	vestigial organs
palenotology	recapitulation
fossils	serology
petrifaction	industrial melanism

II. MASTERING THE CHAPTER

A. LEARNING OBJECTIVES

When you have mastered the material in this chapter, you should be able to:

1. Define all key terms.

2. Explain why <u>The Origin of Species</u> was such an influential work.

3. Discuss the importance of the theory of natural selection.

4. Cite the evidence for evolution provided by fossils, paying particular attention to the "missing link" argument.

5. State the case for evolution as evidenced by a study of comparative anatomy with special reference to homologous and vestigial organs.

6. Explain the theory of recapitulation and its relevance for evolutionary theory.

7. Offer examples from biochemisty and genetics that support evolutionary theory.

8. Explain why the phenomenon of industrial melanism provides such dramatic evidence for evolution.

9. Cite some of the evidence for evolution based on the geographical distribution of animals.

10. State, in general terms, why the theories of evolution and natural selection are held in such high regard by biologists.

11. How does the breeding of domestic animals and plants shed light on the theory of natural selection. Cite two examples of each.

B. REVIEWING TERMS AND CONCEPTS

1. Darwin's book (a) _____ was significant in that it presented a large body of evidence for evolution as well as suggesting a mechanism by which evolution occurs, namely the theory of (b) _____ _____.

313

2. Paleontology, the study of (a) _____, provides an evolutionary record whose breaks gave rise to the term (b) _____, referring to evidence lacking in the history of life. However, the finding of (c) _____ (more/less) complex organisms in deeper (older) layers lends support to the theory of (d) _____. The process by which an organism is copied in stone is (e) _____. Other kinds of fossils include (f) _____, _____, _____, and (g) _____ bodies.

3. Organs that have the same basic structure, the same relationship to other structures, and the same (a) _____ development are said to be (b) _____. Organs with no apparent function, such as the fused vertebrae of humans, are said to be (c) _____ organs. The presence of such organs argues (d) _____ (for/against) special creation, since it seems like (e) _____ (poor/good) design to produce something nonfunctional.

4. The theory of (a) _____ states that embryonic development repeats that of the species's ancestors and that this repetition provides evidence for the theory of (b) _____. These similarities exist not only in embryonic anatomy, but in embryonic (c) _____ as well.

5. Evidence from biochemistry, such as that from the enzyme (a) _____, suggests that the more similar the species being compared, the greater the similarity in the (b) _____ of amino acids in the chain. Such similarities are also noted in the (c) _____ of mammals and other organisms. The similarities of amino-acid sequences between species also suggests similarities of (d) _____ sequences in their DNAs. This can be interpreted to mean similarities in (e) _____ and therefore common (f) _____ _____.

6. With regard to chromosome structure, the more similar the (a) _____ of the two species, the more closely related they are likely to be. Curiously, when two (b) _____ (closely/distantly) related species can be mated, they produce sterile offspring. The sterility is probably due to the failure of the (c) _____ to match precisely during (d) _____ formation. However, the fact that portions of the (e) _____ do match is again evidence of common ancestors.

7. Dramatic evidence for evolution is provided by the example in England of the (a) _____ _____, whose change in coloration is an example of (b) _____ _____. The example demonstrates how (c) _____ against a certain trait, light color, occurred in nature. In the future, if the trees are to be less darkened by soot, you would predict selection (d) _____ (for/against) the dark moths.

8. The strongest geographical support for evolution is provided by the distribution of plants and animals, especially on oceanic (a) _____.

9. By selective (a) _____, humans (b) _____ (have/ have not) created new species. They have developed, however, drastic changes within a species and thus given support to the theory of natural (c) _____.

C. TESTING TERMS AND CONCEPTS

1. The more closely related two species of vertebrates are, the
    a) more likely they are to possess vestigial organs
    b) more likely they are to share identical antibodies
    c) longer their embryonic development proceeds in a parallel fashion
    d) longer they have been separated

2. Strong evidence to show that two species are related would come from
    a) serological tests
    b) analogous structures
    c) a comparison of gene numbers
    d) analogous functions.

3. Which of the following is not an accurate statement about The Origin of Species?
    a) stated that all living things derive from a common ancestor
    b) was the first to propose evidence in support of evolutionary change
    c) proposed a mechanism for evolutionary change
    d) has had great impact on western thinking

4. The 13 species of finches that Darwin found in the Galapagos Islands
    a) were all seed eaters.
    b) were different from all other species of birds elsewhere on the earth
    c) arrived in the islands independently of each other.
    d) were brought by earlier visitors.

5. Generally, the lower the rock strata
    a) the more complex the organisms found fossilized.
    b) the less varied the life forms entombed.
    c) the younger the fossils.
    d) the greater the number of vestigial organs.

6. Which of the following does not supply evidence for the theory of evolution?
    a) amino-acid sequence
    b) vestigial organs
    c) recapitulation
    d) study of analogous structures

7. Darwin's theory of evolution is an attractive one because it
    a) is so easily proven.
    b) has been accepted for over 100 years.
    c) continues to fit the facts.
    d) fits with everyones' religious beliefs.

8. Recapitulation
   a) may be biochemical as well as anatomical.
   b) is restricted to vertebrates.
   c) accounts for the production of nonfertile offspring.
   d) has been proven wrong.

9. Petrifaction results in the preserving of
   a) the soft parts of the organism.
   b) the hard parts of the organism.
   c) footprints.
   d) a copy of the structures of the organism.

10. The more closely related two species are on the basis of homologous organs, then
   a) the less likely they are to have a common ancestor.
   b) the more similar are their karyotypes.
   c) the greater will be their blood serum differences.
   d) the greater the number of vestigial organs they will possess.

11. The most valuable contribution of Darwin's book, The Origin of Species, to evolutionary theory is
   a) volume of evidence presented to support the theories.
   b) theory of convergent evolution.
   c) fact that it has withstood 100 years of criticism.
   d) it was so easily read by laymen.

12. Chromosome pairing in interspecific hybrids
   a) results in mismatched chromosomal regions.
   b) never results in viable offspring.
   c) results in increased fertility.
   d) has never been observed.

13. Homologous organs are those which
   a) are virtually all vestigial.
   b) are used for similar activities.
   c) have the same basic structure.
   d) disappear during embryonic development.

14. Which is most likely to be the guiding factor in evolution?
   a) mutations
   b) variation
   c) genetic recombination
   d) natural selection

15. Vestigial structures support the theory of evolution because they
   a) are newly evolved.
   b) are no longer functional, but once were
   c) are found only in highly evolved organisms.
   d) will help as future conditions change.

D. UNDERSTANDING AND APPLYING TERMS AND CONCEPTS

Decide whether items 1 through 5 are true (T) or false (F).

_____ 1. For Darwin, the most compelling evidence for evolution came from comparative embryology.

_____ 2. Gaps in the fossil record have been used to discredit the theory of evolution.

_____ 3. Through the years, the complexity of life has increased while its diversity has decreased.

_____ 4. Oak trees grow universally in the same habitat throughout the world.

_____ 5. Correlation of amino-acid sequences can be used to estimate relatedness among species.

6. Why do fossils provide such strong support for the theory of evolution? _____

_____
_____
_____

7. What kind of evidence to support the theory of evolution would have been least likely used by Darwin? _____

_____

8. Which two embryos would you expect to look most alike?
   a) fish at 1 week
   b) human at 1 month
   c) monkey at 7 months

_____

9. Which two of the above embryos would you expect to be most similar in their DNA? _____

_____

10. Let the letters A through T represent the twenty amino acids. Here is an hypothetical peptide from a portion of a human enzyme:

B-C-F-M-M-R-A-S-B-B-N-E

Here are homologous peptides from three different animals:

   I.   B-C-F-N-M-R-A-A-B-B-N-R

   II. B-C-F-N-M-S-S-A-B-A-M-E

   III. B-C-F-M-M-R-A-A-B-B-N-F

a) Which animal is probably most closely related to human?

_____

b) Which animal is probably least closely related to the human? _____

317

ANSWERS TO CHAPTER EXERCISES

Reviewing Terms and Concepts

1. a) <u>The Origin of Species</u> b) natural selection

2. a) fossils b) missing links c) less d) evolution e) petrifi-
    cation f) footprints/bones/shells g) whole

3. a) embryonic b) homologous c) vestigial d) against e) poor

4. a) recapitulation b) evolution c) biochemistry

5. a) cytochrome b) sequence c) antigens d) nucleotides e) genes
    f) ancestors

6. a) karyotype b) closely c) chromosomes d) gamete e) chromosomes

7. a) peppered moth b) industrial melanism c) selection d) against

8. a) islands

9. a) breeding b) have not c) selection

Testing Terms and Concepts

1.	c	6.	d	11.	a
2.	a	7.	c	12.	a
3.	b	8.	a	13.	c
4.	b	9.	d	14.	d
5.	b	10.	b	15.	b

Understanding and Applying Terms and Concepts

1. F

2. T

3. F

4. F

5. T

6. They provide visual evidence of ancestral forms.

7. biochemical

8. a and b

9. b and c

10. a) III b) II

EVOLUTION: THE MECHANISMS

I. REVIEWING THE CHAPTER

A. CHAPTER HIGHLIGHTS

33.1 Inheritable Variation: the Raw Material of Evolution

Through observation of the efforts of animal breeders, Darwin
came to realize that inheritable variations were used in a selective
breeding process to develop new, pure-breeding lines of animals. These
inheritable variations take two forms: continuous variation, wherein
the trait is displayed over a continuum of forms, and discontinuous
variation, wherein the trait expression must fall into one of a limited
number of categories.

33.2 Natural Selection

Drawing upon the essay of Thomas Malthus, On Population, and his
own observations, Darwin reasoned that a process similar to selective
breeding was occurring in nature by which the fittest survived. This
process has been labeled natural selection.

33.3 The Measure of "Fitness"

The measure of "fitness" in the biological sense involves several
criteria. These include: potential for survival through the reproductive years, sexual desirability, ability to produce larger mature
families.

33.4 The Genetic Source of Variability

Upon the rediscovery of Mendel's work, the source and mechanism
by which variability was transmitted became known and another criticism
of Darwin's theory was laid to rest. The sources of variability were
seen to be the processes by which genes were reshuffled and the process of mutation by which new "cards" were introduced into the "deck."
When considering evolution in genetic terms, one must be conscious of
the significance of diploidy and haploidy and look at the genetics of
the population rather than the individuals in the population.

33.5 The Hardy-Weinberg Law

Of considerable use in studying population genetics is the Hardy-
Weinberg law, which helps explain why recessive alleles are retained
in the population, thereby providing a reservoir of variability from
which future changes in the genetic constitution of a population can

be drawn. The Hardy-Weinberg law does not explain evolution, however.

33.6 When the Hardy-Weinberg Law Fails to Apply

It is when the Hardy-Weinberg law fails to apply that evolutionary change occurs. The five factors which bring about this change are: mutation, drift, gene migration, nonrandom mating, and differential reproduction.

33.7 Kin Selection

The concept of kin selection provides an understanding of how altruistic behavior can be a genetic endowment that is operated upon by natural selection working at the family level. In some supposed examples, such selection may simply be another example of one of Darwin's criteria of fitness, the ability to produce larger, mature families.

33.8 The Effects of Selection on Population

The pressure of natural selection can affect the distribution of phenotypes in a population in several ways. Stabilizing selection enhances the reproductive success of those phenotypes near the mean, thereby working to maintain the status quo, whereas directional selection favors those removed from the mean at one end of the spectrum. When both extremes are favored, disruptive selection results.

33.9 Can Humans Direct their Own Evolution?

It is possible for humans to direct their own evolution. However, many of the techniques employed by animal breeders would be considered reprehensible for application to humans. Eugenics can influence the improvement of the human gene pool by supporting measures such as sperm banks and by counseling against certain individuals' producing offspring, as in the case of PKU carriers.

33.10 The Effects of Increased Selection Pressure/33.11 The Effects of Relaxed Selection Pressure

Selection pressure and its effects can vary. Increased pressure tends to stabilize gene pools, reducing variability and increasing efficiency, eventually leading to the brink of extinction. Decreased selection pressure yields greater variability within the population.

33.12 What Is a Species/33.13 The Role of Isolation in Speciation

A species may be defined as an actually or potentially interbreeding natural population that does not interbreed with other such populations. Evolution provides answers to the question of how speciation occurs. A requirement for speciation is reduced selection pressure and its concomitant intraspecific variability. Geographic isolation of subpopulations then allows natural selection and/or genetic drift to produce new species.

33.14 Reunion

Reunion, the process by which two formerly isolated subpopulations are brought back together, results in a test of speciation. If the subpopulations fail to interbreed, then speciation has occurred. The factors preventing interbreeding may be varied and numerous. Reunion without interbreeding increases selection pressure, thereby hastening the evolutionary process. Speciation that results in a number of diverse species from a single ancestor is referred to as adaptive radiation.

33.15 Speciation by Polyploid/33.16 Convergent Evolution

Speciation is a gradual process with one exception, that being when polyploidy occurs. Polyploidy provides instant reproductive isolation and has occurred often in plants and very rarely in animals. Sometimes, due to similar selective forces in the process of evolution, two species of different genealogy come to resemble one another. This is termed convergent evolution and in no sense should it be construed as the opposite speciation.

33.17 Summary

Evolution involves adoptive changes in the genotypes and pheno-types of a population. When the changes lead to subpopulations that can no longer interbreed and produce fertile offspring, speciation has occurred.

B. KEY TERMS

inheritable variation	founder effect
selective breeding	gene migration
continuous variation	nonrandom mating
discontinuous variation	assortative mating
polymorphism	differential reproduction
heritability	balanced polymorphism
natural selection	differential fecundity
struggle for existence	kin selection
survival of the fittest	stabilizing selection
fitness	directional selection
adaptation	disruptive selection
crossing-over	eugenics
random assortment	selection pressure

mutation	speciation
multiple factors	reunion
population genetics	polyploidy
Hardy-Weinberg Law	chromosome homology
gene pool	convergent evolution
drift	adoptive radiation

II. MASTERING THE CHAPTER

A. LEARNING OBJECTIVES

When you have mastered the material in this chapter, you should be able to:

1. Define all key terms.

2. Distinguish between continuous and discontinuous variation. Explain the importance of such variations in the processes of selective breeding and evolution.

3. Explain the connection between the works of Malthus and Darwin. Reconstruct the logical process by which Darwin arrived at the theory of natural selection.

4. List and give examples of the criteria by which biological "fitness" is measured in nature.

5. Discuss the genetic sources of variability, distinguishing between those that reshuffle genes, and mutation, the source of new genes.

6. Explain the evolutionary significance of diploidy and haploidy as well as the importance of the study of population genetics.

7. Discuss the Hardy-Weinberg law, particularly as it can be used to explain the maintenance of a reservoir of recessive genes in a population.

8. Offer a definition of evolutionary change in terms of the failure of the Hardy-Weinberg law to operate. Enumerate and explain the factors which bring about such change and the law's failure.

9. Explain kin selection and how it conforms to the survival test of fitness, particularly with regard to altruistic behavior.

10. Discuss the three varieties of selection and how they affect the distribution of phenotypes in a population.

11. Indicate how humans can direct their own evolution.

12. Contrast increased and relaxed selection, noting the effects upon variability, efficiency, and gene pool stability.

13. Explain how and why isolation is a primary requirement for speciation.

14. Discuss the concept of reunion as it provides evidence for speciation and affects selection pressure.

15. List the steps that lead to speciation. Note how polyploidy leads to speciation.

16. Compare and contrast speciation and convergent evolution.

B. REVIEWING TERMS AND CONCEPTS

1. Only (a) _____ variations can be used to develop new breeds through the process of (b) _____ breeding. Variation occurs in two forms: (c) _____ and _____.

2. Thomas Malthus' work, (a) _____, influenced Darwin's thinking, contributing to his development of the theory of (b) _____ _____.

3. In evolutionary terms, those individuals in a population who leave the most mature offspring are said to be the (a) _____. The mechanisms that play a role in fitness are (b) _____ selection, (c) _____, and (d) _____ size.

4. New genes are the result of (a) _____, whereas new combinations result from (b) _____ assortment, (c) _____ _____, and (d) _____.

5. The (a) _____ _____ law offers a mathematical explanation for why (b) _____ alleles remain in a random-mating population. It also allows us to calculate (c) _____ frequencies.

6. The above mentioned law may fail to apply in circumstances where (a) _____, _____, (b) _____ mating, (c) _____ reproduction and (d) _____ migration occur. Such forces bring about (e) _____ change.

7. The reproductive advantage of certain phenotypes can stem from the cause of (a) _____ mortality, (b) _____ _____, and differential (c) _____.

8. Natural selection working at the family level is called (a) _____ _____.

9. Natural selection may affect the distribution of population phenotypes in one of three ways, called (a) _____, _____, and _____ selection. Individuals whose phenotype is close to the mean are favored when (b) _____

selection. Individuals whose phenotype is close to the mean are
favored when (b) _____ selection is operating, whereas (c)
_____ selection favors both extremes.

10. The application of genetic knowledge to improving the human
gene pool is called (a) _____.

11. Gene pools are stabilized by (a) _____ (increased/
decreased) selection pressure which leads to (b) _____ (greater/
lesser) specialization and increased efficiency. Increased variability
in the population is evidence of (c) _____ (increased/decreased)
selection pressure.

12. The formation of many species from a few is referred to as (a)
_____ and has (b) _____ (increased/reduced)
selection pressure as its first requirement. In addition to the increased variability, speciation requires (c) _____ isolation.
Only then can (d) _____ _____ and/or _____
_____ produce subspecies.

13. If two previously separated subpopulations cannot interbreed
during reunion, then (a) _____ has occurred. Such reunions
generally increase (b) _____ pressure. On the other hand, such
pressure may lead to character (c) _____ and so lessen the
competition between them.

14 Instant speciation results from the condition of (a) _____
and is common among (b) _____.

15. If two species of different genealogy resemble one another,
(a) _____ _____ is said to have taken place.
It should not be interpreted as the opposite of (b) _____,
since species multiplication has resulted in both instances.

C. TESTING TERMS AND CONCEPTS

1. Which of the following does not cause a reshuffling of genes?
   a) sexual reproduction.
   b) crossing over.
   c) random assortment.
   d) mutation.

2. The Hardy-Weinberg law does not hold
   a) in cases of incomplete dominance.
   b) in cases of nonrandom mating.
   c) when discussing an entire population.
   d) in cases of very large populations.

3. Evolutionary change
   a) is a product solely of mutation.
   b) occurs where the Hardy-Weinberg law fails to apply.
   c) will not occur in cases where differential reproduction takes
      place.
   d) does not involve changes in gene frequencies.

4. Stabilizing selection does not
   a) favor variations at opposite ends of the variation spectrum.
   b) maintain the status quo.
   c) favor those individuals whose characteristics are close to the mean.
   d) select against those at the extremes.

5. A shift in the Hardy-Weinberg numerical values indicates that
   a) natural selection may be occurring.
   b) mortality is increasing.
   c) convergent evolution is occurring.
   d) fecundity is increasing.

6. Decreased selection pressure
   a) leads to increased efficiency.
   b) permits increased variability in the population.
   c) decreases variation in a population.
   d) is theoretical only.

7. Speciation is
   a) the direct result of cline formation.
   b) the result of geographical isolation of subpopulations.
   c) the opposite of convergent evolution.
   d) always followed by reunion.

8. When pushed to the extreme, specialization may result in
   a) polyploidy.
   b) extinction.
   c) evolution.
   d) reunion.

9. Speciation
   a) is the result of natural selection.
   b) requires increased selection pressure.
   c) is caused by the process of reunion.
   d) means genetic equilibrium.

10. Balanced polymorphism
    a) is a direct result of mutation.
    b) can be explained by the phenomenon of diploidy.
    c) occurs in situations where the heterozygous genotype exists.
    d) leads to polyploidy.

11. Eugenics
    a) is a means of introducing new genes into a population.
    b) can shift the Hardy-Weinberg equilibrium.
    c) attempts to maintain the gene pool variability.
    d) is a measure of "fitness."

12. Genetic variability is the direct result of
    a) selective breeding.
    b) random assortment.
    c) phenotypes.
    d) inbreeding.

13. Body weight, height, and coat color are examples of
    a) continuous variation.
    b) speciation.
    c) discontinuous variation.
    d) convergent evolution.

14. A factor not affecting fitness is
    a) sexual selection.
    b) survival.
    c) average lifespan.
    d) family size.

15. Kin selection is
    a) natural selection operating at the family level.
    b) a form of inbreeding.
    c) found only among primates.
    d) found only in plants.

16. The founder effect explains
    a) the functioning of the Hardy-Weinberg law.
    b) differential reproduction.
    c) how kin selection operates in animal populations.
    d) how gene pools of new populations can be different from those
       of old.

17. Increased selection pressure
    a) increases population variability.
    b) affects the Hardy-Weinberg equilibrium.
    c) is a form of stabilizing selection.
    d) increases convergent evolution.

18. Assortive mating can
    a) upset the Hardy-Weinberg equilibrium.
    b) occur only when near-relatives mate.
    c) be used to explain polyploidy.
    d) occur only if genes on separate chromosomes.

19. Increased variability in a population
    a) results from directional selection.
    b) is made possible by virtue of the Hardy-Weinberg law.
    c) is an effect of relaxed selection pressure.
    d) occurs along with reunion.

20. Stabilizing selection is
    a) synonymous with directional selection.
    b) a means of conserving those members of a population found at
       the lower extreme of a variable characteristic.
    c) a form of natural selection.
    d) a result of polyploidy.

D. UNDERSTANDING AND APPLYING TERMS AND CONCEPTS

Decide whether statements 1 through 15 are true (T) or false (F).

_____ 1. Kin selection provides a basis for arguing that some forms of altruistic behavior are genetically determined.

_____ 2. The greater the number of offspring produced, the greater fitness those family members possess.

_____ 3. A trait that increases an organism's chance of survival during its reproductive years is referred to as an adaptation.

_____ 4. The Hardy-Weinberg law is a mathematical summation of the forces behind speciation.

_____ 5. In the struggle for survival, aggression is the only criterion of success.

_____ 6. Mendel's work on genetics strongly influenced Darwin.

_____ 7. By knowing the percentage of individuals with a recessive phenotype, it is possible to calculate the frequency of the dominant allele in a population.

_____ 8. The more specialized a species becomes, the greater its chances of becoming extinct.

_____ 9. For a given protein, there can be as much variation within a species as between two different species.

_____ 10. If living conditions improve for a certain species, there is likely to be greater variation within the population.

_____ 11. As medical practices improve, we are likely to see greater variation in our species.

_____ 12. Incest is an example of assortative mating.

_____ 13. In-breeding often leads to "defective" offspring. This is because relatives share the same environment.

_____ 14. Selection is always in one direction.

_____ 15. In times of rapid environmental change, one would predict greater variability within a species.

16. Suppose that, in a dessert of light sand, there is a species of hawk that preys upon mice. In the mice, there is variation of color from gray to black. With time, what will happen to the color of the mouse population? (a)_____

What will happen to the gene pool with respect to the alleles for color? (b)_____

If the mice are the prey, will there be any selection for or against the predators (hawks)? (c)_____ Cite a possible example. (d)_____

Suppose, as mice become fewer and fewer, that some of the hawks start to catch fish from a nearby lake. Cite the conditions under which two species of hawks might arise. (e)

_____

_____

_____

_____

17. Horses and donkeys probably appear more alike than do toy poodles and Great Danes. The former, however, belong to different species while the latter belong to the same species. Explain. (a)_____

_____

What condition(s) would need to occur before the poodles and Danes became separate species? (b)_____

_____

_____

ANSWERS TO CHAPTER EXERCISES

Reviewing Terms and Concepts

1. a) inheritable b) selective c) continuous, discontinuous

2. a) Essay on Population b) natural selection

3. a) fittest b) sexual c) survival d) family

4. a) mutation b) random c) crossing-over d) outbreeding

5. a) Hardy-Weinberg b) recessive c) gene

6. a) mutation, drift b) nonrandom c) differential d) gene
    e) evolution

7. a) differential b) nonrandom mating c) fecundity

8. a) kin selection

9. a) stabilizing, directional, disruptive b) stabilizing c) disruptive

10. a) eugenics

11. a) increased b) greater c) decreased

12. a) speciation b) reduced c) geographical d) natural selection, genetic drift

13. a) speciation b) selection c) displacement

14. a) polyploidy b) plants

15. a) convergent evolution b) speciation

Testing Terms and Concepts

1.	d	6.	b	11.	b	16.	d
2.	b	7.	b	12.	b	17.	b
3.	b	8.	b	13.	a	18.	a
4.	a	9.	a	14.	c	19.	c
5.	a	10.	c	15.	a	20.	c

Understanding and Applying Terms and Concepts

1.	T	6.	F	11.	T
2.	F	7.	T	12.	T
3.	T	8.	T	13.	F
4.	F	9.	T	14.	F
5.	F	10.	T	15.	F

16. a) There will be a greater percentage of light mice as the dark
   ones are more likely to be eaten.
   b) The genes for black color will decrease in frequency.
   c) Yes d) There will be selection for those with better eyes.
   e) The hawks more suited for catching fish will mate with others
   that "fish", and those that eat mice will stay in the dessert
   and mate with "likes." Eventually a genetic or chromosomal change
   will occur and render the two varieties incapable of producing
   fertile offspring when cross-mated.

17. a) The two dogs, when mated, will produce fertilie offspring; the
   horse and donkey will not.

   b) When the two varieties of dogs become genetically or chromo-
   somally different enough so that they no longer produce fertile
   offspring, then speciation will have occurred.

THE ORIGIN OF LIFE

I. REVIEWING THE CHAPTER

A. CHAPTER HIGHLIGHTS

34.1 Early Theories of the Origin of Life

The theory of evolution explains the diversity of life on earth today by the descent of each and every species from common ancestors. But what was the first form of life? At best we can make intelligent guesses about its characteristics and when it arose and how it arose.

There are two basic approaches to account for the origin of life: (1) a theory of divine creation, as described in the Bible, for instance, and (2) a scientific approach embodying the belief that the forces governing the world can be known, that they act uniformly at all times and in all places, and that their effects can be measured and predicted. One theory, the cosmozoa theory, explains that life arose here as the result of the arrival of material from outer space. Some early theories stated that life arose from nonliving matter-spontaneous generation. The experiments of Redi, Spallanzani, and Pasteur showed that living things were the products of other living things.

34.2 Oparin's Theory

A. I. Oparin proposed, in 1936, that living matter arose spontaneously sometime during the first two billion years of earth's history. It arose, supposedly, out of a rich supply of organic molecules synthesized out of methane, ammonia, and water vapor, the synthesis energized by lightning, radioactivity, and other energy sources. Stanley Miller showed that in the laboratory amino acids and other organic molecules can be synthesized under primitive earth conditions now believed to have existed some four billion or so years ago. The results of Miller's experiments become more significant when one considers that similar organic compounds have come to earth in meteorites.

The early "protocells" would have needed a boundary layer, complex polymers within, and a means of reproduction. Experimental support for the spontaneous formation of cells from primitive organic molecules has not been successful.

34.3 The Age of the Earth

The earth's age has been estimated at about five billion years, the estimate having been arrived at by radioactive dating. Radioactive dating methods depend on the fact that the atoms of certain radioactive elements, such as uranium-238, break down into the atoms of another element, in this case lead-206. This "decay" process occurs at a definite, measurable rate. For example, it takes 4.51 billion years for half the atoms in any sample of U-238 to decay, this period of time being called the half-life of that element. Often two "clocks" are used to confirm the accuracy of the dating. This method has also been used to date meteorites and moon samples.

34.4 The Dawn of Life

Complex life forms (trilobites, mollusks, etc.) have been dated back 600 million years. More primitive forms of life have been found in rocks that are over three billion years old. Nourishment for the first form of life was no problem since that life form must have been heterotrophic and was surrounded by the same nutrient soup of organic molecules from which it arose. This heterotrophic way of life could not continue indefinitely, however, and eventually autotrophic organisms evolved. Such organisms were capable of synthesizing new organic molecules from inorganic substances in the environment. It also is possible that the autotrophic organisms evolved first. Hence, this "spontaneous generation" was probably a onetime occurrence.

34.5 Summary

There is evidence that all the small organic molecules from which living things are constructed may have originated spontaneously on the primitive earth. Some living forms appear to have existed as far back as 3.5 billion years ago.

B. KEY TERMS

cosmozoa theory   heterotrophic

spontaneous generation   autotrophic

protocell   chemoautotrophic

transmutation   greenhouse effect

half-life   abiotic

stromatolites

II. MASTERING THE CHAPTER

A. LEARNING OBJECTIVES

When you have mastered the material in this chapter, you should be able to:

1. Define all key terms.

2. Distinguish between attempts to account for the origin of life on

a scientific basis as opposed to those based on a theory of divine creation.

3. Cite the essence of the cosmozoa theory.

4. Cite the essence of the theory of spontaneous generation.

5. State four characteristics of a complex of organic molecules qualifying as living matter.

6. Cite three methods used to estimate the earth's age and specify the most reliable method.

7. State the significant aspect of radioactive transmutation that makes it reliable as a geologic "clock."

8. Distinguish between heterotrophy and autotrophy, citing why hetero-trophic living matter was at a disadvantage compared with autotro-phic living matter after life arose on earth.

9. Outline the experiments of Redi, Spallanzani, and Pasteur

B. REVIEWING TERMS AND CONCEPTS

1. The theory of (a) _____ explains the diversity of life on earth today by the descent of each and every species from common (b) _____.

2. The two basic approaches to account for the origin of life are (1) (a) _____ creation and (2) a (b) _____ ap-proach hypothesizing (c) _____ creation of living matter out of a soup of organic molecules over (d) _____ billion years ago.

3. Stanley Miller showed that in the laboratory (a) _____ acids could be synthesized out of (b) _____, _____, and water vapor energized by an electric spark.

4. The earth's age has been estimated at about (a) _____ billion years through a process called (b) _____ dating. Because radioactive elements decay at known (c) _____, they can be used as geologic clocks. For example, uranium-238 has a (d) _____-life of (e) _____ years and decays into the element (f) _____ _____.

5. The first living matter to arise on earth must have carried on a (a) _____ way of life since it presumably took ready-made nutrient molecules directly from the environment. Later, some forms of living matter evolved into an (b) _____ way of life and were able to synthesize their own organic nutrient molecules out of (c) _____ molecules of the environment.

6. Such a spontaneous generation of living matter could not occur today because of the lack of (a) _____ _____ in the

earth's waters and because of the presence of already existing forms of
(b) _____.

C. TESTING TERMS AND CONCEPTS

1. Which of the following is the most reliable method of estimating
   the earth's age?
   a) using the decay rate of lead-206 into uranium-238
   b) measuring the total thickness of sedimentary rock deposits and
      dividing this by the estimated annual increase in the thickness
      of ocean sediments.
   c) using the decay rate of uranium-238 into lead-206
   d) using an isotope with a half-life of less than five years

2. What makes radioactive clocks so reliable is that they
   a) never run down.
   b) all have different rates.
   c) have characteristic and constant rates.
   d) are universally accepted by scientists.

3. The proposal that meteorites brought life to the earth is known as
   a) adaptive evolution.
   b) the cosmozoa theory.
   c) spontaneous creation.
   d) creationist theory.

4. The first life was characterized as having
   a) some sort of membrane to separate it from its environment.
   b) the ability to take in molecules from the environment and dis-
      charge waste molecules back to the environment.
   c) both a and b.
   d) neither a nor b.

5. The length of time it takes 50 percent of the atoms of a radioactive parent material to change into atoms of a daughter product
   is called the element's
   a) transmutation rate.
   b) half-life.
   c) clock rate.
   d) atomic-life.

6. In Pasteur's experiments with spontaneous generation, fresh air
   was
   a) allowed to enter the flask freely through an S-shaped tube.
   b) filtered through cotton before entering the flask.
   c) purified before entering the flask.
   d) kept out of the flask.

7. Spontaneously generated living matter arising today most likely
   would not survive because of the presence of
   a) inorganic materials in the oceans.
   b) already existing forms of life.
   c) carbon dioxide in the atmosphere.
   d) radioactive fallout in the environment.

333

8. Methane and ammonia are presumed to have been in the earth's primitive atmosphere because we today observe these gases in the atmosphere of
   a) Jupiter and Saturn.
   b) Mercury.
   c) the earth.
   d) the moon.

9. Francesco Redi showed that maggots
   a) arise spontaneously in decaying meat.
   b) are produced from the eggs of flies.
   c) decompose meat through protein action.
   d) only develop in the presence of open air.

10. The constancy of radioactive clocks is affected by
    a) extreme temperature changes.
    b) changes in gravity coupled with magnetic changes.
    c) whether the atoms are in living organisms.
    d) nothing that we know.

11. In Miller's apparatus, organic molecules were
    a) synthesized.
    b) broken down.
    c) converted into living molecules.
    d) not observed.

12. Organisms capable of synthesizing nutrient organic molecules out of inorganic molecules of the environment are said to be
    a) autotrophic.
    b) mesotrophic.
    c) heterotrophic.
    d) paratrophic.

13. One would expect that one of the first developments in "protocells" was the evolution of a(n)
    a) DNA repair mechanism.
    b) mitochondrion-like organelle.
    c) a boundary layer or membrane.
    d) means of sexually reproducing.

14. The man who first proposed a theory on how chemicals in the primitive earth could have evolved into life was
    a) Darwin.
    b) an early Greek scholar.
    c) Pasteur.
    d) Oparin.

15. Energy for the synthesis of primitive organic molecules could have been provided by all of the following, except
    a) the sun.
    b) electrical discharges.
    c) ATP.
    d) heat from volcanos.

D. UNDERSTANDING AND APPLYING TERMS AND CONCEPTS

1. Glycine, H-C-COOH, is one of the

$$NH_2$$

$$H$$

amino acids. If it formed as Oparin theorized, what was the source of the following elements?

C _____

N _____

O _____

2. A major assumption in using radioactive materials to "date" objects is: _____
_____
_____

3. Does Miller's experiment prove that life arose spontaneously during the early phases of the earth's development? _____
_____

4. What is possible to conclude from the results of Miller's experiment? _____
_____
_____

5. If we found organic molecules or living things in other parts of the universe, which theory would be supported, 1) special creation 2) "spontaneous creation"? _____
_____

6. If we found primitive life forms on another planet, what properties would you expect it to have? _____
_____
_____
_____

7. Why is reproduction not an absolute requirement for determining if something is alive? _____
_____

8. From a scientific point of view, why are attempts to explain the origin of life on a supernatural basis, i.e., divine creation, unacceptable? _____
_____

ANSWERS TO CHAPTER EXERCISES

Reviewing Terms and Concepts

1. a) evolution b) ancestors

2. a) divine (special) b) scientific c) spontaneous d) three

3. a) amino b) methane, ammonia

4. a) five b) radioactive c) rates d) half e) 4.51 billion
   f) lead-206

5. a) heterotrophic b) autotrophic c) inorganic

6. a) organic, molecules b) life

Testing Terms and Concepts

1.	c	6.	a	11.	a
2.	c	7.	b	12.	a
3.	b	8.	a	13.	c
4.	c	9.	b	14.	d
5.	b	10.	d	15.	c

Understanding and Applying Terms and Concepts

1. C: methane

    N: ammonia

    O: water

2. the rate of decay was the same in the past as it is now.

3. No

4. It is possible for the "building blocks" of life to form without any special creative forces.

5. Either one or both

6. a definite arrangement or organization and the requirement for energy to maintain its organization

7. Things can be sterile and still alive. (Reproduction is a requirement for species propagation—not life.)

8. They are not testable.

35

THE CLASSIFICATION OF LIFE

I. REVIEWING THE CHAPTER

A. CHAPTER HIGHLIGHTS

35.1 Importance

In order to bring some order to a study of the large number of organisms that inhabit the earth, various attempts have been made at classifying those organisms according to shared similarities. The discovery of microorganisms that displayed both plant and animal characteristics led Haeckel to suggest setting up a third kingdom, the Protista. Later discoveries have led to the establishment of a fourth kingdom, the Monera.

35.2 The Principles of Classification

Classification consists of placing together in categories those things that resemble each other. One of the earliest schemes utilized analogous organs as a criterion. Inceases in anatomical knowledge revealed the superficiality of such a scheme and led to the development of a system based upon homologous structures. This improved system was first proposed by Carolus Linnaeus in 1753 and is essentially the one in use today. Utilizing homology as a criterion for classification is especially valid since homologous organs reflect the existence of an evolutionary kinship.

35.3 An Example

The fundamental unit of classification is the species. Similar species are placed in groups called genera. These in turn are placed in families. Then, families constitute orders, which are grouped into classes. Classes constitute a phylum, and finally, all related phyla make up a kingdom.

35.4 Evolutionary Implications of Modern Taxonomy

Since the presently accepted classification scheme reflects degree of kinship, a family tree can be set up that reflects the evolutionary history of the group. It is important to note that this implies

1. that all the species in one group have shared a common ancestor more recently than they have with species in any other group and

2. that no living organism is the ancestor of other organisms.

35.5 Scientific Names

In order to avoid confusion, a system of scientific naming was established by Linnaeus in which an organism was given two names—the genus name and species name. Today this naming procedure, with modifications, follows an internationally agreed upon set of rules.

35.6 Higher Categories

The species is a unit that exists in nature by virtue of the reproductive-capacity criterion. All other taxonomic categories are human constructs. Although the goal of taxonomy is to develop a "natural" classification, a lack of evidence for some groups of organisms sometimes causes taxonomists to resort to convenient criteria.

B. KEY TERMS

Protista	species
Monera	genus
analogous organs	family
taxonomy	order
homology	class
convergent evolution	phylum

II. MASTERING THE CHAPTER

A. LEARNING OBJECTIVES

When you have mastered the material in this chapter, you should be able to:

1. Define all key terms.

2. Discuss the importance of classifying living organisms.

3. Cite the criteria employed by earlier and present-day classification systems. Explain the significance of the system developed by Linnaeus and why it is in use today.

4. List the units of classification in use today, and name the five kingdoms.

5. Discuss the evolutionary implications of modern taxonomy, particularly with regard to the subject of ancestry.

6. Explain the naming system developed by Linnaeus and the modifications employed by present taxonomists.

7. Distinguish between a natural classification and one based on convenience, noting why the former is preferred but the latter is sometimes employed.

B. REVIEWING TERMS AND CONCEPTS

1. The classification of organisms is referred to as the science of (a) _____, of which (b) _____ _____ is considered to be the father.

2. When two kingdoms proved insufficient for classifying all organisms, the German biologist (a) _____ proposed setting up a third kingdom to be named (b) _____. This solution was prompted by the confusion resulting from the invention of the microscope and the discovery of (c) _____.

3. Early classification schemes were based on the principle that creatures possessing (a) _____ organs should be grouped together. Linnaeus's system, however, groups organisms according to (b) _____ and is based upon evolutionary (c) _____.

4. A genus is made up of various (a) _____, whereas related families comprise (b) _____. A kingdom is subdivided into various (c) _____.

5. If the genealogy of a particular classification is correct, then all the species in one group have, in the past, shared a common (a) _____. Also, this implies that no living organism is the (b) _____ of other organisms.

6. In naming organisms, Linnaeus gave each two (a) _____ (language) names. The first identified the (b) _____ of the animal while the second specified its (c) _____.

7. Of all the taxonomic categories, only the (a) _____ actually exists in nature. It consists of organisms that can successfully (b) _____. All other categories are formed on the basis of human (c) _____.

C. TESTING TERMS AND CONCEPTS

1. A classification system based upon the principle of homology is significant because it
   a) is a more modern approach and utilizes advanced biochemical techniques.
   b) allows for the division of life into five categories.
   c) leads to a relatively permanent classification.
   d) is a system which establishes genealogical relationships.

2. Taxonomists would be more likely to resort to superficial, convenient criteria in classifying which of the following types of

organisms?
a) microorganisms
b) birds
c) primates
d) the races of humans

3. Two unrelated species that come to resemble each other closely because of similar adaptations to the same habitat are an example of
a) homologous evolution.
b) divergent evolution.
c) parallel evolution.
d) convergent evolution.

4. The taxonomic system in use today
a) gives each organism two latinized names, the family name and the species name.
b) gives two and sometimes three latinized names to an organism.
c) groups analogously similar organisms.
d) is essentially that proposed by Haeckel.

5. The study of classification
a) is called taxonomy.
b) is called zoology.
c) finds its origins in the work of Haeckel.
d) owes its beginning to van Leeuwenhoek.

6. If two birds shared membership in one of the following groups, which group membership would imply the closest relationship?
a) phylum
b) kingdom
c) suborder
d) family

7. In the designation Canis familiaris L., the letter L
a) indicates the organism is a subspecies.
b) refers to the phylum in which the organism is classified.
c) stands for the individual who first named the organism.
d) indicates the part of the world from which the type specimen was collected.

8. If a newly discovered organism were given the name Canis tigris, it would be most closely related to which of the following?
a) Felis trigris
b) Canis familiaris
c) Felis domestica
d) Tigris domestica

9. Which of the following have most recently descended from a common ancestor?
a) two birds in the same class
b) two snakes in the same family
c) three vertebrates in the same genus
d) three worms in the same species

10. The kingdom Monera contains the
    a) bacteria.
    b) bacteria and blue-green algae.
    c) blue-green algae.
    d) all algae.

11. The German biologist Haeckel proposed
    a) the kingdom Protista.
    b) the kingdom Monera.
    c) giving organisms two-part names.
    d) utilizing Latin names for organisms.

12. The modern system of classification
    a) is based upon the work of Linnaeus.
    b) is less than 100 years old.
    c) utilizes both the principles of homology and those of analogy.
    d) has not been able to deal with the problems created by con-
       vergent evolution.

13. All creatures sharing homologous organs are
    a) placed in the same genus.
    b) related to one another.
    c) from the same habitat.
    d) about the same size.

14. Originally, life was divided into
    a) the plant and animal groups.
    b) the plant, animal, and Monera groups.
    c) the plant, animal, and protist groups.
    d) plants, animals, protists, and Monera.

15. The best characters on which to base a taxonomic classification
    are
    a) characters that interact intimately with the environment, like
       color and thickness of the fur or pattern of root growth through
       the soil.
    b) characters not in direct association with the environment, like
       bone structure or shape of the liver.
    c) behavioral characters, like how a nest is made or when the
       flowers open.
    d) hard parts, like teeth and nail structure.

D. UNDERSTANDING AND APPLYING TERMS AND CONCEPTS

1. Arrange the terms on the right in descending order, i.e., from the
   group containing the greatest number of organisms to the group
   containing the least

       a)_____   A. family

       b)_____   B. genus

       c)_____   C. class

d)_____	D. order
e)_____	E. species
f)_____	F. phylum
g)_____	G. kingdom

2. Consider the five kingdoms used by many taxonomists: plant, animal, Protista, Funqi, and Monera. The viruses are extremely simple in their structure and are fundamentally unlike any other group of organisms. Their evolutionary affinities are almost completely unknown. What kingdom would you place them in, and why? _____

_____

3. Suppose you are a "splitter", not a "lumper", and interested in the taxonomy of dogs. Might you establish several species of dogs where now there is only one? Explain. _____

_____
_____
_____

4. The following diagram illustrates the evolutionary history of five animals.

"A" and "B" are water dwellers, are similar in body shape, and are placed in one class. "C" and "D" are terrestrial and are placed in a second class, and "E" is aerial and a member of a third class. Criticize this taxonomic scheme. _____

_____
_____
_____

ANSWERS TO CHAPTER EXERCISES

Reviewing Terms and Concepts

1. a) taxonomy b) Carolus Linnaeus

2. a) Haeckel b) Protista c) microorqanisms

3. a) analogous b) homology c) kinship

4. a) species b) orders c) phyla

5. a) ancestor b) ancestor

6. a) Latin b) genus c) species

7. a) species b) interbreed c) judgment

Testing Terms and Concepts

1. d	6. d	11. a
2. a	7. c	12. a
3. d	8. b	13. b
4. b	9. d	14. a
5. a	10. b	15. b

Understanding and Applying Terms and Concepts

1. a) G b) F c) C d) D e) A f) B g) E

2. The viruses could be put in the Monera, the simplest of the five kingdoms listed, but this would be based on superficial, "convenient" characteristics. This would indicate nothing about their kinship. It would be more informative to establish a sixth kingdom, the kingdom Virus.

3. A splitter does tend to increase the number of taxonomic categories, but not of species. A species is the one category that is objectively defined: a group of individuals potentially capable of interbreeding. So, dogs comprise one species, whether you are a splitter or a lumper.

4. A classification should not be based on similar habitats or superficial anatomical resemblance (body shape) but on kinship. Clearly, "D" and "E" are most closely related and should be in the same category, perhaps the same order. "C" is almost as closely related and should perhaps be included in that same order or in a different order in the same class as "D" and "E." "B" might be placed in a third order in that one class or in a second class, and "A" is the least closely related to the others, maybe in a completely separate phylum.

THE PROKARYOTES (KINGDOM MONERA)

I. REVIEWING THE CHAPTER

A. CHAPTER HIGHLIGHTS

36.1 The Nature of Prokaryotes

Prokaryotes ("prenuclear"), unlike eukaryotes ("truly nuclear"),
do not possess organelles. Also, they have only a single chromosome
with no histones attached. They lack microtubules, centrioles, spindles,
and basal bodies. Some have flagella, but these are not made of microtubules. The ribosomes are different than those in eukaryotes. There
is no meiosis in prokaryotes, and therefore most reproduction is asexual.

36.2 The Bacterial Cell/36.3 The Classification of Bacteria

Bacteria possess cell walls and occur in three major shapes by
which they can be grouped: bacilli (rod-shaped), cocci (spherical),
and spirilla (curved). The bacterial wall is composed of a complex
polymeric substance called peptidoglycan and is extremely strong,
owing to its elaborate covalent structure. Some bacteria have "pili."
The function of pili is largely unknown, but some are involved in conjugation (sexual reproduction). Mesosomes are folds of membranes that
serve a variety of functions, particularly those requiring a reserve
supply of membranes. Although no membrane-bounded nucleus is found
within bacteria, a loop of DNA is folded together in a "nuclear body"
within the cell. Many bacteria form spores that are extremely resistant
to environmental extremes. The classification of bacteria is based
upon several criteria but, at present, does not reflect evolutionary
affinities. A commonly used method is the so-called gram stain, which
divides bacteria on the basis of whether they retain a purple dye.

36.4 The Photosynthetic Bacteria/36.5 The Chemoautotrophic Bacteria

Photosynthetic bacteria utilize the sun as a source of energy
yet, unlike plants, never use water as a source of electrons. Their
photosynthetic pigments are called bacteriochlorophylls and are incorporated in the mesosome membranes. Most bacteria in this group
are obligately anaerobic and therefore restricted in their habitat.
In addition to being able to produce carbohydrates, photosynthetic
bacteria can also "fix" nitrogen. The chemoautotrophic bacteria also
manufacture carbohydrates but, lacking chlorophyll, they must rely on
the energy derived from the oxidation of certain compounds in their
environment. The important nitrifying bacteria are members of this
group.

36.6 Gram-Positive Rods/36.7 Gram-Positive Cocci

Several gram-positive, rod-shaped bacteria are of particular interest because they are producers of especially virulent conditions in humans. These bacteria are members of the genus Clostridium (tetanus, botulism, etc.) and the genus Bacillus (anthrax, etc.). Gram-positive cocci of special note are those belonging to the genera Staphylococcus (occasional minor infections, common food poisoning), Streptococcus (strep throat, impetigo, and, rarely, rheumatic fever), and Pneumococcus (bacterial pneumonia).

36.8 Gram-Negative Rods/36.9 Gram-Negative Cocci/36.10 Spirilla

The gram-negative bacilli include an enormous number of species, with Escherichia coli being the most thoroughly studied. Some of the more notorious members of the group are those that cause plague, cholera, and typhoid fever. Among the gram-negative cocci two are of particular note. They belong to the genus Neisseria and cause meningococcal meningitis and gonorrhea. The spirilla are gram-negative with a rigid cell wall that gives them their shape.

36.11 Actinomycetes and their Relatives

The actinomycetes grow as thin filaments rather than as single cells and, although similar to fungi in appearance, are prokaryotic rather than eukaryotic. Some actinomycetes are pathogenic yet their significance comes from the fact that, as dominant members of the microbial population in soil, they play a major part in the decay of organic wastes. Also, they are an important source of antibiotics. Two close relatives of the actinomycetes are the mycobacteria and the corynebacteria. The former include species responsible for tuberculosis and leprosy and the latter include the agent that causes diphtheria.

36.12 Spirochetes/36.13 Mycoplasmas

Spirochetes are long, thin, helix-shaped bacteria with cell walls that are less rigid than those of spirilla. Some members of the group are serious parasites, e.g. the spirochete that causes syphilis. Mycoplasmas are tiny, nonmotile bacteria without cell walls. Some are free-living while others are parasitic. This group includes the smallest free-living organisms known and is responsible for some forms of pneumonia.

36.14 Rickettsias and Chlamydiae/36.15 The Gliding Bacteria

The rickettsias are small like the mycoplasmas yet differ from them in that almost all of them are obligate intracellular parasites. Typhus and Rocky Mountain spotted fever are caused by members of this group. The chlamydiae resemble rickettsias in many ways and cause psittacosis and trachoma. The gliding bacteria are named for their means of locomotion. Most are heterotrophic but a few are chemoautotrophic. Their close resemblance to blue-green algae has led to the suggestion that they are algae that have lost their photosynthetic ability.

36.16 The Blue-Green Algae (Phylum Cyanophyta)

The blue-green algae are photosynthetic and prokaryotic. Some blue-green algae can fix atmosphere nitrogen in special colorless cells called heterocysts. Because of their many bacteria-like properties, many prefer to call them cyanobacteria. They possess chlorophyll a, use water as a source of electrons, and utilize photosystems I and II. They also possess a blue pigment, phycocyanin, and/or a red pigment, phycoerythrin, as well as B-carotene. Like gliding bacteria, they are encased in a wall of peptidoglycan surrounded by a gummy sheath. As a result, they are extremely hardy. Very old fossils resembling modern blue-green algae have suggested a key evolutionary role for these algae in preparing the way for heterotrophic organisms.

36.17 The Prochlorophyta

Recently some organisms classified as cyanophyta were found not to be blue-green algae. They contain no phycocyanin or phycoerythrin, but are prokaryotes in other respects. They contain chlorophyll b and have been considered as prime candidates for having evolved into the chloroplasts of higher plants.

36.18 The Archaebacteria

The archaebacteria are prokaryotes that look like bacteria, but do not have peptidoglycan in their cell walls. Their protein synthesizing machinery is also different. Some members of this group are methanogens, and many live in rather harsh environments.

36.19 The Viruses

Viruses are not prokaryotes, possessing almost none of their characteristics. They are incapable of metabolism and self-reproduction. It is problematical as to whether or not they are living organisms. Their importance derives from the fact that they are pathogenic. The first virus discovered was TMV, tobacco mosaic virus. Further work with TMV showed that, in a crystalline form, it retained its infectivity indefinitely. Virus particles (virions) consist of a core of either DNA or RNA surrounded by a protein sheath called a capsid. Other ingredients can also be found in some capsids.

The majority of DNA viruses contain a core in the form of a double helix. The infection cycle of the DNA bacteriophage consists of (1) attachment of virion to host cell and injection of DNA core, (2) transcription and translation of the early genes, (3) transcription and translation of the late genes, (4) assembling of complete virions, (5) lysozyme production followed by rupture of cell wall and release of virions. The RNA infection cycle is a similar process.

In some instances, the intracellular events of the lytic cycle are not completed following infection by a DNA bacteriophage. In this disappearing-virus act the virus can be found in the cell's descendants in a stable relationship called lysogeny. In some instances, viruses "freed" from this relationship can carry and insert bacterial genes into a new host in the process known as transduction. Sometimes hidden virus infections can transform call metabolism and cell division in an oncogenic (cancer-producing) fashion.

36.20 Summary

Prokaryotes are organisms without nuclei, mitochondria, plastids, Golgi apparatus, or microtubules. Most are heterotrophic, although some are parasites and cause severe diseases. There are also some that are photosynthetic. Viruses are incapable of metabolism and self-reproduction, but are pathogenic.

B. KEY TERMS

Monera	spirochetes
prokaryotes	mycoplasms
eukaryotes	rickettsia
peptidoglycan	chlamydiae
pili	Cyanophyta
mesosomes	heterocysts
spores	cyanobacteria
gram-positive/negative	prochlorophyta
purple sulfur bacteria	archaebacteria
green sulfur bacteria	TMV
chemoautotrophic	filterable viruses
nitrifying bacteria	methanogens
bacilli	capsids
cocci	virions
spirilli	disappearing viruses
streptococci	lytic cycles
pneumococci	prophage
actinomycetes	lysogeny
mycobacteria	oncogenic
corynebacteria	

II. MASTERING THE CHAPTER

A. LEARNING OBJECTIVES

When you have mastered the material in this chapter, you should be able to:

1. Define all key terms.

2. Discuss the differences between eukaryotes and prokaryotes with reference to structure, organelles, and reproduction.

3. Discuss the bacterial cell, paying particular attention to the cell wall. Cite the various methods that are combined to form a technique for classifying bacteria. Note various other structures and their functions, especially pili, mesosomes, flagella, and spores. explain the most promising approach to making the bacterial classification system a more "natural" one.

5. Explain how the chemoautotrophic bacteria convert carbon dioxide into carbohydrate by means of "dark" reactions. In particular, give the energy sources of the sulfur, iron, and nitrifying bacteria.

6. Discuss the gram-positive rods and cocci that are of special interest because of their disease-producing ability. Explain the mechanisms involved in tetanus, botulism, anthrax, staph and strep infections, and bacterial pneumonia. Display an awareness of the main groups to which these pathogenic bacteria belong.

7. Discuss the gram-negative rods and cocci that are of interest because of their harmful and beneficial roles. Cite the organisms and mechanisms involved in typhoid fever, cholera, plague, meningococcal meningitis, and gonorrhea. Note the relative unimportance of the spirilla in human activities.

8. Discuss the actinomycetes with special regard for their structure and functions. Cite examples of closely related mycobacteria and corynebacteria.

9. Describe the structure of the spirochetes and mycoplasmas and give examples of their roles in causing disease.

10. Distinguish among the mycoplasmas, the rickettsias, and the chlamydiae. Give examples of their pathogenic roles.

11. Explain the relationship between the gliding bacteria and the blue-green algae. Discuss the photosynthetic mechanism of the blue-green algae, their hardiness, and the probable role of their ancestors in the origin of early forms of life.

12. Cite some possible criteria to be used when deciding whether to place an organism in the Cyanophyta or Prochlorophyta.

13. Give reasons for and against placing members of the Archaebacteria in a third major division.

14. Offer reasons why viruses fail to meet many of the criteria for living things. Indicate why they are of importance in biological studies, noting not only their pathogenic roles, but also their ability, in some cases, to produce genetic changes. Cite the importance of TMV to viral studies.

15. Compare and contrast the structure of the DNA and RNA viruses. List the stages involved in the infection cycle of bacteriophages. Discuss the disappearing viruses with regard to the lytic cycle, lysogeny, and their function in the process of transduction. Explain the oncogenic properties associated with the disappearing viruses.

B. REVIEWING TERMS AND CONCEPTS

1. Cells which do not have organelles are classified as (a) _____ and include the (b) _____ and the (c) _____ (algae). Reproduction in this group is most commonly (d) _____.

2. Bacteria display three general shapes, scientifically referred to as (a) _____, _____, and _____. The bacterial cell walls are composed of a complex polymeric material called (b) _____. Many bacteria are able to swim by means of (c) _____ and some have a second set of protein filaments called (d) _____, whose function is unknown. Portions of the cell membrane fold into the cytoplasm, forming (e) _____, which serve a variety of functions. When their food supply runs low, many bacteria have the ability to produce (f) _____.

3. A method of classifying bacteria that is based upon whether the (a) _____ _____ are colorized or decolorized by chemicals is referred to as the (b) _____ _____ method. In the future, a classification system based upon (c) _____ makeup may result in a more natural classification system.

4. Photosynthetic bacteria never use water as a source of (a) _____ and most cannot tolerate free oxygen; they are therefore referred to as (b) _____. Also, these bacteria are able to fix (c) _____.

5. Tetanus and botulism are produced by (a) _____ positive bacilli belonging to the genus (b)_____. Those organisms that cause plague and cholera are gram-negative (c) _____, whereas those producing gonorrhea are gram-negative (d) _____ belonging to the genus (e) _____.

6. A major role in the decay of organic wastes is played by the (a) _____. These soil inhabitants are also an important source of (b) _____. Diphtheria is the result of the toxin of a microorganism belonging to the group (c) _____, a close relative of the actinomycetes.

7. Syphilis is caused by a (a) _____. The (b)

8. The (a) _____ are obligate intracellular parasites responsible for such diseases as Rocky Mountain spotted fever and typhus. A close relative to the above group are the (b) _____, which are responsible for (c) _____ (parrot fever) and the eye infection (d) _____.

9. Blue-green algae belong to the phylum (a) _____ and are able to use (b) _____ as a source of electrons. Like the gliding bacteria, their cell walls are composed of (c) _____. Thanks to colorless cells known as (d) _____, some are able to fix atmospheric (e) _____.

10. Certain unicellular algae that were classified as (a) _____ were not blue-green algae because they lacked (b) _____ and phycoerythrin, but did contain chlorophyll (c) _____. These organisms have now been placed in a new (d) _____, called Prochlorophyta.

11. A group of bacteria that could have been among the first forms of life on earth are called (a) _____. Many members of the group produce (b) _____ from hydrogen and water. They are (c) _____ (difficult/easy) to study in the laboratory, but enough is known about them to cause some people to consider them a (d) _____ major division of life.

12. The first virus discovered was that which has become known by the initials (a) _____. Virus particles, called (b) _____, consist of an interior core of (c) _____ acid surrounded by a coat of protein known as a (d) _____.

13. Viruses containing a double-helix core belong to the (a) _____ virus group. A virus that can infect bacteria is referred to as a (b) _____ and goes through a (c) _____- step infection cycle.

14. If lysis occurs, the infection cycles are referred to as (a) _____ cycles. This does not occur in the case of infection by (b) _____ viruses. A stable relationship called (c) _____ develops and the phage is called a (d) _____. Later, when these virions infect new hosts, they are capable of (e) _____, a process of genetic transfer.

C. TESTING TERMS AND CONCEPTS

1. The bacterial wall
   a) when formed, indicates the bacterium is going into its spore stage.
   b) owes its strength to its covalent structure.
   c) causes bacteria to be placed in the group Prokaryotes.
   d) is more like a membrane than a wall.

2. Photosynthetic bacteria are characterized by the fact that
   a) water is never used as a source of electrons for their biochemical processes.

b) they are unable to fix nitrogen.
c) they are obligately aerobic.
d) all their pigments are unique.

3. The genus Clostridium is notable because it is
   a) a good host for viruses.
   b) the largest of the gram-negative cocci genera.
   c) among the best-studied group, owing to its being an inhabitant of the gastrointestinal tract of humans.
   d) responsible for especially virulent conditions in humans.

4. A virus-mediated genetic transfer is
   a) known as lysogeny.
   b) produced by the action of PPLOs.
   c) called transduction.
   d) never found in bacteria.

5. Important agents in the breakdown of organic wastes are the
   a) spirochetes.
   b) actinomycetes.
   c) rickettsias and chlamydiae.
   d) Archaebacteria.

6. The blue-green algae differ from photosynthetic bacteria in that the algae
   a) contain chlorophyll a.
   b) are unable to use water as a source of electrons.
   c) can utilize only photosystem II.
   d) have a nuclear membrane.

7. Dipphtheria is caused by
   a) a member of the spirochete group.
   b) the effect of lysis upon body cells.
   c) the toxin released by the microorganisms.
   d) a virus.

8. Which of the following descriptions does not apply to lysogeny?
   a) is a stable relationship
   b) is the bursting of the cell wall
   c) precedes transduction
   d) involves a prophage

9. Many bacteria form spores
   a) as a first step in asexual reproduction
   b) by discarding the cell wall and rearranging the molecules of the cell membrane.
   c) in response to a decreasing food supply.
   d) following invasion by a virus.

10. Viruses do not possess which of the following
    a) capsid.
    b) protein.
    c) prokaryotic structure.
    d) a nucleic acid.

11. The prokaryotes are characterized by
    a) asexual reproduction and no microtubules.
    b) organelles and a single chromosome.
    c) nuclei and plastids.
    d) being universally saprophytic.

12. A group long thought to be fungi because of their growth as thin, moldlike filaments
    a) provides an important source of antibiotics.
    b) is significant because it includes many pathogenic species.
    c) is responsible for several varieties of venereal disease.
    d) are not fungi because the group contains nuclei.

13. The iron and sulfur bacteria
    a) utilize sunlight as a source of energy.
    b) are members of the Cyanophyta group.
    c) oxidize reduced substances in their environment.
    d) are eukaryotic.

14. Hidden virus infections in animal cells
    a) can be oncogenic.
    b) prevent cell transformation.
    c) are generally caused by RNA viruses.
    d) are exemplified by diseases like polio.

15. The rickettsias differ from the mycoplasmas primarily in that
    a) rickettsias are almost all obligate intracellular parasites.
    b) mycoplasmas can grow and reproduce only within the living cells of their host.
    c) rickettsias are far larger than the mycoplasmas.
    d) rickettsias form spores.

16. Portions of the bacterial cell membrane may fold into the cytoplasm forming
    a) pili.
    b) stromatolites.
    c) flagella.
    d) mesosomes.

17. The gram-stain technique
    a) provides a method of classifying anaerobes but not aerobes.
    b) is the most promising method of making the bacterial classification system more natural.
    c) distinguishes between two fundamentally different types of cell walls.
    d) reveals if a bacterium is infected with a virus.

18. Organisms classified as prokaryotes
    a) reproduce most often by sexual means.
    b) are always bacteria.
    c) lack organelles.
    d) are probably the most recently evolved.

19. All of the following are true for members belonging to the

Prochlorophyta <u>except</u>
a) they are prokaryotic.
b) they have phycocyanin.
c) they lack mitochondria.
d) they have ribosomes like prokaryotes.

20. Which of the following is <u>not</u> true of the Archaebacteria?
a) Many are methanogens.
b) They are difficult to grow in the laboratory.
c) Their protein-synthesizing machinery resembles eukaryotes.
d) Their cell walls are peptidoglycan.

D. UNDERSTANDING AND APPLYING TERMS AND CONCEPTS

Decide whether the following are true (T) or false (F)

_____ 1. Bacteria do not use water as an electron source during photosynthesis.

_____ 2. Without reproductive organs, bacteria can no exchange genetic material.

_____ 3. Many bacteria are harmful because of the toxins they produce.

_____ 4. Superficially, actinomycetes look like fungi, i.e., they are filamentous.

_____ 5. Rickettsias are most commonly found as free-living saprophytes.

_____ 6. Stromatolites cause various infectious diseases.

_____ 7. Viruses which induce cancers are said to be oncogenic.

_____ 8. Movement in bacteria is facilitated by flagella.

_____ 9. Heterocysts are caused when bacteria invade skin.

_____ 10. Transformation and transduction are synonymous.

_____ 11. In the absence of many morphologically distinguishing features, how does one classify the hundreds of species of bacteria?
_____
_____

12. In general, a cut is less dangerous than a puncture wound. Why? _____
_____
_____

13. Although rabies is caused by a virus, people are often given a "tetnus" shot after a dog or cat bite. Why? _____
_____
_____

14. The bubonic plague was caused by a bacterium, but rats were implicated. Why? _____

_____

_____

15. Why do antibiotics kill bacteria and not the humans who take them? _____

_____

_____

_____

16. Why have phosphate detergents been implicated in the "fouling" of lakes? _____

_____

_____

_____

17. Living things require constant energy inputs. Do viruses?
(a)_____ How are viruses more like salt than cells?
(b)_____

What component of a cell is a virus most similar? (c) _____

_____

ANSWERS TO CHAPTER EXERCISES

Reviewing Terms and Concepts

1. a) prokaryotes b) bacteria c) blue-green d) asexual

2. a) bacilli, cocci, spirilla b) peptidoglycan c) flagella d) pili
   e) mesosomes f) spores

3. a) cell walls b) gram stain c) genetic

4. a) electrons b) obligately anaerobic c) nitrogen

5. a) gram b) Clostridium c) bacilli d) cocci e) Neisseria

6. a) actinomycetes b) antibiotics c) corynebacteria

7. a) spirochete b) mycoplasmas

8. a) rickettsias b) chlamydiae c) psittacosis d) trachoma

9. a) Cyanophyta b) water c) peptioglycan d) heterocysts
   e) nitrogen

10. a) Cyanophyta b) phycocyanin c) b d) phylum

11. a) Archaebacteria b) methane c) difficult d) third

12. a) TMV b) virions c) nucleic d) capsid

13. a) DNA b) bacteriophage c) five

14. a) lytic b) disappearing c) lysogeny d) prophage e) transduction

Testing Terms and Concepts

1.	b	6.	a	11.	a	16.	d
2.	a	7.	c	12.	a	17.	c
3.	d	8.	b	13.	c	18.	c
4.	c	9.	c	14.	a	19.	b
5.	b	10.	c	15.	a	20.	d

Understanding and Applying Terms and Concepts

1. T

2. F

3. T

4. T

5. F

6. F

7. T

8. T

9. F

10. F

11. biochemical criteria (staining properties, metabolic reactions, etc.)

12. Cuts leave the wound open to the air, and thus do not allow harmful "anaerobes" (particularly Clostrium tetani) to grow.

13. Spores of C. tetani might be introduced along with the bite (a puncture wound).

14. The organism was transmitted to humans by fleas which had bitten rats.

15. Bacteria have certain biochemical properties which we do not share. Thus, antibiotics usually interfere with one of those processes.

16. Certain microbes thrive on the increased phosphate, and therefore they multiply rapidly.

17. a) No b) They can exist as crystals. c) chromosome

355

THE PROTISTS AND FUNGI

I. REVIEWING THE CHAPTER

A. CHAPTER HIGHLIGHTS

37.1 The Kingdom Protista: Characteristics

The Kingdom Protista is made up of phyla whose members are largely unicellular. Although their evolutionary relationships are obscure, most of these phyla appeared on earth before the plants or animals.

37.2 The Evolution of Eukaryotes

It has been suggested that the eukaryotes evolved by the symbiotic association of two or more kinds of prokaryotes. This view is supported by the facts that mitochondria, chloroplasts, and basal bodies are semiautonomous genetically, i.e. are capable of independent self-duplication, and further, that the genetic mechanism is very similar to that of prokaryotes. This parallel also applies to the protein synthesis occurring within mitochondria and chloroplasts. Additional evidence is provided by the manner in which some heterotrophic organisms exploit photosynthetic endosymbionts. Their obligatory relationship may represent an intermediate evolutionary state.

37.3 The Rhizopods (Phylum Sarcodina)/37.4 The Flagellates (Phylum Mastigophora)

The protozoa have now been divided into four phyla. The first of these are known as the rhizopods or phylum Sarcodina. Its members move by means of pseudopodia. The amoeba is the classic example of the group. This phylum also includes the foraminifera, radiolaria, and heliozoans. Organisms in a second phylum, the Mastigophora or flagellates, move by means of one or more flagella. These are constructed from microtubules with the "9 + 2" pattern. Members of their group are responsible for African sleeping sickness. On occasion, some flagellates also possess pseudopodia. As a result, some biologists have suggested that, although lacking in chloroplasts, the flagellates have all the other components of eukaryotes and may therefore have been the ancestral stock from which eukaryotes evolved.

37.5 The Ciliates (Phylum Ciliophora)

A third phylum belonging to the Protozoa is the Ciliophora or ciliates. They move by the rhythmic beating of cilia—also built on the "9 + 2" pattern. Ciliates have one or more micronuclei and a large macronucleus that controls the general metabolic activity of the cell. They reproduce asexually by fission but may, on occasion, reproduce sexually by a process known as conjugation, which involves the exchange of the genetic material within the micronuclei. No new daughter cells result from this process but genetic recombination does occur, with the result that the "offspring" contribute jointly to the formation of a single clone when asexual reproduction is resumed. The high level of complexity and large size of many ciliated protozoans have led some biologists to suggest that they are probably acellular rather than unicellular. Their complexity exceeds that of many cells found in multicellular organisms, and has led to the speculation that they reached an "evolutionary dead-end."

37.6 The Sporozoans (Phylum Sporozoa)

Sporozoans are parasitic. Members of the genus Plasmodium produce malaria, most forms of which are chronic and insidious in that the sporozoites can change their antigenic determinants and avoid attack by antibodies. The complex life cycle requires a second host, mosquitoes of the genus Anopheles.

37.7 The Eukaryotic Algae/37.8 The Red Algae (Phylum Rhodophyta)/37.9 The Dinoflagellates (Phylum Pyrrophyta)/37.10 The Euglenophytes (Phylum Euglenophyta)

The eukaryotic algae are photosynthetic and often are included in the plant kingdom by biologists, even though some bear slight resemblance to plants. They can be unicellular, colonial, or truly multicellular (with very little cell differentiation). The red algae, phylum Rhodophyta, are almost exclusively marine and mostly multicellular. They lack chlorophyll b and their shared similarities with the prokaryotic blue-green algae suggest they are a primitive group. In contrast to other eukaryotes, the red algae never have the "9 + 2" flagella. The dinoflagellates (phylum Pyrrophyta) are mostly unicellular, displaying the eukaryotic type ("9 + 2") of flagellum and, when reproduction is explosive, producing poisonous red tides. The euglenophytes (phylum Euglenophyta) are photosynthetic flagellates which lack a cell wall. Euglena is a typical member of this group which, if it lacked its chloroplasts, could be labeled a protozoan, suggesting that many of this latter group are evolutionary outcomes of spontaneous chloroplast loss by their ancestors.

37.11 The Green Algae (Phylum Chlorophyta)

The green algae (phylum Chlorophyta) resemble the euglenophytes in the possession of chlorophylls a and b; however, they possess a rigid cell wall of cellulose. They range from unicellular (Chlamydomonas) to colonial (Spirogyra) to multicellular (Volvox). Members of this group may or may not possess flagella, with the nonmotile ones being plantlike in appearance and leading many biologists to suggest that the plant kingdom evolved from early ancestors of this phylum.

37.12 The Golden Algae (Phylum Chrysophyta)/37.13 The Brown Algae (Phylum Phaeophyta)

The golden algae (phylum Chrysophyta) contain chlorophylls a and c and derive their color from a carotenoid called fucoxanthin. Most are unicellular and many are flagellated. The diatoms comprise the vast majority of the group and are crucial to the economy of nature in that they accomplish a major portion of all occurring photosynthesis. The brown algae (phylum Phaeophyta) also contain chlorophylls a and c as well as fucoxanthin; however, they are multicellular and almost exclusively restricted to salt water. Kelp and rockweeds are examples of this group, whose members are used for food and as sources of iodine and fertilizer.

37.14 The Slime Molds (Phylum Myxomycetes)

This group is so named because at one stage in their life cycle they appear as a slimy mass. This plasmodium has thousands of nuclei that are ultimately incorporated into spores. The cellular slime molds are aggregates of thousands of individual amoebalike cells. Both types are of great scientific but little economic interest.

37.15 The Kingdom Fungi: Characteristics

Most fungi grow as tubular filaments called hyphae, which are coenocytic. Mycelium is the term applied to an interwoven mass of hyphae. The walls of hyphae contain a polymer known as chitin, which provides strength. Owing to a lack of chlorophyll, fungi are heterotrophic, deriving food molecules from the soil, manufactured foods, and living and dead organisms. Many fungi cause disease yet many more engage in the essential role of decomposing dead organisms into their component nutrients.

37.16 The Phycomycetes

The Phycomycetes are made up of nonaquatic and aquatic types, the latter being called water molds. A number of water molds are economically important, especially those parasitizing fish and causing "late blight" in potatoes. Terrestrial molds lack the motile spores and gametes of the water molds and must rely on air currents for dispersal of their spores. All members of this phylum form their spores within a sporangium and their hyphae lack septa.

37.17 The Phylum Ascomycetes

The members of the phylum Ascomycetes produce one type of spore (conidia) asexually. Sexual reproduction produces ascospores that develop inside a saclike structure known as an ascus. Among the members of this phylum are the yeasts and those that cause mildew and chestnut blight, produce penicillin, and are used in the production of some cheeses.

37.18 The Phylum Basidiomycetes/37.19 The Fungi Imperfecti (Phylum Deuteromycetes)

The Basidiomycetes bear spores at the tips of club-shaped structures known as basidia. Mushrooms and toadstools are masses of inter-

woven hyphae sent up by the main part of the mycelium growing below
ground level. The Basidiomycetes are of economic importance, but
largely of a negative nature. The smuts, and the rusts, which can have
complex life cycles, account for a great deal of crop damage. The
fungi imperfecti (phylum Deuteromycetes) is a catch-all category into
which are placed fungi whose reproductive picture is incomplete and
that therefore cannot be assigned to either the ascomycetes or
basidiomycetes.

37.20 The Lichens

        Lichens are a composite of a fungus mycelium within which algal
cells are imbedded. The fungus portion can be a basidiomycete, a member of the fungi imperfecti or, more usually, an ascomycete. The unicellular alga is of either the green or blue-green type. Although most
fungi exist only in areas of abundant food, those found in lichens inhabit even hostile rock surfaces thanks to their symbiotic partners,
the algae, which contribute sugars and, in some cases, fix nitrogen.
Lichen dispersal methods are unknown but essential since lichens are
among the first to colonize harsh, new environments.

37.21 Summary

        The protists and fungi are both eukaryotes. The former group
consists of four phyla of protozoans and six phyla of algae. All fungi
are saprophytes or parasites. Lichens are organisms composed of a
fungus and an alga.

B. KEY TERMS

Protista	red algae
fungi	Rhodophyta
plants	phycocyanin
animals	phycoerythrin
symbiotic	dinoflagellates
endosymbiosis	Pyrrophyta
protozoa	Euglenophyta
rhizopods	green algae
Sarcodina	Chlorophyta
foraminifera	golden algae
radiolaria	Chrysophyta
heliozoans	diatoms

pseudopodia	brown algae
flagellates	Phaeophyta
Mastigophora	slime molds
trypanosomes	Myxomycetes
ciliates	plasmodial
Ciliophora	cellular
conjugation	hyphae
macro and micronuclei	mycelium
Paramecium	Phycomycetes
clone	Ascomycetes
sporozoans	ascospores
Sporozoa	Basidiomycetes
Plasmodium	basidia
sporozoites	fungi imperfecti
merozoites	Deuteromycetes
	lichens
	conidia

II. MASTERING THE CHAPTER

A. LEARNING OBJECTIVES

When you have mastered the material in this chapter, you should be able to:

1. Define all key terms.

2. List the characteristics of the Protista.

3. Discuss the evolution of the eukaryotes, noting the evidence available to support the hypothesis that the symbiotic association of prokaryotes led to the appearance of eukaryotes.

4. Discuss the criterion for membership in the phylum Sarcodina and list the three groups of aquatic protozoa commonly included in this phylum.

5. Explain the means of locomotion by which the Mastigophora are classified and give examples of some of their pathogenic members.

6. Discuss the Ciliophora with regard to their locomotion, size, and various nuclei. Explain their two methods of reproduction, especially in terms of their effect upon genetic makeup. Cite the evidence for classifying some ciliates as acellular.

7. Outline the life cycle of Plasmodium.

8. List the general characteristics of the eukaryotic algae.

9. Discuss the characteristics of the red algae, noting their similarity to the blue-green, prokaryotic algae.

10. Compare and contrast the dinoflagellates and euglenophytes with special attention to their cell walls. Cite the importance that the former may have for humans, and the close relationship between the latter and the Mastigophora.

11. Discuss the characteristics of the green algae, giving examples of the various genera. Explain the evolutionary significance of the phylum as well as its role as a food source.

12. Compare the golden and brown algae with regard to their chemical and structural makeup. Cite the important ecological and economic roles of their various members, with special attention being paid to diatoms.

13. Explain why the Myxomycetes are commonly known as slime molds. Distinguish between the cellular slime molds and the plasmoidal. List the traits that cause these microorganisms to be of great scientific interest.

14. List the major characteristics of the fungi, paying particular attention to hyphae, chitin, lack of chlorophyll, and dispersal.

15. Compare the water molds and the terrestrial molds which constitute the Phycomycetes. Give examples of their economic impact.

16. Discuss the ascomycetes with special reference to their sexual and asexual methods of spore production. Cite the important economic role played by this group and give examples.

17. Discuss the characteristics of the basidiomycetes, also noting their economic importance and the complex life histories of some of the rusts.

18. Explain the reason for creating the phylum Deuteromycetes and why certain fungi are assigned to this group.

19. Discuss the lichens with regard to their components. Explain the unique sumbiotic relationship that prevails and why the fungi appear to have the better part of the relationship. Cite why lichens

are of importance in the economy of nature.

B. REVIEWING TERMS AND CONCEPTS

1. Eukaryotes are divided into the plants, animals, (a) _____, and _____. The eukaryotic protists may have evolved by the (b) _____ association of two or more kinds of (c) _____. Both (d) _____ and _____ (also basal bodies) can duplicate themselves independently of the cell in which they exist.

2. The rhizopods, phylum (a) _____, move by means of (b) _____ and include two large marine groups, the (c) _____ and _____, and the freshwater (d) _____. Members of the phylum Mastigophora, or (e) _____, move by means of (f) _____ and include a form pathogenic to humans that causes African (g) _____ _____.

3. The ciliates, phylum (a) _____, possess a large, polyploid (b) _____ that controls metabolic activity. Asexual reproduction in this group is by means of (c) _____ and sexual reproduction by means of (d) _____. The fact that some ciliates are large and possess many organelles has led some biologists to suggest that they are (e) _____ rather than unicellular.

4. The sporozoans, which are all (a) _____ feeders, are most infamous for including the genus (b) _____, the members of which produce various types of (c) _____. The complex life cycle of _Plasmodium_ requires two separate (d) _____, with the (e) _____ mosquito injecting (f) _____ into the bloodstream, from which they travel to the (g) _____.

5. Most algae live in the (a) _____ and, although some are truly multicellular, there is very little (b) _____ of cells.

6. The red algae, phylum (a) _____, are almost exclusively (b) _____ (habitat), and although some are unicellular, most are (c) _____. They utilize chlorophyll a but lack chlorophyll (d) _____ and bear many similarities to the prokaryotic (e) _____ algae.

7. The (a) _____, phylum Pyrrophyta, possess the eukaryotic type ("9 + 2") (b) _____ and can be responsible for the poisonous (c) _____ _____. Members of the phylum (d) _____ are photosynthetic flagellates that lack a (e) _____ cell wall. With their plastids removed, the euglenophytes might be assigned to the group (f) _____.

8. The green algae, phylum (a) _____, possess rigid cell walls of (b) _____ and chlorophylls (c) _____ and _____. Most biologists believe that the (d) _____ kingdom evolved from early members of this phylum.

9. The (a) _____ _____, phylum Chrysophyta, are
comprised mostly of (b) _____ whose shells are impregnated with
(c) _____. The brown algae, phylum (d) _____, are
all (e) _____, somewhat plantlike forms, found almost exclusively in (f) _____ _____.

10. The phylum (a) _____ contains two types of slime
molds, the (b) _____ and the _____. During their
life cycle they are (c) _____, multicellular, protozoa-like
and (d) _____-like.

11. Most of the fungi grow as tubular filaments called (a) _____,
which are sometimes interwoven into a structure referred to as a (b)
_____. The tubular filaments are not compartmentalized in-
to separate cells and are therefore said to be (c) _____.
The walls of the filaments are strengthened by the presence of (d)
_____.

12. The water molds and terrestrial molds compose the phylum (a)
_____. The water molds have (b) _____ spores.
A species of this group caused the (c) _____ famine in
Ireland.

13. Ascomycetes produce two kinds of spores, one formed asexually,
called (a) _____, and one formed sexually, called (b)
_____, which develop inside a structure known as a(n) (c)
_____.

14. Fungi dispersed by spores borne at the tip of clubshaped
structures called (a) _____ belong to the phylum (b)
_____. In addition to the economically positive
members of this phylum, the (c) _____, with their complex life
histories, cause serious crop losses. If the category assignment of a
fungus is in doubt, it is generally placed in the phylum (d) _____
_____, also called the fungi (e) _____.

15. Lichens are a composite of a fungus (a) _____ in
which are embedded (b) _____ cells. These cells may be members of either the (c) _____ or _____ groups. The
relationship existing between the two types of organisms is referred
to as (d) _____.

C. TESTING TERMS AND CONCEPTS

1. The plant kingdom may have evolved from early ancestors of the
    a) Euglenophyta.
    b) Chrysophyta.
    c) Chlorophyta.
    d) Rhodophyta.

2. Kelp and other large seaweeds belong to the phylum
    a) Chlorophyta.
    b) Rhodophyta.
    c) Pyrrophyta.
    d) Phaeophyta.

363

3. The Chrysophyta are composed mainly of
    a) multicellular forms.
    b) photosynthetic flagellates lacking a rigid cell wall.
    c) diatoms.
    d) algae with both chlorophyll $\underline{b}$ and $\underline{d}$.

4. Although some eukaryotic algae are multicellular, they all lack
    a) chlorophyll $\underline{c}$. phycocyanin.
    b) extensive cell differentiation.
    c) organelles.
    d) chitinous cell wall.

5. Which of the following does not characterize the Chlorophyta?
    a) lack of chlorophyll $\underline{b}$
    b) cell wall containing cellulose
    c) unicellular, colonial, and multicellular forms
    d) possess a well-defined nucleus

6. Fungi do not possess chlorophyll and are therefore
    a) eoenocytic.
    b) heterotrophic.
    c) parasitic.
    d) anaerobes.

7. Which of the following is not a characteristic of the phylum
    Ascomycetes?
    a) produce two kinds of spores
    b) contains parasitic members
    c) of little economic importance
    d) have septate hyphae

8. The algal partner in the lichen composite is
    a) multicellular.
    b) a provider of free electrons.
    c) in some instances, a fixer of nitrogen.
    d) a member of the Phylum, Masigophora.

9. The basidiomycetes
    a) include mushrooms and rusts.
    b) are, for the most part, beneficial to humans.
    c) lack mycelia.
    d) form no sexual spores.

10. Which of the following does not characterize the dinoflagellates?
    a) cause "red tides"
    b) lack the "9 + 2" type of flagellum
    c) mostly unicellular
    d) lack histones on their chromosomes

11. Which of the following does not apply to malaria?
    a) is caused by members of the phylum Mastigophora
    b) is produced by an organism with a complex life cycle requiring
       two hosts
    c) can be caused by any one of several species
    d) can change antigenic determinants

12. Which of the following does <u>not</u> provide support for the theory that eukaryotes evolved from prokaryotes?
    a) prokaryotes are the more complex of the two groups
    b) protein synthesis of prokaryotes is similar to that within chloroplasts and mitochondria
    c) genetic replication within chloroplasts occurs independently of the cell's nucleus
    d) fossils of eukaryotes are not as old

13. The slime molds, at some point in their life cycle, are both
    a) photosynthetic and nonphotosynthetic.
    b) cellular and plasmoidal types.
    c) of great economic importance.
    d) spore producing and multicellular.

14. The symbiotic relationship in lichens is one in which the
    a) fungi provide nutrients for the algae.
    b) algae could not grow separately from the fungi.
    c) fungi appear to benefit more than the algae.
    d) two organisms can actually exchange genetic material.

15. Which of the following does <u>not</u> apply to modern ciliates?
    a) may represent an evolutionary dead end
    b) possess few organelles
    c) metazoans may have evolved from their early ancestors
    d) none of these apply

16. Of the following sequences, which one probably is the most accurate?
    a) eukaryotes ➝ symbiotic association ➝ prokaryotes
    b) prokaryotes ➝ symbiotic association ➝ eukaryotes
    c) obligate dependency ➝ symbiotic association ➝ prokaryotes
    d) prokaryotes ➝ eukaryotes ➝ symbiotic association

17. The phylum Sarcodina includes the
    a) foraminifera, rhizopoda, and Mastigophora.
    b) radiolaria, heliozoans, and protozoa.
    c) foraminifera, radiolaria, and heliozoans.
    d) protozoa, foraminifera, and heliozoans.

18. The lack of septa within hyphae and the formation of spores within a sporangium are two requirements for membership in the phylum
    a) Basidiomycetes.
    b) Ascomycetes.
    c) Myxomycetes.
    d) Phycomycetes.

19. Protists are characterized by the fact that
    a) none possess specialized tissues.
    b) they are unicellular.
    c) they are prokaryotic.
    d) evolved before the prokaryotes.

20. Which of the following is <u>not</u> a characteristic of the red algae?
    a) lack chlorophyll b
    b) a relatively advanced group

c) members of the Rhodophyta
d) contain pigments found in the blue-green algae

21. The euglenophytes
    a) are photosynthetic flagellates.
    b) possess a rigid cell wall.
    c) lack chloroplasts.
    d) do not have chlorophyll b.

22. Members of the phylum Ciliaphora can reproduce
    a) by means of macronuclei.
    b) asexually, as well as by fission.
    c) by conjugation, as well as asexually.
    d) only when in the proper host.

23. The flagellates are not characterized by the fact that they
    a) possess flagella.
    b) are pathogenic in some instances.
    c) can change their antigenic determinants.
    d) lack pseudopodia in all instances.

24. All of the following are true of members of the fungi imperfect except:
    a) some cause ringworm.
    b) they have well defined sexual cycles.
    c) they are placed here because they can not be placed in one of the other groups.
    d) when a "home" is found for them, it is usually in the ascomycetes.

D. UNDERSTANDING AND APPLYING TERMS AND CONCEPTS

Decide whether the following statements are true (T) or false (F).

_____ 1. Protists reproduce only asexually.

_____ 2. Endosymbiosis is synonymous with parasitism.

_____ 3. The four phyla of protozoa are divided on a very natural basis.

_____ 4. of the algae, the Chlorophyta contains members most similar to higher plants.

_____ 5. Some brown algae are larger than corn plants.

_____ 6. Fungi are classified accoring to their habitat.

_____ 7. Members of the Deuteromycetes (fungi imperfecti) could, with as much justification, be classified with the protozoa or algae.

_____ 8. Because they contain many nuclei in a common cytoplasm plasmodial slime molds should probably not be considered as eukaryotes.

9. Why has the use of DDT (a pesticide) greatly reduced malaria (caused by a protozoan)? _____
_____
_____

10. Why does <u>Euglena</u> stand on a taxonomic border between plants and animals? _____
_____
_____

11. If you discovered a new algal species, you would assign it to a phylum on what basis? _____

12. What major organelle is found in all species of protozoans as well as algae? _____
_____

13. What organelle is found in all algae, but no protozoans?
_____

14. What organelle is found in some algae and some protozoans?
_____

15. Which organelle is found in plants but not fungi? _____
_____

16. Why are fungi considered plants when they do not undergo photosynthesis? _____
_____

17. List five commercially important products of fungi. _____
_____

18. What do many fungi do that is sometimes good and sometimes bad? _____
_____
_____

ANSWERS TO CHAPTER EXERCISES

Reviewing Terms and Concepts

1. a) protists, fungi b) symbiotic c) prokaryotes d) mitochondria, chloroplasts

2. a) Sarcodina b) pseudopodia c) foraminifera, radiolaria
   d) heliozoans e) flagellates f) flagella g) sleeping sickness

3. a) Ciliophora b) macronucleus c) fission d) conjugation
   e) acellular

4. a) parasitic b) <u>Plasmodium</u> c) malaria d) hosts e) <u>Anopheles</u>
   f) sporozoites g) liver

367

5. a) ocean b) differentiation

6. a) Rhodophyta b) marine c) multicellular d) b e) blue-green

7. a) dinoflagellates b) flagellum c) red tides d) Euglenophyta
   e) rigid f) Mastigophora

8. a) Chlorophyta b) cellulose c) a, b d) plant

9. a) golden algae b) diatoms c) silica d) Phaeophyta e) multicellular f) salt water

10. a) Myxomycetes b) plasmodial, cellular c) unicellular d) fungus

11. a) hyphae b) mycelium c) coenocytic d) chitin

12. a) Phycomycetes b) motile c) potato

13. a) conidia b) ascospores c) ascus

14. a) basidia b) Basidiomycetes c) rusts d) Deuteromycetes
    e) imperfecti

15. a) mycelium b) algal c) green, blue-green d) symbiosis

Testing Terms and Concepts

1.	c	6.	b	11.	a	16.	b	21.	a
2.	d	7.	c	12.	a	17.	c	22.	c
3.	c	8.	c	13.	d	18.	d	23.	d
4.	b	9.	a	14.	c	19.	a	24.	b
5.	a	10.	b	15.	b	20.	b		

Understanding and Applying Terms and Concepts

1. F 2. F 3. F 4. T 5. T 6. F 7. F. 8. F

9. It kills the mosquitoes, which are intermediate (alternate) hosts
   for the protozoan.

10. It contains plastids (therefore plant), but no cell wall (therefore
    animal).

11. Pigment(s)

12. Nucleus

13. Plasids

14. flagellum

15. plastids

16. They have cell walls.

17. mushrooms, antibiotics (particularly penicillin), cheeses, beer and wine, bakery products.

18. Cause decay.

38

THE PLANT KINGDOM

I. REVIEWING THE CHAPTER

A. CHAPTER HIGHLIGHTS

38.1 The Geological Eras

Although there are fossils of prokaryotes dating from three billion years ago, the majority of fossils, and hence a detailed fossil
record, extend back only 600 million years. The geological and biological history of the earth is divided into segments in which the
correlation of geological and biological changes is not a matter of
chance but, rather, of cause and effect.

38.2 The Evolution of Plants

During the Cambrian and Ordovician periods, the land contained
few, if any, autotrophs and hence no heterotrophs. By the end of the
Silurian, however, algae living near water-land interfaces evolved into the first land plants.

Organisms are included in the plant kingdom on the basis of their
(1) containing chlorophylls a and b, (2) lacking means of locomotion,
(3) possessing differentiated cells, (4) having sexual reproductive
systems, and (5) undergoing partial embryonic development within the
parent organism. With modification of the above in particular cases,
those plants lacking chlorophyll, as well as the algae, can be included in the kingdom. However, the author chooses to place the algae
in the kingdom Protista.

38.3 The Mosses and Liverworts (Phylum Bryophyta)

If one excludes the algae, the plant kingdom divides into two
phyla, the Bryophyta and Tracheophyta. The bryophytes consist of the
mosses and liverworts, with the latter being the more primitive of the
two. Neither possesses woody tissues or a specialized system for
vascular transport. Because of their reproductive processes, they are
restricted to habitats which, at least periodically, have abundant
moisture.

The mosses are important pioneer plants in that they can grow in
barren areas and, by their decomposition, produce humus. The Bryophyta
either represent a second unsuccessful attempt by aquatic plants at
colonizing the land or are degenerate descendants of land plants.

38.4 The Vascular Plants (Phylum Tracheophyta)

Earliest fossils indicate that the first plants were vascular and therefore legitimate members of the phylum Tracheophyta. By the close of the Devonian period, four distinct subphyla had evolved, each of which has descendants living today. The subphylum Psilopsida includes four species with no roots or leaves but with a rhizome and an erect aerial stem containing xylem and phloem. The Lycopsida, commonly known as club mosses, bear spores in clublike structures called strobili. They are not mosses at all but vascular plants with xylem and phloem, roots, and small leaves. The Sphenopsida, or horsetails, develop each season from an underground stem and incorporate large amounts of silica in their stems. About 25 species exist today.

The subphylum Pteropsida is divided into three classes: ferns, gymnosperms, and angiosperms. This subphylum differs from the other three members of the Tracheophyta in possessing large leaves with many, often branching, veins. The ferns (class Filicinae) are homosporous, with vascular roots and stem located beneath the ground. The gymnosperms descended from now-extinct seed ferns and today are most evident in the conifers. The angiosperms are the most dominant plant type, making up about seven-eighths of the living species in the plant kingdom. The angiosperms are divided into the monocots and dicots, the latter being the older and larger of the two groups. The monocots are characterized by one cotyledon in the seed, leaves with parallel veins, vascular bundles scattered, and flower parts in threes or multiples of threes. The dicots have two cotyledons, network veination, a radial arrangement of vascular bundles, and flower parts in fours or fives, or multiples thereof.

38.5 Adaptions of Angiosperms

Angiosperms have been successful from an evolutionary point of view and are essential to human existence. They are composed of three main organs, (1) roots for anchorage and absorption, (2) stem for leaf support and flower bud production, for connecting link leaves and roots, and, occasionally, for storage of food, (3) leaves for food production and, in some instances, additional functions. Growth patterns vary widely among the angiosperms. Their great success is attributable to their varied and efficient adaptations to living on dry land and their efficient methods of sexual reproduction and seed dispersal. These adaptations include (1) roots for the extraction of water and minerals from the soil, (2) cambium layer, which produces wood for support, (3) xylem and phloem for transport, (4) retention of the female gametophyte within the megasporangium and the production of pollen, thereby eliminating the need for surface water to transport sperm between plants, (5) the seed, an embryonic plant with a built-in food supply and, most importantly, (6) the production of fruits and flowers, events unique to the angiosperms that make pollination and seed dispersal more efficient.

38.6 Summary

The plant kingdom can be divided into two major groups: (1) the Bryophyta, which has species without a well-defined vascular system, and swimming sperm, and (2) the Tracheophyta, which contains species with well-defined vascular systems. The angiosperms, particularly those with specialized organs, culminate the evolution of plants.

B. KEY TERMS

Precambrian	rhizome
Paleozoic	class Gymnospermae
Mesozoic	heterosporus
Cenozoic	conifers
mosses	class Angiospermae
liverworts	monocots
phylum Bryophyta	dicots
vascular plants	roots
phylum Tracheophyta	stems
subphylum Psilopsida	leaves
subphylum Lycopsida	simple leaves
strobili	compound leaves
subphylum Sphenopsida	perennial
subphylum Pteropsida	biennial
class Filicinae	fruit
homosphorus	tendrils

II. MASTERING THE CHAPTER

A. LEARNING OBJECTIVES

When you have mastered the material in this chapter, you should be able to:

1. Define all key terms.

2. Discuss the fossil record with regard to its detail and evidence for marked geological and biological changes.

3. Discuss the evolution of plants with regard to the appearance of the land plants.

4. List the criteria used by the author to determine members of the

plant kingdom.

5. Discuss the phylum Bryophyta, giving the characteristics of its members and providing examples. Cite the possible evolutionary lines of the Bryophyta.

6. Discuss the evolution of the vascular plants. Give the characteristics of the subphyla Psilopsida, Lycopsida, and Sphenopsida and provide examples of each.

7. Give the characteristics of the three classes of the subphylum Pteropsida and provide examples of each. Discuss the relative efficiency of their reproductive methods, paying particular attention to the angiosperms.

8. Explain the great success of the angiosperms with regard to their various adaptations. List the main organs of angiosperms and give their various functions.

B. REVIEWING TERMS AND CONCEPTS

1. The earliest land plants were members of the (a) _____ group and made their appearance by the end of the (b) _____ period. The author includes in the plant kingdom all plants that have chlorophylls (c) _____ and do not move by means of contracting (d) _____.

2. The mosses and liverworts belong to the phylum (a) _____ and have no specialized (b) _____ system for the transport of water. Mosses are important as pioneer vegetation because they form decomposed plant material called (c) _____.

3. The (a) _____ have no roots or leaves, yet possess an underground stem called a (b) _____. The Lycopsida bear spores in clublike structures called (c) _____. Some of the members of this subphylum produce two kinds of spores, (d) _____ and _____. Scouring rushes belong to the subphylum (e) _____ and owe their name to the fact that their stems contain (f) _____.

4. Plants possessing large leaves and many, often branching, veins belong to the subphylum (a) _____. This subphylum subdivides into (b) _____ classes, with the Filicinae, or (c) _____, containing plants that possess an underground stem called a (d) _____. Some early ferns were (e) _____, i.e. they produced both microspores and megaspores, and were the earliest of the (f) _____. Today, the most abundant of the gymnosperms are the (g) _____.

5. Toward the end of the Mesozoic era, the dominant plants belonged to the class (a) _____. This class is divided into the (b) _____ and _____. The larger of the two groups are the (c) _____ and the older of the two are the (d) _____.

6. The plant body of the angiosperms consists of three main organs:
(a) _____, _____, and _____. The
leaves may be (b) _____ or _____. Angiosperms
which require two years to complete their life cycle are called (c)
_____, whereas those needing only one year are called (d)
_____. The angiosperms are more efficient than other groups
in pollination and seed dispersal because they produce (e) _____
and _____.

C. TESTING TERMS AND CONCEPTS

1. Members of the subphylum Pteropsida
   a) are commonly called horsetails or scouring rushes.
   b) possess large leaves with many veins.
   c) are all homosporous
   d) could have been included with the algae.

2. The majority of fossils date from
   a) 600 million years ago.
   b) 3 billion years ago.
   c) the early Precambrian.
   d) the past several thousand years.

3. Which of the following is not typical of the angiosperms?
   a) veins in the leaves may lie parallel to one another
   b) may be monocots
   c) are the smallest class within the Pteropsida
   d) have highly specialized reproductive structures

4. A change in the geology of the earth
   a) is preceded by a change in life forms.
   b) indicates a change in the forces of natural selection.
   c) complicates the uranium-dating process.
   d) is not likely to have biological repercussions.

5. Bryophytes are not typified by which of the following?
   a) relatively small size
   b) lack of wood tissue
   c) lack of flowers
   d) specialized vascular system

6. The author divides the plant kingdom into the
   a) mosses and liverworts.
   b) Psilopsida and Lycopsida.
   c) Tracheophyta and Bryophyta.
   d) mosses and vasculars.

7. By the close of the Devonian period the
   a) bryophytes had all but disappeared.
   b) tracheophytes were making their appearance.
   c) four distinct subphyla of vascular plants had appeared.
   d) only plants present were algae.

8. Which of the following is not true of at least some angiosperm
   stems?

a) produce leaves
b) store food
c) lack a vascular system
d) play a support role

9. Which of the following is not an adaptation of the angiosperms?
a) early expulsion of the female gametophyte from the megasporangium
b) a cambium layer producing wood for support
c) both xylem and phloem for transport
d) waxy coverings to prevent water loss

10. The first plants which populated the land
a) were descendants of the bacteria.
b) probably were derived from the green algae.
c) were, of necessity, heterotrophs.
d) had roots for obtaining water.

11. Two unique characteristics of angiosperms are
a) xylem and phloem.
b) roots and stems.
c) leaves and seeds.
d) flowers and fruits.

12. Which of the following is not an absolute requirement for inclusion in the plant kingdom?
a) possession of chlorophyll
b) a lack of the power of locomotion by means of contracting fibers
c) possession of cells differentiated to form tissues and organs
d) possession of complex sex organs

13. Mosses
a) are more important to nature than to the economy.
b) are more important to the economy than to nature.
c) are mostly found in the dessert.
d) have elaborate, but tiny flowers.

14. Which of the following probably did not contribute to coal formation?
a) ferns
b) lycopsids
c) sphenopsids
d) dicots

15. An underground stem is called a
a) root.
b) heterospore.
c) strobilus.
d) rhizome.

D. UNDERSTANDING AND APPLYING TERMS AND CONCEPTS

Decide whether the following statements are true (T) or false (F).

_____ 1. Plants probably appeared on land before in the water.

_____ 2. A major criterion for inclusion in the plant kingdom is the production of flowers.

_____ 3. Species with swimming sperm are restricted to moist habitats, at least part of the time.

_____ 4. When you get an "order of fries," you are getting stems.

_____ 5. The onions on the burger that you got with the fries are leaves.

_____ 6. If you want tomato slices on your burger, you are really ordering tomato "fruit."

_____ 7. The beef in the burger comes from a cow that was most likely fed with plants belonging to the Bryophyta.

_____ 8. A bouquet from the florist, in addition to having flowers, will likely have a member from this class of plants.

9. Why do people often work moss into potting soil or their gardens? _____
   _____
   _____

10. Give a possible reason for the Sphenopsida incorporating silica in their stems—the stems often have the texture of sandpaper? _____
    _____
    _____

11. Peanuts are easily broken into two _____. Incidently, break a peanut open and look for the embryonic leaves.

12. The wood in your pencil is what kind of plant tissue? _____
    _____

13. What prevents land plants from drying out? _____
    _____

14. A pair of cotton jeans is made from a part of which plant organ? _____

15. Why should you—THANK A LYCOPSIDA TODAY!? _____
    _____

ANSWERS TO CHAPTER EXERCISES

Reviewing Terms and Concepts

1. a) tracheophyte b) Silurian c) a and b d) fibers

2. a) Bryophyta b) vascular c) humus

3. a) Psilopsida b) rhizome c) strobili d) microspores, megaspores
    e) Sphenopsida f) silica

4. a) Pteropsida b) three c) ferns d) rhizome e) heterosporous
    f) gymnosperms g) conifers

5. a) Angiospermae b) monocots, dicots c) dicots d) dicots

6. a) roots, stem, leaves b) simple, compound c) biennials
    d) annuals e) flowers, fruit

Testing Terms and Concepts

1. b	6. c	11. d
2. a	7. c	12. a
3. c	8. c	13. a
4. b	9. a	14. d
5. d	10. b	15. d

Understanding and Applying Terms and Concepts

1. F 2. F 3. T 4. T 5. T 6. T 7. F

8. Filicinae (ferns)

9. 1) aids water holding capacity, 2) loosens soils 3) better
    aeration

10. protection—Animals would probably not want to chew on these.

11. Cotyledons

12. Xylem (stem)

13. a waxy cuticle or covering

14. flowers

15. Much of our fossil fuel comes from their ancestors.

39

THE INVERTEBRATES

I. REVIEWING THE CHAPTER

A. CHAPTER HIGHLIGHTS

39.1 Introduction

In general, animals are organisms that do not have chlorophyll,
are capable of locomotion, and are multicellular. There are 25 to 30
different animal phyla, all of which are invertebrate except for one
that is mostly vertebrate and one that is wholly vertebrate.

39.2 The Sponges (Phylum Porifera)

The sponges are among the most primitive invertebrates and possess a body consisting of two layers of cells with a jellylike layer,
called the mesoglea, between them. They secure food and oxygen from
water drawn continuously through their pores. Also, they possess a
skeleton composed of spicules and disperse their offspring by producing
free-swimming larvae.

39.3 The Cnidarians (Phylum Cnidaria)

Cnidarians all possess stinging cells called cnidoblasts, which
contain poison-filled nematocysts. The mesoglea has cells scattered
throughout it and is considered by some as a third cell layer. The
body is a hollow cylinder with a single opening at one end, the mouth.
The phylum is divided into three classes: (1) Hydrozoa—most produce
both polyp and medusa forms, (2) Scyphozoa—main body is the medusa,
(3) Anthozoa—polyp stage only. These classes contain the Portugese
man-of-war, jellyfish, and corals, respectively.

39.4 The Flatworms (Phylum Platyhelminthes)

The flatworms are bilaterally symmetrical and can achieve locomotion by means of cilia and undulation of the entire body. The
freshwater planarians are examples of this group. Like the cnidarians,
the flatworms digest food in a gastrovascular cavity with a single
opening. Their sense receptors are concentrated at their anterior end,
thereby displaying cephalization. The phylum subdivides into the
classes (1) Turbellaria, containing the most primitive members of this
phylum, (2) Trematoda, which are exclusively parasitic and of which
the flukes are an example, and (3) Cestoda, also exclusively parasitic and represented by tapeworms.

39.5 The Origin of Animals

The theories proposed to account for the origin of animals fall into two categories: (1) the multicellular characteristic of animals arose as a consequence of the formation of colonies of cells that then later became multicellular because of increased interdependence and specialization, and (2) animals arose by the cellularization of ciliated protozoans. Evidence is available to support both theories.

39.6 The Roundworms (Phylum Nematoda)

The roundworms possess two evolutionary advances over the flatworms from which they may have evolved: (1) a one-way digestive tract, and (2) a definite body cavity which, because it develops from the blastocoel, is called a pseudocoel. The majority of nematodes are small, cylindrical animals. They are found practically everywhere and exist in some parasitic forms, although most are free-living. Parasitic forms that infect humans can produce hookworm, trichinosis, and elephantiasis.

39.7 The Annelid Worms (Phylum Annelida)

The annelid worms are segmented, with the units repeating some body structures such as the excretory organs, but not others, such as the digestive tract. Other characteristics include: (1) an efficient, closed circulatory system, (2) a fairly elaborate ventral nervous system, (3) a large, fluid-containing body cavity called a coelom that permits extensive body movements and is lined with mesoderm. The Annelida are divided into three classes, Polychaeta, Oligochaeta (earthworms), and Hirudinea (leaches).

39.8 The Mollusks (Phylum Mollusca)

The mollusks are a large group consisting of soft-bodied, unsegmented (with one exception) animals, many protected by one or more shells of calcium carbonate. The phylum consists of three major and several minor classes. The major ones include: (1) the Bivalvia, possessing two shells, bilateral symmetry, and a filtering method of feeding, (2) the Gastropoda, which possess one shell, lack a plane of symmetry altogether, and possess a distinct head, and (3) the Cephalopoda, characterized by a large, well-developed head bearing prominent eyes, and surrounded by a ring of tentacles, and exclusively salt-water habitats. Lesser classes include the Scaphopoda, Polyplacophora, and Monoplacophora.

39.9 The Arthropods (Phylum Arthropoda)

The arthropods contain the largest number of species of all the phyla, outnumbering all other species added together. They possess a segmented body enclosed by an exoskeleton containing chitin. Their symmetry is bilateral and they possess pairs of jointed appendages used for various functions. Their segments display a variety of structures and are combined into the head, thorax, and abdomen regions. Their circulatory system is open and the main part of the nervous system is found on the organism's ventral side.

The phylum consists of several subphyla. The Trilobitomorpha includes the trilobites of the Cambrian period. Members of the Chelicerata have a head and thorax that are fused into a cephalothorax, and the first pair of appendages (chelicerae) are adapted for feeding. The Chelicerata include the classes: (1) Merostomata—extinct eurypterids and living horseshoe crabs, and (2) Arachnida—descended from eurypterids, including ticks, scorpions, spiders. Members of the subphylum Mandibulata possess mandibles and antennae and include four major classes: (1) Crustacea—cephalothorax, mostly aquatic, includes crayfish, lobsters, shrimp, (2) Chilopoda—the centipedes, with one pair of legs per body segment, carnivorous, (3) Diplopoda—the milli- pedes, with two pairs of legs per segment, herbivorous, and (4) Insecta—over one-half of all living species and the dominant class of arthropods everywhere, excepting in salt water. The thorax has three distinct segments with a pair of legs on each and many adult types. are winged. The efficiency of insect structure and function causes them to be one of the most successful groups on earth.

39.10 The Phylum Onychophora

The phylum Onychophora contains one living animal, Peripatus, which has features of both the annelids and insects. It provides the clue that a primative segmented worm may have been ancestral to both these groups.

39.11 The Echinoderms (Phylum Echinodermata)

The members of the phylum Echinodermata all live in salt water, possess spiny skin, radial symmetry, and a water vascular system that is used to extend their many tube feet during locomotion. The phylum contains five classes: (1) Crinoidea—sea lilies, (2) Asteroidea—star= fish, (3) Ophiuroidea—brittle stars, (4) Echinoidea—sea urchins and sand dollars, (5) Holothuroidea—sea cucumbers.

39.12 The Chordates (Phylum Chordata)

The chordates possess bilateral symmetry, some segmentation, and an internal skeleton. Further, they possess the unique features of a notochord, gill pouches, at some developmental stage, and a hollow, dorsal nerve cord that enlarges to form the brain at the anterior end. The phylum subdivides into three subphyla. The subphylum Vertebrata is the subject of the following chapter. The remaining subphyla are the Cephalochordata (amphioxus) and the Tunicata or Urochordata (sea squirts).

39.13 Deuterostomia and Protostomia

Chordates and echinoderms share a number of characteristics, particularly in their embryonic development, that are not possessed by other phyla. These include pattern of egg cell cleavage, occasional identical twin production, site in initial gastrulation, mechanism of coelom development, and the compound creatine phosphate. These shared similarities and sharp contrasts with other phyla have led some taxo- nomists to establish two superphyla—the protostomia and the deuterostomia. It has been suggested that the phylum Nemertina may have been ancestral to the echinoderms and chordates. Vertebrates may have evolved from ancestral tunicates whose larva abandoned metamorphosis in freshwater estuaries.

B. KEY TERMS

mesoglea	Turbellaria
cnidoblasts	flukes
gastrovascular cavity	Trematoda
Coelenterata	Cestoda
Hydrozoa	tapeworms
Scyphozoa	trichinosis
Anthozoa	coelom
polyp	bivalves
medusa	snails
Polychaeta	slugs
Oligochaeta	univalves
Hirudinea	Cephalopoda
mantle	Scaphopoda
Bivalvia	Polyplacophora
Gastropoda	Monoplacophora
hemocoel	chitin
planarians	head
cephalization	thorax
abdomen	starfish
Trilobitomorpha	Crinoidea
trilobites	Asteroidea
Chelicerata	Ophiuroidea
cephalothorax	Echinoidea
Merostomata	Holothuroidea

eurypterids	notochord
Arachnida	gill pouches
Mandibulata	gill slits
Crustacea	Vertebrata
Chilopoda	Cephalochordata
centipedes	amphioxus
Diplopoda	Tunicata
Millipedes	Urochordata
Insecta	Nemertina
Onychophora	deuterostomia
tube feet	protostomia

II. MASTERING THE CHAPTER

A. LEARNING OBJECTIVES

When you have mastered the material in this chapter, you should be able to:

1. Define all key terms.

2. Explain how the animals are distinguished from the other kingdoms.

3. List the characteristics of the members of the phylum Porifera.

4. Discuss the phylum Cnidaria and the characteristics of the three classes into which it is subdivided. Give examples of each.

5. Discuss the phylum Platyhelminthes, its general characteristics, and the particular ones of three classes that belong to the group. Explain the pathogenic importance of the parasitic classes and give examples.

6. Explain two theories advanced to account for the origin of animals. Cite the evidence for both theories.

7. Discuss the phylum Nematoda, noting the evolutionary improvements over the phylum Platyhelminthes. Give some of the diseases produced by the parasitic members of the phylum.

8. Discuss the characteristics of the Annelida, paying particular attention to the circulatory and nervous systems and the implications of a coelom. List the three classes that make up this

9. Characterize the phylum Mollusca. Give the particular characteristics of the major classes (Bivalvia, Gastropoda, Cephalopoda) and list the minor classes.

10. Discuss the general characteristics of the phylum Arthropoda, noting especially its segmented body and chitinous exoskeleton. List the three main body sections of an arthropod.

11. Discuss the subphylum Chelicerata and its important classes, giving characteristics and examples of each.

12. Discuss the subphylum Mandibulata and its important classes, providing characteristics and examples of each. Explain why the class Insecta is of such paramount importance and provide examples of the efficiency of some of the insect structures and functions.

13. Explain why the single-membered phylum Onychophora is of such interest to biologists.

14. Discuss the phylum Echinodermata with regards to its characteristics and habitat. List the five classes into which it is subdivided, giving an example of each.

15. Explain the three unique features that characterize the phylum Chordata. Distinguish among the subphyla, providing examples of each.

16. Give reasons for the creation of two superphyla by taxonomists. Suggest the possible evolutionary descent of the deuterostoma. Explain the way by which ancestral tunicates may have given rise to the vertebrates.

B. REVIEWING TERMS AND CONCEPTS

1. In general, animals are organisms that do not have (a) _____, are capable of (b) _____, and are (c) _____. Of the 25 to 30 animal phyla, all but one are (d) _____.

2. The sponges belong to the phylum (a) _____ and consist of two layers of cells with a separating layer called the (b) _____ between them. Their offspring are dispersed in the form of (c) _____.

3. The Cnidaria possess an inner cavity known as the (a) _____ cavity and also possess stinging cells called (b) _____. Formerly, this phylum was known as the (c) _____. Included in this phylum are the classes (d) _____ (jellyfish), whose main body form is the (e) _____, and the (f) _____ (corals), which consist of a (g) _____ stage only.

4. The flatworms, phylum (a) _____, possess (b) _____ symmetry and concentrate their sensory equipment in the head, a condition called (c) _____. The most primitive flatworms

belong to the class (d) _____.

5. Roundworms, phylum (a) _____, include the organism which, when ingested with undercooked pork, results in the disease (b) _____.

6. Segmented worms belong to the phylum (a) _____ and are characterized by the possession of a major (b) _____ trunk along the (c) _____ side. Also, they possess a cavity called a (d) _____, which is lined throughout with (e) _____.

7. The phylum (a) _____ consists of animals with shells made of (b) _____ (limestone). These shells are formed by a structure called the (c) _____. If the organism possesses two shells, it belongs to the class (d) _____, whereas an organism possessing a coiled single shell would belong to the class (e) _____. Tentacled members of this phylum belong to the class (f) _____.

8. Members of the phylum (a) _____ possess a segmented body with a jointed exoskeleton composed of (b) _____. The segments are combined into three regions, the (c) _____, _____, and _____. The members of the subphylum Chelicerata have a head and thorax fused into what is called a (d) _____.

9. Crayfish belong to the subphylum (a) _____ and the class (b) _____. Centipedes belong to the class (c) _____ and have (d) _____ pairs of legs per body segment. The largest class of animals, the (e) _____, contains members whose (f) _____ consists of three segments with a pair of legs on each.

10. The phylum Onychophora contains (a) _____, which has both (b) _____ and _____ characteristics. Its ancestors may have given rise to the phylum (c) _____.

11. The phylum (a) _____ consists of all saltwater animals with spiny skins and a water vascular system used to operate their (b) _____ feet. Starfish belong to the class (c) _____ while sea lilies belong to the class (d) _____.

12. At some point in their development, chordates possess a (a) _____ and pairs of (b) _____ _____. The largest subphylum of this group is the (c) _____. Other subphyla include the (d) _____ (Urochordata), to which the (e) _____ belong, and the subphylum (f) _____, of which the amphioxus is representative.

13. The (a) _____ includes the echinoderms and (b) _____. Members of this group can produce (c) _____ twins and all possess the high-energy compound (d) _____ phosphate.

C. TESTING TERMS AND CONCEPTS

1. Which of the following represents the correct order of complexity,
    from least to greatest?
    a) Porifera, Nematoda, Annelida
    b) Annelida, Nematoda, Mollusca
    c) Annelida, Platyhelminthes, Nematoda
    d) Platyhelminthes, Cnidaria, Onychophora

2. Multicellularity in animals may have derived from
    a) a symbiotic relationship between algal and prokaryotic cells.
    b) a symbiotic relationship between various eukaryotic organisms.
    c) colonies of cells of a single kind of organism.
    d) cellularization of flagellated or amoeboid protozoans.

3. Which of the following is not characteristic of the jellyfish?
    a) belong to the Scyphozoa
    b) radially symmetrical
    c) medusa the main body form
    d) lack a mesoglea

4. Earthworms
    a) belong to the class Hirudinea.
    b) possess a ventrally located nerve cord.
    c) possess a mantle.
    d) have a complete gastrovascular cavity.

5. Arthropods possess
    a) a chitinous endoskeleton.
    b) a closed circulatory system.
    c) three main body regions.
    d) biradial symmetry.

6. Which of the following is not characteristic of the deuterostomia?
    a) spiral cleavage of fertilized egg cell
    b) identical twins
    c) coelom
    d) gastrulation beginning at the part of the embryo that will
       become the anus

7. The class containing the largest number of species is the
    a) Crustacea.
    b) Insecta.
    c) Mandibulata.
    d) Vertebrata.

8. Which of the following is not characteristic of chordates?
    a) gill pouches
    b) notochord
    c) ventral nerve cord with a brain at one end
    d) tail

9. The most primitive bilaterally symmetrical animals belong to the
    class

a) Platyhelminthes.
b) Turbellaria.
c) Cestoda.
d) Gastropoda.

10. The body of a Poriferan is
a) bilaterally symmetrical.
b) radially symmetrical.
c) three cell layers thick.
d) possessed of a layer called the mesoglea.

11. An important difference between Annelida and Nematoda is that
a) annelids have a coelom that permits greater freedom of movement of both internal organs and of the body as a whole.
b) annelids probably evolved first and gave rise to the more specialized nematodes.
c) nematodes are primitive, with no body cavity and a simple thrashing form of locomotion.
d) nematodes have an open circulatory system with a dorsal heart, while the annelids have a closed system and anterior hearts.

12. Characteristics of Onychophora that make it a "missing link" do not include
a) an open circulatory system.
b) malpighian tubules associated with the hindgut.
c) worm-like legs.
d) body segmentation.

13. Which of the following is not characteristic of insects?
a) ability to excrete uric acid
b) one pair of legs for each of the four thorax segments
c) tracheal tubes
d) internal fertilization

14. One of the features of the annelids not found in more primitive phyla is
a) bilateral symmetry.
b) cephalization.
c) nerves.
d) a coelom.

15. The Cephalopoda possess
a) biradial symmetry.
b) eyes homologous to those of humans.
c) a protective shell.
d) a rapid form of locomotion.

D. UNDERSTANDING AND APPLYING TERMS AND CONCEPTS

1. Match each physical characteristic in column A with the appropriate taxonomic category in Column B.

A   B

a) six jointed legs _____

b) tube feet _____

c) gill pouches _____

d) nematocyst _____

e) bony exoskeleton _____

f) chitinous exoskeleton _____

g) mantle with tentacles _____

h) endoskeleton of spiricles _____

i) highly branched gastrovascular cavity _____

j) pseudocoelom _____

k) eight jointed legs _____

l) calcareous shell _____

m) segmented coelom _____

n) radula _____

A. Arthropoda

B. Nematoda

C. Gastropoda

D. Cephalopoda

E. Porifera

F. Arachnida

G. Annelida

H. Platyhelminthes

I. Chordata

J. Vertebrata

K. Echinodermata

L. Mollusca

M. Insecta

N. Cnidaria

2. In each set of terms below, one does not belong. Circle that term and explain why it is out place.

a) collar cell, cniboblast, spicule, mesoglea _____

_____

b) Polychaeta, Gastropoda, Bivalvia, Cephalopoda _____

_____
_____

c) cnidoblast, radula, chelicera, mantle _____

_____
_____

d) Annelida, Echinodermata, Arthropoda, Mollusca _____

_____
_____

e) notocord, backbone, gill slits, dorsal nerve cord _____

_____
_____

f) uric acid, chitinous exoskeleton, internal fertilization, tracheal tubes _____

_____

g) mandibles, antennae, chelicerae, separate head and thorax

_____

_____

h) tube foot, muscular foot, notochord, gill slit _____

_____

_____

i) gastrovascular cavity, open system, closed system, pharynx

_____

_____

j) Cnidaria, Annelida, Nematoda, Porifera _____

_____

_____

3. A sponge and a cnidarian polyp both have a body wall surrounding a central cavity, with a single large opening at top. These organisms are only distantly related, however, and their central cavities are used for completely different purposes. Explain.

_____

_____

_____

4. Animals of several phyla have adopted terrestrial habitats. What adaptations have allowed them to do it?

a) earthworm _____

_____

b) snail _____

c) grasshopper _____

_____

5. Listed below are some important animal characteristics. Number them from 1 to 6 to show the order in which they evolved.

_____ a) segmentation
_____ b) bilateral symmetry
_____ c) jointed exoskeleton
_____ d) multicellularity
_____ e) gastrovascular cavity
_____ f) pseudocoelom

ANSWERS TO CHAPTER EXERCISES

Reviewing Terms and Concepts

1. a) chlorophyll   b) locomotion   c) multicellular   d) invertebrate

2. a) Porifera   b) mesoglea   c) larvae

3. a) gastrovascular   b) cnidoblasts   c) Coelenterata   d) Scyphozoa
    e) medusa   f) Anthozoa   g) polyp

4. a) Platyhelminthes b) bilateral c) cephalization d) Turbellaria

5. a) Nematoda b) trichinosis

6. a) Annelida b) nerve c) ventral d) coelom e) mesoderm

7. a) Mollusca b) calcium carbonate c) mantle d) Bivalvia
   e) Gastropoda f) Cephalopoda

8. a) Arthropoda b) chitin c) head, thorax, abdomen d) cephalothorax

9. a) Mandibulata b) Crustacea c) Chilopoda d) one e) Insecta
   f) thorax

10. a) <u>Peripatus</u> b) annelid, insect c) Arthropoda

11. a) Echinodermata b) tube c) Asteroidea d) Crinoidea

12. a) notochord b) gill pouches c) Vertebrata d) Tunicata
    e) tunicates f) Cephalochordata

13. a) deuterostomia b) chordates c) identical d) creatine

Testing Terms and Concepts

1.	a	6.	a	11.	a
2.	c	7.	b	12.	b
3.	d	8.	c	13.	b
4.	b	9.	b	14.	d
5.	c	10.	d	15.	d

Understanding and Applying Terms and Concepts

1. a) M b) K c) I d) N e) J f) A g) D h) E i) H j) B
   k) F l) L m) A n) C

2. a) The cnidoblast is a characteristic of Cnidaria, while the
   others are characteristic of Porifera. b) The Polychaeta is an
   annelid class; the others are mollusk classes. c) The mantle is
   a protective covering; the others are food gathering structures.
   d) The echinoderms are deuterostomes; the others are protostomes.
   e) A backbone is found only in the vertebrates; the others are
   found in all chordates. f) Tracheal tubes are used in respiration;
   the others reduce dehydration. g) Chelicerae are found in the
   Chelicerata; the others are found in the Mandibulata. h) The gill
   slit aids in respiration, the others in locomotion. i) A pharynx
   <u>acquires</u> food; the others <u>distribute</u> nutrients throughout the body.
   j) Annelida is a coelomate phylum; the others are not.

3. In the sponge, it is simply an exit for water that has been
   filtered of food by individual cells. In the cnidarian, it is a
   digestive cavity for large pieces of food, and the opening is a
   mouth and anus.

4. a) restrict habitat to moist soil b) shell and mucus coverings,
   lung c) chitinous exoskeleton, trachea, conversion of nitrogenous
   wastes to uric acid.

5. a) 5

   b) 3

   c) 6

   d) 1

   e) 2

   f) 4

THE VERTEBRATES

I. REVIEWING THE CHAPTER

A. CHAPTER HIGHLIGHTS

40.1 The Jawless Fishes (Class Agnatha)

The first vertebrate fossils are those of the ostracoderms or
jawless fishes, which first appeared in the Ordovician and lacked both
jaws and paired fins. They possessed armored plates and what evolved
into the first kidneys, structures primarily operated as a device for
maintaining water balance in their freshwater environment. The only
jawless fishes today are the lamprey and hagfish, the most primitive
of living vertebrates.

40.2 The Placoderms

The placoderms appeared early in the Devonian and possessed both
jaws and paired fins. Their lines of descent, after considerable
adaptive radiation, led to the two classes of modern fish, the cartilaginous and bony fish. Drying and warming climates during the
Devonian led to increased selection pressure, such that some of these
fishes evolved lungs for temporary periods of diminished water, and
others retreated to the oceans.

40.3 The Cartilaginous Fishes (Class Chondrichthyes)

The cartilaginous fishes were the group that retreated to the
oceans. Their skeletons were made of cartilage. They went from a
hypotonic environment to a hypertonic one, accomplishing water balance
by allowing urea to accumulate in the bloodstream until it was isotonic
to sea water. This hypertonic environment also led this group to be
the first vertebrates to develop internal fertilization. Sharks,
skates, and rays are modern-day members of this group.

40.4 The Bony Fishes (Class Osteichthyes)

The bony fishes were those which developed primitive lungs to
withstand periodic drought. This group rapidly divided into three
subgroups: (1) the paleoniscoids, which had ray fins and many of which
migrated to the seas, where their lungs became air bladders--most modern
fish are descended from this group; (2) lungfishes, which evolved the
ability to breathe with their mouths closed as well as improved circulatory efficiency through heart changes, and an enzyme that could convert ammonia into less toxic urea; and (3) crossopterygians, also with
internal nostril openings, and with lobed fins that allowed them to

"walk" from one pool to another—it was this group that was to evolve into the amphibians by the end of the Devonian. The Age of Fishes had given rise to the first of the four-legged vertebrates or tetrapods.

40.5   The Amphibians (Class Amphibia)

The amphibians were the vertebrate pioneers of the land with their lungs, bony limbs, more efficient, three-chambered heart, and simple ears. Their method of reproduction and inability to withstand long exposure to the air forced them to shift from land to water and back again, a situation that placed different demands on the specialized amphibian kidney. The Carboniferous, with its vast tropical swamps, has been called the Age of Amphibians. The drier and colder Permian produced a decline that leaves but three orders extant today: (1) Anura—frogs and toads, (2) Urodela—salamanders and newts, and (3) Apoda—caecilians.

40.6   The Reptiles (Class Reptilia)

The Permian's climate favored the reptiles, who first appeared during the Carboniferous. The major advance of the earliest reptiles (cotylosaurs) was the development of a shelled, yolk-filled egg that necessitated internal fertilization. As the Permian became increasingly dry, reptiles developed a dry skin, improved lungs, and a partial septum in the ventricle. Adaptive radiation resulted in five lines of descent. These were: (1) pelycosaurs—legs underneath the body rather than at the side, gave rise to the therapsids from which, 100 million years later, the first mamals evolved; (2) turtles (order Chelonia)— retained their association with the water for the last 200 million years, (3) and (4) plesiosaurs and icthyosaurs—marine reptiles with finlike appendages, and (5) diapsids. This last group, it is thought, developed the ability to excrete nitrogenous wastes as uric acid, al- most completely eliminating their need for drinking water. The water produced by cellular respiration and food ingestion was usually suf- ficient. Diapsids diverged into two separate lines. One line produced the orders Rhynchocephalia and Squamata, the former with one species extant (the tuatara) and the latter comprising lizards and snakes. The second line produced the ruling reptiles of the Mesozoic Era, starting with the thecodonts. There is evidence to suggest that some of the thecodonts were able to maintain a relatively high and well-regulated body temperature. The thecodonts evolved into five orders of reptiles. These included the Crocodilia (crocodiles and alligators), two orders of dinosaurs, and two groups of flying reptiles, one of which gave rise to the modern birds.

40.7   The Birds (Class Aves)

The birds evolved from a feathered reptile called Archeopteryx. These feathers provided a strong, light wing surface as well as suf- ficient heat insulation to allow the birds to maintain a relatively high body temperature even in cold climates. Although they were well established by the end of the Mesozoic, it was the Cenozoic, with its great geological changes, that saw the extensive adaptive radiation of this group. Their four-chambered heart and light and powerful bodies helped them to adapt successfully, whereas many of their reptile cousins became extinct at the close of the Mesozoic.

40.8 Continental Drift

Continental drift began early in the Mesozoic and continues today. This phenomenon provides an explanation for many formerly puzzling facts of geology and biology. At the start of the Mesozoic, all the present continents constituted a single land mass that has been called Pangaea. This then separated into Laurasia in the northern hemisphere and Gondwana in the southern hemisphere. Further separation produced our present continents. Extensive geological evidence exists to support the theory of continental drift. A mechanism (plate tectonics) has been postulated for such movement.

40.9 The Mammals (Class Mammalia)

The first mammals arose late in the Triassic from therapsid ancestors. They were small insectivores that, like the birds, possessed a four-chambered heart and completely separated systemic and oxygenating circuits. In addition, hair conserved body heat, milk was provided for the young, teeth became specialized, and of paramount importance, their cerebrum increased in size. The three surviving groups of mammals are distinguished by their method of providing for their embryonic young. They are the: (1) monotremes—egg-laying, (2) marsupials—pouched, and (3) placental mammals—almost complete intrauterine development. The placental mammals were the most successful, radiating into fourteen different orders after the extinction of the ruling reptiles.

40.10 The Evolution of Homo sapiens

About 60 million years ago, an order of arboreal placental mammals known as primates evolved. By 20 million years ago they had split into three separate stocks: "New World" monkeys, "Old World" monkeys, and hominoids. By 14 million years ago, the hominoids had given rise to the hominids, who came down from the trees and took to walking, and the pongids (true apes), who remained in the trees. By 3 million years ago, a hominid (Australopithecus) with an upright posture had appeared, and by 1.5 million years ago, there were three distinct species. One of these (Homo habilis) made and used tools and had a brain volume of approximately 750 ml., two-thirds again as large as that of Australopithicus. Homo habilis evolved into Homo erectus, Australopithicus died out, and H. erectus spread throughout the Old World. By 100,000 years ago Homo sapiens, whose brain was as large as ours, had appeared and gradually spread to Australia and throughout the New World. This spread of man created isolated gene pools, which initiated processes of subspeciation that produced the races of today.

B. KEY TERMS

ostracoderms   icthyosaurs

paleoniscoid   Rynchocephalia

lungfishes	Squamata
crossopterygians	lizards
Anura	snakes
Urodela	thecodonts
Apoda	crocodiles
pelycosaurs	alligators
therapsids	turtles
Chelonia	diapsids
plesiosaurs	flying reptiles
Crocodilia	feathers
ruling reptiles	plate tectonics
dinosaurs	hair
cotylosaurs	milk
amnion	monotremes
yolk sac	Prototheria
chorion	marsupials
allantois	Metatheria
pterosaurs	placental Mammals
Eutheria	pongids
primates	homonids

II. MASTERING THE CHAPTER

A. LEARNING OBJECTIVES

When you have mastered the material in this chapter, you should be able to:

1. Define all key terms.

2. Discuss the first vertebrates (ostracoderms), paying particular attention to the evolution and function of the kidney.

3. Explain the evolutionary significance of the placoderms and the climate changes of the Devonian.

4. Discuss the cartilaginous fishes and the evolutionary modifications which allowed them to survive in the oceans.

5. Discuss the bony fishes and how they evolved into three main groups. Give the characteristics of each group and its evolutionary importance.

6. Cite the evolutionary improvements of the amphibians. Explain the unique demand placed upon the amphibian kidney. Give reasons for the amphibians' decline to their present three orders.

7. Explain why the reptiles became such a successful group. List the five orders of reptiles, noting the significance of the development of uric acid excretion. Discuss the orders of reptiles that evolved from the thecodonts, noting the characteristics of each.

8. Explain the successful adaptations of birds in their evolution from reptilian ancestors, paying particular attention to their heart and means of achieving lightness.

9. Explain what is meant by continental drift and how it solves many biological and geological puzzles.

10. Distinguish among the three classes of mammals. Offer reasons why the mammals were an "improvement" over their therapsid ancestors. Explain why the placental mammals blossomed into fourteen orders during the Cenozoic.

11. Trace the evolution of humans beginning with the arboreal primates of 60 million years ago. Cite the importance of Homo erectus and offer a possible explanation for the development of the races, noting that speciation did not occur.

B. REVIEWING TERMS AND CONCEPTS

1. The first vertebrates were the (a) _____ fishes or (b) _____, which first appeared during the (c) _____ period. These creatures evolved the first (d) _____ for the purpose of maintaining themselves in a hypotonic environment. During the Devonian, the (e) _____ appeared with their jaws and paired fins. These gave rise to the class Chondrichthyes, or (f) _____ fishes, and the class Osteichthyes, or (g) _____ fishes. As adaptations that helped them cope with their seawater environment, the former evolved (h) _____ fertilization and the ability to allow (i) _____ to build up in the blood until it was isotonic. The latter group, to withstand periodic drought, developed a pair of primitive (j) _____ and quickly radiated into three groups, the (k) _____-finned fishes, the (l) _____, and the crossopterygians. The last group had pectoral and pelvic fins that were (m) _____.

2. The (a) _____ were the vertebrate land pioneers. They had developed a (b) _____-chambered heart and simple (c) _____ out of their spiracles.

395

3. During the Carboniferous, the (a) _____ evolved from the amphibians. The earliest of this new group were the (b) _____, their chief advance being the development of a (c) _____, yolk-filled egg. The most important of the five major lines of reptiles were the (d) _____, with their legs positioned under them. This group was to give rise to the (e) _____, from which the mammals would eventually evolve. The snakes and lizards both possess the ability to excrete (f) _____ wastes in the form of (g) _____ _____, thereby conserving water. Another group that possessed the above ability were the (h) _____, who were to evolve into the ruling reptiles or (i) _____ and the (j) _____.

4. The reptile-like ancestor of the birds was (a) _____. Birds possess (b) _____ bones and a (c) _____-chambered heart and are able to withstand cold thanks to their insulation and high (d) _____ rate.

5. The theory of (a) _____ _____ suggests that, at one time, all continents were part of the same land mass called (b) _____. This mass broke into two areas with (c) _____ in the northern hemisphere and (d) _____ in the southern hemisphere. The mechanism proposed for this movement is called (e) _____ _____.

6. Late in the Triassic, the (a) _____ gave rise to the mammals, whose diagnostic traits are (b) _____, and (c) _____ production. The mammals are divided into three classes on the basis of how they care for their (d) _____. The most advanced class are the (e) _____ mammals and the most primitive are the egg-laying mammals or (f) _____.

7. Humans are of an order of placental mammals called (a) _____, which, some 20 million years ago, split into (b) _____ (how many) groups. One of these groups, the hominoids, divided into two branches, the (c) _____ or true apes and the (d) _____. This latter group eventually gave rise to the human species's ancestor, Homo (e) _____. This last group's genetic isolation probably gave rise to the (f) _____ of humans, or Homo (g) _____, all of which are members of the same (h) _____.

C. TESTING TERMS AND CONCEPTS

1. A notable evolutionary achievement of the earliest reptiles was
   a) nitrate excretion.
   b) a three-chambered heart.
   c) a shelled egg.
   d) homeothermy.

2. The evolutionary modification first noted in Archeopteryx was
   a) the presence of feathers.
   b) a four-chambered heart.

c) shelled eggs.
d) the presence of a bill without teeth.

3. The first vertebrates belonged to the class
   a) Osteichthyes.
   b) Placoderms.
   c) Chondrichthyes.
   d) Agnatha.

4. Marsupials
   a) lack body hair.
   b) are the most common mammals in Australia and South America.
   c) fared poorly when confronted with placental competition.
   d) lay eggs into an abdominal pouch.

5. A recent human ancestor
   a) was Homo erectus, who lived throughout the Old World.
   b) was Homo sapiens neanderthalensis, who lived throughout the Old
      and New Worlds.
   c) were the hominoids, with their very upright posture.
   d) belonged to the genus Australopithicus and used stone tools.

6. Which of the following was not characteristic of the crossopterygians?
   a) bony skeleton
   b) lobed fins
   c) internal nostrils
   d) swim bladder

7. The placoderms were the
   a) first fish to evolve kidneys.
   b) first to evolve a hinged jaw.
   c) evolutionary line that gave rise to the ostracoderms.
   d) evolutionary line that gave rise to the lampreys and hagfish.

8. The therapsids are notable because they
   a) developed finlike limbs for swimming.
   b) produced Archeopteryx.
   c) eventually evolved into mammals.
   d) gave rise to the lizards and snakes.

9. The carboniferous period saw the rise of
   a) amphibians.
   b) reptiles.
   c) mammals.
   d) birds.

10. In which group is internal fertilization not required?
    a) cartilaginous fishes
    b) amphibians
    c) reptiles
    d) birds

11. Uric acid excretion

a) must occur rapidly because uric acid is particularly poisonous.
b) was first developed by the amphibians, as they colonized land.
c) is a means of converting uric acid into nitrogenous wastes,
   which are then excreted.
d) aids in conserving body water.

12. The races of humans
    a) are distinct species belonging to the same genus.
    b) are the result of gene pool isolation.
    c) formed only in the last 20,000 years.
    d) number five, one on each major continent of the world.

13. Cartilaginous fishes
    a) possess both bone and cartilage in their skeletons.
    b) have blood that is isotonic to seawater.
    c) lack a spiracle.
    d) do not exhibit live birth.

14. The tetrapods evolved from
    a) paleoniscoids.
    b) Devonian rayfins.
    c) crossopterygians.
    d) Cartilaginous fishes.

15. Mammals are divided into three groups on the basis of their
    a) manner of caring for their embryonic young.
    b) manner of caring for their postembryonic young.
    c) method of milk production.
    d) oxygenating and systemic circuits.

D. UNDERSTANDING AND APPLYING TERMS AND CONCEPTS

1. The early jawless fishes were small, sedentary filter-feeders,
   very unlike the active, predatory fishes, reptiles, birds, and
   mammals that are among their decendants. What key characteristic
   did these jawless fish have that set them apart from their con-
   temporaries and permitted them to give rise to active predators?
   (a)_____

   What characters are seen in the placoderms and in the later
   bony fishes that allowed them to take the role of active predator?
   (b)_____

   The crossopterygians developed two features that preadapted
   them for a terrestrial existence. What were they? (c)_____

   What habitat and selection pressures caused their development?
   (d)_____
   _____
   _____

2. Listed below are some characteristics that have permitted great
   success of vertebrates on land. For each, name the vertebrate class

in which it first appeared.

a) shelled, amniotic egg _____

b) lungs _____

c) feathers _____

d) fur _____

e) enlarged cerebrum _____

f) dry, waterproof skin _____

g) four-chambered heart _____

h) limbs of muscle and bone _____

i) ear _____

j) inflation of lungs by rib cage _____

k) three-chambered heart _____

l) ability to breathe through nostrils _____

m) excretion of nitrogen as uric acid _____

n) excretion of nitrogen as urea _____

o) teeth specialized for different tasks _____

3. Comparison of the invertebrates and vertebrates reveals, in each, a group of animals that adapted to a terrestrial life in similar ways. Listed below are some areas in which they are similar. Describe how they are similar.

a) water retention _____

b) nitrogenous excretion _____

c) reproduction _____

d) temperature regulation _____

e) limbs _____

What two classes are referred to here?

f) _____

g) _____

of course there are differences between these two classes in major areas. What are they:

h) gas exchange _____

i) circulation of nutrients _____

j) skeleton _____

4. Rank these ancestors of man in the order of their appearance by numbering them from the earliest (beginning with 1) to the most recent. Place a <u>zero</u> before any animal that is <u>not</u> a direct ancestor of man.

a) _____ New World monkey

b) _____ Homo <u>erectus</u>

c) _____ pongid

d) _____ hominid

e) _____ <u>Australopithicus</u>

f) _____ Homo <u>habilis</u>

g) _____ hominoid

h) _____ Homo <u>sapiens</u> <u>neanderthalensis</u>

i) _____ Homo <u>sapiens</u> <u>sapiens</u>

5. It has been suggested that the key event in human evolution was coming down from the trees and moving out onto the plains; it led to our upright posture, tool use, and uniquely enlarged brain. Why did our upright posture arise in an open habitat, rather than in the forested habitat of most primates? (a)_____
_____
_____

What features of this new habitat might have led to tool use?
(b)_____
_____

How might these developments lead to increased intelligence?
(c)_____
_____

What was unusual about these early hominids that caused them to take this evolutionary route, when other plains animals did not?
(d)_____
_____

ANSWERS TO CHAPTER EXERCISES

Reviewing Terms and Concepts

1. a) jawless b) ostracoderms c) Ordovician d) kidneys
    e) placoderms f) cartilaginous g) bony h) internal i) urea
    j) lungs k) ray l) lungfishes m) lobed

2. a) amphibians b) three c) ears

3. a) reptiles b) cotylosaurs c) shelled d) pelycosaurs
    e) therapsids f) nitrogenous g) uric acid h) thecodonts
    i) dinosaurs j) birds

4. a) Archeopteryx b) hollow c) four d) metabolic

5. a) continental drift b) Pangaea c) Laurasia d) Gondwana
    e) plate tectonics

6. a) therapsids b) hair c) milk d) embryonic young e) placental
    f) monotremes

7. a) primates b) three c) pongids d) hominids e) erectus
    f) races g) sapiens h) species

Testing Terms and Concepts

1.	c	6.	d	11.	d
2.	a	7.	b	12.	b
3.	d	8.	c	13.	b
4.	c	9.	a	14.	c
5.	a	10.	b	15.	a

Understanding and Applying Terms and Concepts

1. a) a bony skeleton moved by various skeletal muscles b) movable
    jaw and paired fins c) lobe-fins and lungs d) The early bony
    fish lived in freshwater habitats during periods of drying cli-
    mate. Lungs allowed them to get oxygen in crowded, stagnant pools,
    and muscular fins allowed them to waddle from one pool to another,
    over land.

2. a) Reptilia   f) Reptilia   k) Osteichthyes

    b) Osteichthyes   g) Mammalia   l) Osteichthyes

    c) Aves   h) Osteichthyes   m) Reptilia

    d) Mammalia   i) Amphibia   n) Osteichthyes

    e) Mammalia   j) Reptilia   o) Mammalia

3. a) waterproof skin b) uric acid c) internal fertilization and
    waterproof egg d) poikelothermic e) positioned under the body
    for agile locomotion f) Insecta g) Reptilia h) trachea versus
    lungs i) open versus closed j) exoskeleton versus endoskeleton

4. a) 0 b) 5 c) 0 d) 2 e) 3 f) 4 g) 1 h) 0 i) 6

5. a) Locomotion through the trees <u>requires</u> the use of the hands. b)
With the hands available, tool use becomes possible, and without
trees as refuges from predators, tool use (i.e. rock or club) becomes decidedly advantageous. c) Reliance on tools puts a premium
on mental skills, like being able to see a use for a fallen limb,
being able to remember to carry such a club even when it is not
needed at that moment, and being able to see how modifications
could make a useful thing out of something now useless (i.e.
breaking a too-large limb in half). When mental skills become
important to one's life and reproductive output, then they become
selected for. d) The early hominids had a relatively large brain
to begin with, and they did <u>not</u> have any other good defense against
predators (i.e. fangs, claws, hooves).

ENERGY FLOW THROUGH THE BIOSPHERE

I. REVIEWING THE CHAPTER

A. CHAPTER HIGHLIGHTS

41.1 The Input of Energy

A community of plants, animals, and microorganisms, along with the nonliving features of the environment, constitutes an ecosystem. All the world's ecosystems make up the biosphere. The ecosystems utilize the sun's energy, which varies with season and latitude. The energy received by the tropics throughout the year is equivalent to that received in the temperate regions during the summer: 8000-10,000 $kcal/m^2/day$.

41.2 Ecosystem Productivity

Ecosystem productivity is measured by determining the weight of plant material produced by one square meter of land per year and determining from that how many kilocalories are stored (4.25 kcal/gram of dried plant material). In a termperate forest, approximately 500 $kcal/M^2$ are stored with a photosynthetic efficiency of 1%. The amount of energy stored by a plant community is referred to as its net productivity. Productivity of deserts and oceans is far below that of cultivated land, especially land that is highly fertilized. The manufacture, transport, and application of fertilizer and the other efforts used in agriculture represent energy consumption, all of which should be taken into account when determining net productivity.

41.3 Food Chains

The net productivity of a plant community is utilized by various organisms through various pathways. The total organic matter in an ecosystem is referred to as the biomass or standing crop. When energy passes to a heterotroph and then from animal to animal, the pathway of consumption is referred to as a food chain. The chain consists of the producers (plants), the primary consumers (herbivores), the higher-level consumers (carnivores), and the decomposers, which receive input from all other components. Each consumption level is referred to as a trophic level. Various food chains are interconnected, resulting in food webs of some complexity. The flow of energy through the biosphere is one-way; the materials from which living things are built are re-cycled over and over.

41.4 Energy Flow through Food Chains

As food is transferred from one trophic level to the next, a great deal of energy is "lost" in maintaining the organism, i.e. driving metabolic activity and movement. This energy is ultimately lost in the form of heat. Except in a few controlled situations, conversion of net production from one level to the next is well below 50%, with many carnivores being approximately 5% efficient. A general rule of thumb is that the efficiency of transfer between levels is 10%. This decrease in total available energy from level to level is sometimes referred to as a pyramid of energy.

41.5 The Biomes

Distinctive plant and animal communities are called biomes. Climatic features such as temperature and rainfall interact in the creation and maintenance of a biome. Limiting factors are extremely important in biomes, for example, length of growing season, freezes, total rainfall. Assuming adequate rainfall, four biomes are defined by temperature as one moves northward from the equator: (1) tropical rain forest, (2) temperate deciduous forest, (3) taiga, and (4) tundra. This same temperature effect can be achieved by traveling from a lower to a higher altitude. When rainfall is limiting, three biomes result: (1) grasslands—less than 20 inches of rain per year during the growing season, (2) desert—less than 10 inches of rain per year, and (3) chaparral—20 to 30 inches of rain per year, mostly in the winter.

41.6 Fire

Fire plays an important role in maintaining some biomes. It does this by destroying, and thereby keeping in check, plant species that might overtake and dominate the ones typical of a biome. These typical species recover quickly after a fire. Periodic ground fires keep available fuel scarce, actually preventing the more destructive crown fires.

41.7 Plant Succession

The elimination or alteration of an environmental factor results in the proliferation of a new or previously minor species of plant which may, in turn, alter the environment and bring about the rise of other species. This process is referred to as plant succession and begins as soon as an area is capable of supporting life. In the temperate deciduous forest biome, this proceeds from lichens and mosses to shrubs, to birches and aspens, to pines, to hardwoods. The final, self-sustaining stage is referred to as a climax forest. A similar process occurs in ponds, which eventually fill with organic matter to become low fields, which then become forested. If humans or nature intervenes to destroy a climax area and that area is then left untouched, a secondary succession begins that arrives at the same climax community as the primary form of succession. This tendency for all succession to end in the same climax community within a given region is called convergence. Generally, succession involves increased photosynthetic-efficiency, biomass, and diversity.

41.8 Freshwater Ecosystems

Only 3% of the water on earth is fresh, and 99% of that is permanently frozen or buried in aquifers. However, streams and lakes provide a variety of habitats for biological communities. The shore of a lake is the littoral zone, where light reaches the bottom. In open water, the layer that light penetrates is the limnetic zone. Here microscopic algae are the dominant producers. Below the limnetic zone is the profundal zone and the bottom, or benthos, both of which depend for their calories on organic matter that drifts down from above. Rivers and streams differ from lakes and ponds in being more highly oxygenated and in depending more on surrounding ecosystems for organic input.

41.9 Marine Ecosystems

Oceans are also divided into zones. The margins are the intertidal zones, the relatively shallow ocean out to the edge of the continental shelf is the neritic zone, and the water over the oceanic basin is the oceanic zone. Plaktonic algae are important producers, but primary production is very low. The bottom of the basin is the abyssal plain, which like the lake benthos, depends largely on the downward drift of organic debris. Around rifts, chemoautotrophic bacteria do carry out some primary production.

B. KEY TERMS

ecosystem	nekton
biosphere	secondary consumers
biomass	tertiary consumers
net productivity	trophic level
community	food webs
rivers	pyramid of energy
streams	pyramid of numbers
standing crop	biomes
food chain	tropical rain forest
producers	temperate deciduous forest
primary consumers	taiga
profundal zone	tundra
benthos	grassland
lake	chaparral
pond	desert

littoral zone   climax

limnetic zone   secondary succession

plankton   convergence

II. MASTERING THE CHAPTER

A. LEARNING OBJECTIVES

When you have mastered the material in this chapter, you should be able to:

1. Define all key terms.

2. Explain what constitutes an ecosystem and how ecosystems combine to form the biosphere. Discuss the energy received by the earth in terms of the effects of season and latitude. Compare the energy budget of tropical and temperate latitudes.

3. Explain how ecosystem productivity is determined. Discuss the concept of net productivity and how it relates to cultivated land.

4. Explain what happens to the net productivity of a plant community in terms of the pathways of food consumption, noting the relationship among producers and consumers. Contrast a food chain and a food web. Describe what happens to the net productivity as it passes through trophic levels.

5. Discuss how energy flows through a food chain, paying particular attention to the efficiency of conversion and the particularly low efficiency of carnivores. Distinguish between a pyramid of numbers and a pyramid of energy.

6. Discuss the major factors involved in determining a biome and how latitude and/or altitude is a determining factor. List seven major terrestrial biomes, giving characteristics and examples of each.

7. Explain how fires actually helps maintain a biome and how ground fires prevent larger, more destructive crown fires.

8. Discuss the mechanisms involved in plant succession, contrasting primary and secondary types. Give an example of each type. Discuss what happens to biomass and efficiency as succession proceeds.

9. List and describe the zones of a lake and an ocean ecosystem.

10. Discuss the differences between lake and river ecosystems.

11. Explain why oceans are "biological deserts."

B. REVIEWING TERMS AND CONCEPTS

1. A plant, animal, and microorganism (a) _____ together with the (b) _____ features of its environment constitutes an (c) _____, which in turn is part of the much larger (d) _____.

2. The amount of energy stored by a plant community over a period is its (a) _____ _____. Deserts approximate (b) _____ in terms of productivity.

3. The total organic matter present in an ecosystem is referred to as the (a) _____ or (b) _____. The grass-grasshopper-toad-snake-hawk pathway is called a (c) _____ _____ _____, two or more of which may have inter-connections and constitute a (d) _____ _____, Producers are (e) _____ organisms, where the herbivores that feed upon the producers are (f) _____ _____. Each level of consumption is called a (g) _____ level.

4. Energy flow through a food chain displays an average efficiency of (a) _____% at each level. The energy budget of a (b) _____ ecosystem is in balance. The decrease in available energy as one travels up a food chain is referred to as a (c) _____ _____ _____ and the decrease in individuals at each level reflects the pyramid of (d) _____.

5. A distinctive type of plant and animal community is called a (a) _____, whose nature is controlled by two major determinants, (b) _____ and _____. An increase in (c) _____ has an effect similar to an increase in (d) _____ in determining biomes. An area receiving from 10 to 20 inches of rainfall per year and an adequate growing season is labeled a (e) _____ biome. The biomes subjected to the most severe climate conditions are (f) _____ and _____.

6. (a) _____ fires are often beneficial whereas (b) _____ fires are not. Fires can actually help (c) _____ biomes.

7. A plant (a) _____ that goes from bare rock to eventual hardwood forest is called a (b) _____. One that starts with a burned-over area is called a (c) _____ _____. When succession produces a self-sustaining population, the (d) _____ stage has been reached. The tendency for all plant successions in the same region to end in the same (e) _____ community is called (f) _____.

8. The shore of a lake is the (a) _)_____ zone; here (b) _____ plants can grow. In open water, (c) _____ is the major producer in the (d) _____ zone, and there is no primary production in the (e) _____ zone or (f) _____. Rivers differ from lakes in being more highly (g) _____ and more dependent on surrounding fields or forests for (h) _____ input.

9. The margin of the ocean is the (a) _____ zone, over the

continental shelf is the (b) _____ zone, and over the ocean basin is the (c) _____ zone. The deep-sea bottom is the abyssal plain. Oceans have a very low productivity because (d) _____ is available only at the surface, while (e) _____ are available only near the shores or bottom sediments.

C. TESTING TERMS AND CONCEPTS

1. If a food chain possesses an amount of energy (x) at one trophic level, then two levels above, the amount remaining would be
    a) x.
    b) x/10.
    c) x/20.
    d) x/100.

2. Which of the following characterizes the tundra?
    a) permafrost, adequate rainfall, stunted plants
    b) "spruce-moose" association
    c) 20 to 30 inches of rainfall during winter, plants dormant in summer
    d) only 20 inches of rainfall during the growing season

3. Which of the following is not characteristic of a climax forest?
    a) a large number of organisms but fewer different species—low diversity
    b) low net productivity
    c) a self-sustaining population
    d) a large, stable biomass

4. When fire is excluded from an ecosystem,
    a) crown fires no longer prevent ground fires.
    b) soil nutrient levels fluctuate markedly.
    c) the area frequently returns to chaparral.
    d) the climax community often changes.

5. Which term includes the other three?
    a) community
    b) biosphere
    c) ecosystem
    d) population

6. Each step in a food chain is referred to as a
    a) food web.
    b) trophic level.
    c) standing crop.
    d) consumer.

7. Which of the following does not apply to the temperate regions?
    a) receive an average of 8000-10,000 kcal/m^2/day throughout the year
    b) productivity in summer may match that of the tropics
    c) subjected to more variation in energy received than the tropics
    d) sunlight less intense than in the tropics

8. What portion of the earth's productivity is accounted for by the oceans?
   a) 1/10
   b) 1/3
   c) 1/2
   d) 3/4

9. A cow can be best labeled a
   a) primary consumer.
   b) secondary consumer.
   c) tertiary consumer.
   d) producer.

10. Which of the following terms includes the other three?
    a) producers
    b) consumers
    c) food web
    d) food chain

11. Which of the following best represents an ecosystem?
    a) the fish in an aquarium
    b) all organisms in an aquarium
    c) all organisms in an aquarium plus the light source
    d) all organisms in an aquarium, the light, water, and sand

12. In temperate forests, the efficiency of photosynthesis is
    a) 1%.
    b) 25%.
    c) 50%
    d) 75%

13. An increase in latitude produces an effect similar to an increase in
    a) longitude.
    b) altitude.
    c) rainfall.
    d) temperature.

14. Which of the following is closest to the oceans in terms of its productivity?
    a) temperate forest
    b) temperate grassland
    c) cultivated land
    d) desert

15. As one travels upwards in a pyramid of numbers, the
    a) number of organisms increases.
    b) efficiency increases.
    c) biomass decreases.
    d) individual size decreases.

D. UNDERSTANDING AND APPLYING TERMS AND CONCEPTS

1. A pine beetle (which carries a tree-killing fungus) is currently

doing severe damage to large tracts of ponderosa pine in the Rocky Mountains. Previously, the beetle did scattered, insignificant damage to only weak trees. The beetle requires very old and/or weak trees in order to establish itself in any significant way. All chemical attempts at control have proved futile. Mr. Earl W. Ashley, a private citizen, has suggested a "method" for minimizing the damage potential of the pine beetle—one, in fact, that the U.S. Forestry Service has also suggested. Can you suggest what this proposed cure might be? _____

_____

_____

2. List the four trophic levels found in most ecosystems and give an example of each.

Trophic Level   Example

a) _____   _____

b) _____   _____

c) _____   _____

d) _____   _____

Note that many members of the fourth trophic level really occupy the fourth level only (or fifth or sixth) so rare? e) _____

_____

_____

3. Communities are characterized more by their plant life than by their animal life. For each ecosystem listed below, describe two ways plants are adapted for life there.
a) tropical rain forest _____

_____

b) temperate deciduous forest _____

_____

c) taiga _____

_____

d) tundra _____

_____

e) grasslands _____

_____

f) desert _____

_____

g) chaparral _____

_____

4. What is the ultimate limiting factor that determines how much food can be produced on earth? (a) _____

_____

Once all available land is put to agricultural use, there will be two theoretically possible ways to further increase plant food production on land. What are they? (b) _____

(c) _____

5. In both desert and tundra ecosystems, human destruction is severe
    and long-lasting. Why here more than in other biomes? _____
    _____
    _____

6. In undisturbed ocean ecosystems, the neritic zone and regions of
    upwelling are fairly rich in fish life, compared to the oceanic
    zone. Why? (a) _____
    _____
    Why is the intertidal zone especially rich? (b) _____
    _____

7. The oceans are often described as biological deserts—coastal areas
    produce less than they might and the open ocean produces very
    little indeed. How might each of these regions be made to produce
    more plant and animal life? (a) coastal regions _____
    _____
    (b) open ocean _____
    _____

ANSWERS TO CHAPTER EXERCISES

Reviewing Terms and Concepts

1. a) community b) nonliving c) ecosystem d) biosphere

2. a) net productivity b) oceans

3. a) biomass b) standing crop c) food chain d) food web
    e) photosynthesizing f) primary consumers g) trophic

4. a) 10 b) mature (or climax) c) pyramid of energy d) numbers

5. a) biome b) rainfall, temperature c) altitude d) latitude
    e) grassland f) desert, tundra

6. a) Ground b) crown c) maintain

7. a) succession b) primary succession c) secondary succession
    d) climax e) climax f) convergence

8. a) littoral b) rooted c) phytoplankton d) limnetic e) profundal
    f) benthos g) oxygenated h) energy

9. a) intertidal b) neritic c) oceanic d) sunlight e) minerals

Testing Terms and Concepts

1.	d	6.	b	11.	d
2.	a	7.	a	12.	a
3.	a	8.	b	13.	b
4.	d	9.	a	14.	d
5.	b	10.	c	15.	c

Understanding and Applying Terms and Concepts

1. The forest has been overprotected by man. Let the beetles carry out their natural weeding of the sick and old. When these are gone, the numbers of beetles will fall to very low levels too.

2. a) producer, grass b) herbivore, grasshopper c) primary, carnivore, chicken d) secondary carnivore, human e) About 90% of the energy of one level is lost in the transfer to the next level. After two or three such transfers, little is left to support a consumer population.

3. a) Trees grow tall in competition for light, and vines climb for the same reason. b) Trees drop their leaves as a protection against freezing, and plants of the understory are adapted to lower levels of light. c) Trees rely on needles, rather than broad leaves, because needles lose less water during the long periods when all water is frozen, and they produce "antifreezes" to protect the tissues themselves from freezing. d) Some plants grow very low, and others reproduce quickly, produce seeds, and die to avoid the cold, dry winds. e) Grasses have vast root systems to catch the little available water and underground stems that survive frequent fires. f) Some plants have very deep root systems to tap deep aquifers, and others store water in specialized stems. g) Plants are dormant during dry season and have thick waxy coatings to reduce water loss.

4. a) The amount of solar energy reaching the surface b) Produce more efficient crop plants through artificial selection. c) Bring more energy to earth, using orbiting solar collectors.

5. In these ecosystems, the layer of life over the surface of the earth is very thin and already severely stressed by natural factors, so growth and repair is extremely slow.

6. a) In these two regions, nutrient-rich bottom waters are mixed into sunlit surface layers. b) Here, runoff from the land brings in still more nutrients.

7. a) Stop coastal pollution and dredging and filling. b) The problem here is a lack of inorganic nutrients in the upper, sunlit layers. Bottom waters could be pumped up or artificial fertilizers could be applied.

42

THE CYCLES OF MATTER IN THE
BIOSPHERE

I. REVIEWING THE CHAPTER

A. CHAPTER HIGHLIGHTS

42.1 The Carbon Cycle

The flow of energy through the biosphere is one-way; there is no
"energy cycle." However, the matter of our planet is continuously
being turned into organic molecules by organisms, then broken down into inorganic molecules, over and over again. These are "material
cycles." Carbon, owing to its bonding ability, provides the molecular
size and diversity required for life. Autotrophs take in carbon dioxide and produce organic compounds. Hetertrophs ingest the organic
compounds, releasing a good deal of carbon dioxide, passing some compounds on to the next trophic level, but transferring the majority of
organic matter to the decay organisms. In this way carbon is returned
to the atmosphere to continue in the cycle. Human burning of fossil
fuels has increased the amount of carbon dioxide in the air with perhaps some two-thirds of this stimulating and being incorporated by
photosynthesis. Carbon dioxide also dissolves in seawater to eventually
become stored in the form of calcium carbonate. Whether or not the
increase in atmospheric carbon dioxide has altered the earth's temperature is not known.

42.2 The Oxygen Cycle

Oxygen is utilized by the process of respiration as the final
acceptor of electrons, producing water. In photosynthesis light energy
strips the oxygen atoms in water of their electrons, finally resulting
in molecular oxygen to complete the cycle. For every molecule of oxygen
used in the cellular respiration of glucose, a molecule of carbon dioxide is liberated. The converse applies for photosynthesis. Each
oxygen molecule in the atmosphere today represents one carbon atom that
has so far escaped oxidation in fossil fuel deposits or the bodies of
organisms. However, burning of all available fossil fuels would lower
the concentration of oxygen in the atmosphere only slightly, because
most reduced carbon is too thinly distributed or too deep to serve as
fuel. Aquatic environments are far less stable, because their oxygen
content is less and may be depleted as an ultimate result of pollution.

42.3 The Nitrogen Cycle

In spite of the large reservoir of atmospheric nitrogen, nitrogen
is a limiting ingredient for living things since plants must obtain it

in fixed form. To break the nitrogen molecule apart requires tremendous amounts of energy as evidenced in the industrial production of ferti- lizer. This production accounts for approximately one-third of the nitrogen fixed on earth and can have significant effects upon ecosystems. Nitrogen fixation is accomplished by legumes containing nitrogen-fixing bacteria. Some other prokaryotic organisms can fix nitrogen. How these natural nitrogen-fixers overcome the high energy barriers is not under- stood, but they do require an enzyme and large amounts of ATP. At each trophic level, nitrogen is returned to the environment as excretions and deaths. The microorganisms of decay break organic nitrogen mole- cules down into ammonia that can then be absorbed directly by plants once more. Ammonia is also oxidized to absorbable nitrates in the pro- cess of nitrification. Denitrification, the reverse process, is also carried out by bacteria, thereby maintaining the atmospheric supply of nitrogen.

42.4 The Sulfur Cycle

Sulfur is found in almost all proteins and moves through the bio- sphere in two cycles. An inner cycle involves passing from soil (or water) to plants, to animals and back to soil (or water). Bacteria are crucial to the cycle, breaking down sulfur compounds and transforming them into gases which, after oxidation, return to the earth in rain- water. The "outer" sulfur cycle involves the movement of sulfur com- pounds to the oceans and their conversion to various gases, which are carried over land and returned to the soil in precipitation. Industrial activities have increased the amount of gaseous sulfur in the atmos- phere significantly. This results in acid rain, a definite threat to lakes in many regions.

42.5 The Phosphorus Cycle

Phosphorus plays a very small, yet essential, role in living things, being a constituent of such items as nucleic acids and ATP. Terrestrial ecosystem productivity can be increased by the addition of phosphorus compounds as witnessed by their inclusion in fertilizer. Life sulfur, phosphorus participates in both an inner and a global cycle. In the inner cycle decay makes phosphorus available to roots for reincorporation into organic matter. The global cycle involves a return to the land of sea-borne phosphorus through the excrement of sea birds and the uplifting of marine sediments. The mining of phos- phate rocks for fertilizer and water softeners contributes eventually to the increase of phosphorus in water, thereby producing algal blooms and accelerating eutrophication.

42.6 Other Mineral Requirements

Other minerals are essential to living organisms but in such small amounts that they exist in abundance, with perhaps iodine being a localized exception. Some elements are required in such minute quantities that they are referred to as trace elements and can easily be added to a system.

42.7 Analytical Techniques

Study of material cycles involves careful measurements of inputs, using rain gauges, and outputs, using calibrated settling basins at stream outflows. However, extrapolation of these data to a global scale does introduce enormous uncertainties.

42.8 Water and the Biosphere/42.9 The Properties of Water

Water is the single most abundant molecule in living things. Indeed, all of living chemistry is aqueous chemistry. Also, water is the habitat for many species of organisms. Water is essential to life and all the cycles connect with it. It is unsurpassed as a solvent and has the greatest heat capacity of any common substance. Also, it possesses a high heat of vaporization, thereby allowing it to act as an important temperature regulating device.

42.10 The Water Cycle

Approximately 97% of the earth's water is found in the oceans, with less than 1% found as fresh, liquid water, most of which is present as groundwater. Rainfall causes rivers to bring a significant quantity of salt to the oceans. Water evaporates from the oceans and from the land and from plants. This vapor condenses and falls on the land again. In some instances water passes down to the water table and resides in a zone of saturated rock and/or soil known as an aquifer.

42.11 Soil

Along with temperature and water, soil is a major determinant in the earth's productivity because it is here, through the roots, that most materials enter into food webs. In passing vertically down through the soil, several horizons are encountered: (1) partially decayed organic matter, (2) humus—decayed organic matter, (3) subsoil—inorganic nutrients, and (4) weathered parent material. Soil is formed by the disintegration of parent material, the formation and incorporation of humus, and the movement of dissolved minerals. Thus chemical, organismic, and water action contribute to soil formation. Rainfall influences soil productivity greatly. If more than 30 to 40 inches of rainfall occurs each year, minerals from the soil are carried by the water on its way to the water table, causing the soil to be acidic. About 20 inches of rainfall ensures that little or no water travels to the water table and minerals remain within reach of plant roots. With 10 inches of rainfall, the water remains near the surface and high evaporation rates cause the soil to become alkaline.

42.12 Prospects for Increasing the World's Cropland

Three means of increasing the world's cropland have been explored in an attempt to meet the needs of a growing population. The irrigation of desert soil is difficult because of salt buildup caused by high evaporation, the large amounts of water required per acre, and impervious layers of material beneath shallow soil. Terracing of hills and mountains is of limited success because of the impossibility of large scale use of machinery and the problem of erosion. A third possibility for new cultivated land is the jungle areas of Africa and

South America. The problem here is that high rainfall washes minerals out of the soil very quickly. The minerals that do exist are contained in the jungle plants. When the land is cleared, the soil quickly loses its fertility. The thin humus layer and high iron content of these lateritic soils only further aggravate the problem to the extent that coventional agricultural attempts meet with failure. The "slash and burn" technique of the Tsembaga still is the most successful.

B. KEY TERMS

decay	nitrogen fixation
nitrification	water table
denitrification	aquifer
greenhouse effect	humus
acid rain	subsoil
eutrophication	lateritic

II. MASTERING THE CHAPTER

A. LEARNING OBJECTIVES

When you have mastered the material in this chapter, you should be able to:

1. Define all key terms.

2. Trace the path of carbon through the carbon cycle. Discuss the effects of human activities on the amount of carbon dioxide in the air and its possible effect upon plants and climate.

3. Explain how oxygen is utilized in the processes of respiration and photosynthesis in such a way that the oxygen cycle is maintained. Discuss the effects of the burning of fossil fuels and the effects of organic pollution upon aquatic oxygen levels.

4. Discuss the nitrogen cycle, noting the tremendous amount of energy required in industrial nitrogen-fixing operations. Discuss other ways in which nitrogen is fixed. Explain the processes of nitrification and denitrification.

5. Discuss the sulfur cycle and the important role played by bacteria. Explain what effect industrial activities have had upon the cycle.

6. Explain the importance of phosphorus to living organisms. Discuss the inner and global cycles of phosphorus and how the global differs from other cycles. Note the effects of phosphate-rock mining upon water ecosystems.

416

7. List some of the other mineral requirements of living things and why their supply poses few problems.

8. List the properties of water and how they are important to living things.

9. Discuss the water cycle, paying particular attention to the aquifer zone and the water table.

10. Cite the importance of soil for terrestrial ecosystems. Describe the origin and composition of soil horizons. Explain how rainfall amounts affect the productivity of soil.

11. Discuss the prospects for increasing the world's cropland by examining the techniques that have been tried and their degrees of success.

B. REVIEWING TERMS AND CONCEPTS

1. Although a rare element in nature, (a) _____ accounts for 18% of living matter. The ability of its atoms to (b) _____ to each other is essential to life. At each trophic level, carbon returns to the atmosphere or water as the gas (c) _____ ____ _____. Carbon is unlocked from corpses and debris by the action of (d) _____ and _____.

2. Molecular oxygen comprises (a) _____% of the earth's atmosphere. In the process of (b) _____, oxygen serves as the final acceptor of electrons, producing (c) _____. In the process of (d)_____, light energy removes electrons from the (e) _____ atoms of water, reducing (f) _____ to carbohydrate and releasing molecular (g) _____.

3. In order for plants to manufacture protein, they must secure nitrogen in its (a) _____ forms. Nitrogen atoms unite with oxygen atoms in the air under the influence of (b) _____ and dissolve in rain, forming (c) _____. Agriculture relies heavily upon (d) _____-fixed nitrogen, to the extent that roughly (e) _____ of all fixed nitrogen is produced in this way. The process of decay breaks down organic nitrogen molecules into (f) _____. Most of the (g) _____ from decay is converted into nitrates by the process of nitrification.

4. Sulfur is a component of virtually all (a) _____ and moves through the biosphere in two different cycles. Under both aerobic and anaerobic conditions, (b) _____ play a crucial role in the sulfur cycle. The combustion of fossil fuels adds sulfur to the atmosphere in the form of (c) _____ _____, which results in (d) _____ rain.

5. ATP and nucleic acids depend upon the (a) _____ cycle, which actually consists of two different cycles. The global cycling of this mineral relies upon (b) _____ and (c) _____ of ocean sediments for its return to terrestrial ecosystems.

417

6. Some elements are needed for living in such minute quantities that they are referred to as (a) _____ elements. One example is molybdenum, which is needed for (b) _____ _____.

7. It is water's versatility as a (a) _____ that enables it to serve as a transport medium. Also, it has the greatest (b) _____ capacity of any common substance, allowing it to act as a (c) _____ against temperature fluctuations. Many (d) _____ maintain a stable body temperature by relying upon its high heat of (e) _____.

8. Approximately (a) _____% of the earth's water is found in the oceans and less than (b) _____% exists as fresh, liquid water. The regular cycling of water from ocean, to (c) _____ _____, to land, and back to the ocean again can be interrupted by the movement of water down to water-bearing rock or (d) _____.

9. Soil can be divided into layers called (a) _____, with the layer of decayed organic matter referred to as the (b) _____ layer; the soil below that is (c) _____, derived from the chemical breakdown of the (d) _____ material. Soils high in aluminum and iron are called (e) _____.

10. Three means of increasing the world's cropland have been explored in an attempt to meet the needs of a growing population. The (a) _____ of desert soil is difficult because of (b) _____ buildup caused by high evaporation, the large amounts of water required per acre, and (c) _____ layers of material beneath shallow soil. (d) _____ of hills and mountains is of limited success because of the impossibility of large scale use of machinery and the problem of erosion. A third possibility for new cultivated land is the (e) _____ areas of Africa and South America. The problem here is that high rainfall washes (f) _____ out of the soil very quickly. When the land is cleared, the soil quickly loses its (g) _____. The (h) _____ technique of the Tsembaga still is the most successful.

C. TESTING TERMS AND CONCEPTS

1. Which of the following processes does not make nitrogen available to plants?
   a) lightning
   b) decay
   c) denitrification
   d) nitrification

2. Increased amounts of sulfur dioxide in the air
   a) increase the pH of normally acid rainwater.
   b) produce "acid rains."
   c) increase photosynthesis.
   d) are buffered by the granite underlying such areas, as the N.Y. Adirondacks.

3. In photosynthesis, oxygen
   a) accepts electrons from carbon compounds.
   b) liberates energy from sulfur.
   c) is released from water molecules.
   d) combines with hydrogen to form water.

4. The global cycling of phosphorus differs from that of nitrogen in that phosphorus
   a) forms no volatile compounds.
   b) produces an unstable gaseous compound.
   c) has no mechanism by which it returns to terrestrial ecosystems.
   d) is found in the excrement of marine birds.

5. Much of the organic matter at each trophic level
   a) passes to decomposers.
   b) is utilized by a lower trophic level.
   c) is broken back down to its component elements.
   d) is leached downward through the soil toward the underlying aquifer.

6. The concentration of atmospheric carbon dioxide has increased because of the
   a) burning of coal and oil.
   b) release of carbon dioxide from ocean deposits of calcium carbonate.
   c) increased photosynthesis resulting from agriculture.
   d) human population explosion and this population's combined breathing rate.

7. Bacteria are crucial to the sulfur cycle because they
   a) break down sulfur-containing compounds into gases.
   b) make elemental sulfur available to plant roots.
   c) aid in the return of sulfur to the terrestrial cycle.
   d) produce sulfates.

8. The pool of oxygen in the atmosphere is accounted for by the
   a) fossil fuel deposits and the world-wide biomass.
   b) process of respiration, which splits water.
   c) liberation of oxygen from calcium carbonate.
   d) action of bacteria and other microorganisms.

9. Which of the following does not play an important part in the fixation of nitrogen?
   a) lightning
   b) industrial processes
   c) chemical decomposition of rocks and soil
   d) bacterial metabolism

10. The organisms of decay
    a) break ammonia into atmospheric nitrogen.
    b) break organic nitrogen molecules down to ammonia.
    c) reduce nitrates to nitrogen.
    d) fix nitrogen in leguminous roots.

11. Of the following, the element that is likely to be so scarce in the soil as to endanger human health is
    a) molybdenum.
    b) calcium.
    c) iodine.
    d) sodium.

12. The average temperature of the atmosphere may be increasing because of
    a) increased carbon dioxide from human activity.
    b) increased smoke and soot of human activity.
    c) increased acreage of cleared land and artificially created reservoirs.
    d) increased photosynthesis.

13. A zone of saturated soil or rock is known as
    a) surface water.
    b) fossil water.
    c) the water table.
    d) an aquifer.

14. Plants use more water than animals do because
    a) plant bodies contain more water than animal bodies.
    b) the chemical reactions of plant metabolism are different and require more water than those of animal metabolism.
    c) Plants are unable to recycle water within their bodies as animals can.
    d) Plants, with extensive root systems, have greater access to water than animals have.

15. The high heat capacity of water does not
    a) arise from the polarity and hydrogen bonding of its molecules.
    b) protect individual plant and animals from sudden temperature fluctuations.
    c) explain the effectiveness of sweating and panting as temperature control mechanisms.
    d) delay the onset of spring in coastal regions.

D. UNDERSTANDING AND APPLYING TERMS AND CONCEPTS

1. List the three major abiotic reservoirs of carbon in the biosphere and explain how carbon moves from each into a food chain.

Reservoir	Transfer to Food Chain
a) _____	_____
b) _____	_____
c) _____	_____

2. Enormous quantities of nitrogen fertilizers are used in agriculture today. We are in danger of depleting the world's supply of nitrates?

Explain. _____
_____
_____
_____

3. Describe the steps that must occur between the death of a plant and the incorporation of these elements into the bodies of other plants.

   a) carbon _____
   _____

   b) nitrogen _____
   _____

   c) sulfur _____
   _____

4. Ultimately, it is solar energy that drives the various cycles of matter. Explain how this occurs for the following cycles.

   a) carbon cycle _____
   _____

   b) oxygen cycle _____
   _____

   c) sulfur cycle _____
   _____

   d) water cycle _____
   _____

5. Insectiverous plants, like the Venus's-fly-trap, can carry out photosynthesis as well as any other plant. Why do they digest insects as well? _____
   _____
   _____

6. Hydrogen bonding is responsible for both the heat capacity of water and its high heat of vaporization. Explain this relationship.
   _____
   _____
   _____

7. A moderate amount of household sewage sprayed onto a terrestrial ecosystem will result in a long-term increase in productivity. On the same sized pond ecosystem, it will result in the death of the community. Why? _____
   _____
   _____

ANSWERS TO CHAPTER EXERCISES

Reviewing Terms and Concepts

1. a) carbon b) bond c) carbon dioxide d) bacteria, fungi

2. a) 20 b) respiration c) water d) photosynthesis e) oxygen

f) carbon g) oxygen

3. a) fixed b) lightning c) nitrates d) industrially e) one-third
    f) ammonia g) ammonia

4. a) proteins b) bacteria c) sulfur dioxide d) acid

5. a) phosphorus b) sea birds c) uplifting

6. a) trace b) nitrogen fixation

7. a) solvent b) heat c) buffer d) animals e) vaporization

8. a) 97 b) 1 c) water vapor d) aquifers

9. a) horizons b) humus c) subsoil d) parent e) lateritic

10. a) irrigation b) salt c) impervious d) Terracing e) jungle
    f) minerals g) fertility h) slash-and-burn

Testing Terms and Concepts

1.	c	6.	a	11.	c
2.	b	7.	d	12.	a
3.	c	8.	a	13.	d
4.	a	9.	c	14.	c
5.	a	10.	b	15.	c

Understanding and Applying Terms and Concepts

1. a) atmospheric carbon dioxide; absorption by plants during photo-
    synthesis b) fossil fuels; liberation of carbon dioxide during
    burning and its absorption during photosynthesis c) carbonate
    rocks; liberation of carbon dioxide during weathering

2. On the contrary, industrial fixation of nitrogen and the cultiva-
    tion of vast areas of legumes are significantly increasing the
    world's supply of nitrates.

3. a) respiration by decay organisms and liberation of carbon dioxide,
    absorption of carbon dioxide by plants and incorporation during
    photosynthesis b) metabolism by decay organisms and liberation of
    ammonia, which can be taken up by plants c) metabolism by decay
    organisms and liberation of sulfates, which can be absorbed by
    plants

4. a) Sunlight powers photosynthesis, which involves the absorption
    of carbon dioxide. b) Sunlight powers photosynthesis, which re-
    leases oxygen for heterotrophs. c) Sunlight powers photosynthesis,
    which powers plant absorption of sulfates. d) Sunlight evaporates
    water and so returns it to terrestrial ecosystems as precipitation.

5. The insects are consumed, not for their energy content, but for
    their mineral content. These plants grow in soils that are defi-

cient in nitrates and other nutrients.

6. Both increasing the temperature of a body of water and changing it to the gaseous state involves increasing intermolecular movement. Hydrogen bonding interferes with this movement, so large amounts of heat are required.

7. In both communities, these nutrients stimulate aerobic metabolism. In a terrestrial ecosystem, oxygen is plentiful, but in the pond, it is not, and virtually all available oxygen can be used up. This kills the aerobic organisms.

THE GROWTH OF POPULATIONS

I. REVIEWING THE CHAPTER

A. CHAPTER HIGHLIGHTS

43.1 The Human Population

The human population is increasing very rapidly at the present time. More importantly, this increase has been accelerating, i.e. the growth curve has become increasingly steeper. At the present growth rate, the earth's population will double in 38 years to 8 billion people. How many people the earth can support remains an unresolved question owing to a great number of unknown factors.

43.2 Principles of Population Growth

Growth rate is expressed as the decimal equivalent of the increase in population per 1000 individuals in one year. It is computed by subtracting the death rate ($\underline{d}$) from the birth rate ($\underline{b}$) to obtain the rate of natural increase ($\underline{r}$). Migration also influences population growth. Thir $\underline{r}$ factor can then be used to determine future growth by applying the equation: $\underline{N} = \underline{N}_0 e^{rt}$, where $\underline{N}_0$ is the starting population, $\underline{N}$ is the population after a certain time ($\underline{t}$) has elapsed, and $\underline{e}$ is the constant 2.718. The plot of the resulting curve of this equation is an exponential growth curve. Populations would (and sometimes do) exhibit this growth were it not for various controls.

43.3 Density-independent Checks on Population Growth

Density-independent controls are those that exert their effects irrespective of the size of the population at the time. They are usually abiotic mechanisms (onset of winter, forest fire), and they often cause drastic decreases in numbers.

The value of $\underline{r}$ is also controlled by the birth rate. One means of limiting the reproductive rate of a group is to limit food and other essentials. Some species stabilize their populations by producing fewer offspring even when sufficient food is available. This may be accomplished by inhibitory chemicals, higher infant mortality, and even cannibalism, but the adult density seems to be the primary factor. Another way to limit offspring is to limit the number of parents. Among some mammals and birds, this is accomplished by the necessity of a sufficiently large breeding territory for each mated pair. Humans can accomplish this by tradition, regulation, celibacy, or birth control devices.

43.4 Density-dependent Checks on Population Growth

Density-dependent controls exert their effects with an intensity that grows with the size of the population. They are usually biotic mechanisms (intraspecific or interspecific competition for food or space, predation, parasitism), and they usually cause a leveling off of growth rather than a population crash.

The environment can act as a limiting factor upon population growth. The uppermost limit of a population that can be supported is referred to as the carrying capacity ($\underline{K}$) of the environment. A population far below $\underline{K}$ displays exponential growth until it starts to meet increasing environmental resistance, a relationship that can be expressed as

$$\frac{K-N}{N} \quad .$$

This expression implies that as $\underline{N}$ approaches $\underline{K}$, the exponential growth will slow until zero population growth (ZPG) results. The plot of this curve is called the logistic growth curve. Many factors can contribute to $\underline{K}$. Food supply, interspecific competition, the $\underline{K}$ of predators and parasites, and habitat destruction are some of these factors. In eliminating pests or preventing extinction, the $\underline{K}$ factors of species under scrutiny are of paramount importance. Similar attention should be paid to the harvesting of animals, such that their numbers do not drop below the maximum sustainable yield.

43.5 $\underline{r}$ Strategists and $\underline{K}$ Strategists

Species that rely on rapid reproduction for their evolutionary success are referred to as $\underline{r}$ strategists. In general, they: (1) are found in disturbed and/or transitory habitats, (2) have short life spans, (3) have short generation times, (4) have large numbers of offspring, (5) display high infant mortality, and (6) are controlled by density-independent factors. If a habitat becomes filled with a diverse population, then the competition favors the more efficient $\underline{K}$ strategists. These have populations close to the value of $\underline{K}$ with nothing to gain from a high $\underline{r}$. $\underline{K}$ strategists generally: (1) are found in stable habitats, (2) have long lifespans, (3) have long generation times, (4) have few offspring, (5) display low infant mortality, (6) are closely adapted to their ecological niche, and (7) are controlled by density-dependent factors.

B. KEY TERMS

exponential growth curve	predation
carrying capacity	parasitism
logistic growth curve	ecological niche
maximum sustained yield	reproductive competition

intraspecific competition   migration

interspecific competition

II. MASTERING THE CHAPTER

A. LEARNING OBJECTIVES

When you have mastered the material in this chapter, you should be able to:

1. Define all key terms.

2. Discuss the past and present growth of world population, paying particular attention to the importance of incremental gains and the present doubling time.

3. Explain how the rate of natural increase is calculated. Utilize the formula for determining the size of a population after an interval of time has elapsed. Explain what is meant by an exponential growth curve.

4. Distinguish between density-independent and density-dependent population controls and give examples of each.

5. Explain the different ways in which the birth rate can be lowered by various species, paying particular attention to humans.

6. Discuss the role of migration in affecting population size and when this is likely to occur.

7. Explain what is meant by the carrying capacity (K) of an environment and how it can be used mathematically to derive the logistic growth curve. Give examples of factors that might contribute to K. Suggest the correct approach to pest control, preventing extinction, and animal harvesting in light of these factors.

8. Compare r strategists and K strategists, giving the general characteristics of both. Contrast their evolutionary strategies.

B. REVIEWING TERMS AND CONCEPTS

1. The rate of natural increase is denoted by (a) _____ and is determined by finding the difference between the (b) _____ rate and the _____ rate. Another factor which can influence this figure is (c) _____. With no limiting factors, the growth curve of a population will be (d) _____.

2. Density- (a) _____ controls are usually abiotic, and they often result in a population (b) _____. Density- (c) _____ controls are biotic and usually result in (d) _____ _____ growth.

3. Fruit flies lay fewer eggs under crowded conditions, resulting in a form of (a) _____ control, which, like some other forms, is triggered by adult (b) _____ . Lemming migration is an example of population reduction by (c) _____ .

4. The maximum number of individuals that can be supported by an environment is referred to as the (a) _____ and is designated by the letter (b) _____ . When $N = K$ then the growth curve (c) _____ _____ and the (d) _____ _____ is zero. If a growth curve is plotted for $(K - N)/N$ values of one to zero, then a (e) _____ growth curve results. As the population density of a species increases, the K of the predators (f) _____ . When harvesting organisms, the wisest course is not to exceed the (g) _____ _____ yield.

5. r strategists are found in (a) _____ or _____ habitats. They have short (b) _____ _____ and (c) _____ _____ , and large families to compensate for high (d) _____ mortality. K strategists have stable populations whose numbers are very close to the value (e) _____ . They are as closely adapted to their habitat or (f) _____ _____ as possible.

C. TESTING TERMS AND CONCEPTS

1. Which of the following is not a form of birth control?
   a) need for breeding territories
   b) celibacy
   c) dominance or peck-order
   d) infanticide

2. Which of the following is least likely to be a density-dependent control?
   a) spraying an insecticide on a garden pest
   b) plague among humans
   c) predation on songbirds
   d) lack of water among desert plants

3. Which of the following is not characteristic of an r strategist?
   a) high infant mortality
   b) short lifespans
   c) high birthrate
   d) relatively stable population size

4. An exponential growth curve will level off
   a) if predation lessens.
   b) in a closed environment.
   c) if the death rate decreases.
   d) if faced with a density-independent control mechanism.

5. If the values for K and N are equal, then the population is
   a) at maximum.
   b) growing at its maximum rate.

c) about to display exponential growth.
d) declining rapidly.

6. If a population of 10,000 experienced 120 births and 80 deaths in one year, the r would be
    a) 4.0
    b) .40
    c) .04
    d) .96

7. Without controls, a population would
    a) display an exponential growth curve.
    b) display an arithmetic growth rate.
    c) display sigmoid growth.
    d) display cyclic growth.

8. K strategists
    a) are closely adapted to their environmental niche.
    b) exploit with ever-increasing efficiency an ever-broader slice of the environment.
    c) are found in transitory environments.
    d) are well adapted to counter the effects of high infant mortality.

9. The Grand Canyon deer population decreased because of
    a) its own previous increase.
    b) increased predation.
    c) increased hunting by humans.
    d) a climate change in the area.

10. If a species, such as the whooping crane, is facing extinction, probably the most significant effect upon its numbers and future would be gained by
    a) eliminating its predators.
    b) increasing its habitat.
    c) rearing it in zoos.
    d) providing supplemental food.

11. If an island's inhabitants experienced 80 births, 30 deaths, and 50 immigrants per 1000 population, the r for that year would be
    a) 0.0
    b) .05
    c) .06
    d) .10

12. Humans should be considered
    a) r strategists, because they can colonize almost any habitat successfully.
    b) r strategists, because they have a very high birth rate.
    c) K strategists, because all ages are at risk of being killed by various environmental hazards and infant mortality is relatively high.
    d) K strategists, because of their long gestation time and long period of childhood.

13. The huge increase in numbers of humans over the last 200 years can best be explained by the
    a) increased carrying capacity.
    b) increased birth rate.
    c) increased rate of natural increase.
    d) increased migration.

14. Populations tend to increase
    a) rapidly.
    b) in a simple-interest fashion.
    c) exponentially.
    d) linearly.

15. The world's population will probably
    a) double within 38 years.
    b) reach 4 billion in 38 years.
    c) level off by the year 2000.
    d) reach 10 billion by the year 2000.

D. UNDERSTANDING AND APPLYING TERMS AND CONCEPTS

1. In natural communities, which is the more powerful regulator of population size, intraspecific or interspecific competition? (a) _____ Why? (b) _____
    _____
    _____

2. Explain why each of the following regulates population growth in a density-dependent manner.

    a) cannibalism _____
    _____

    b) migration _____
    _____

    c) parasitism _____
    _____

    d) war _____
    _____

3. Consider these population growth curves

population
  size

a) Which curve describes the early growth of a population introduced into a vast, previously unexploited habitat? _____

b) Which curve is most likely to show a sharp decrease in population size in the near future? _____

c) Which curve shows the early growth of a particularly slowly maturing species, where the offspring do no reproducing in the time period being considered? _____

d) Which curve describes a population regulated by density-dependent controls? _____

e) Which curve describes a population regulated by no controls at all? _____

f) Which curve best describes the pattern of growth you would expect to see in an r strategist? _____

g) Which curve best describes the pattern of growth you would expect to see in a K strategist? _____

h) Which curve best describes the growth of the human population over the last thousand years? _____

4. If industrially caused cancer resulted in everyone dying at age 60 instead of 70, what effect could we expect to see in the current exponential growth curve? _____
    _____
    _____

5. If some density-dependent mechanism gradually controls human population growth, what kind of a growth curve will we see? (a) _____ Ideally, what density-dependent mechanism will it be? (b) _____

6. If the human population continues to grow exponentially, what will likely happen to the growth curve? (a) _____
    Why? (b) _____
    _____
    _____

ANSWERS TO CHAPTER EXERCISES

Reviewing Terms and Concepts

1. a) r b) birth c) death, migration d) exponential

2. a) independent b) crash c) dependent d) sigmoid

3. a) birth b) density c) emigration

4. a) carrying capacity b) K c) levels off d) growth rate
    e) logistic f) increases g) maximum sustainable

5. a) disturbed, transitory b) life spans c) generation times
    d) infant e) K f) ecological niche

Testing Terms and Concepts

1.	d	6.	c	11.	d
2.	a	7.	a	12.	d
3.	d	8.	a	13.	a
4.	b	9.	a	14.	c
5.	a	10.	b	15.	a

Understanding and Applying Terms and Concepts

1. a) intraspecific competition b) Members of the same species are more similar to each other and therefore compete with each other for more resources than members of different species. Furthermore, interspecific competition leads to evolutionary divergence and a gradual lessening of the competition.

2. a) Mothers eat their young more, as the population rises. b) More young are forced to leave an area, as the area becomes more crowded. c) Disease organisms are transmitted faster through crowded populations. d) Wars have often been started to permit expansion of a crowded state or to make economic opportunities for a growing population.

3. a) B b) B c) A d) D e) B f) B g) D h) B

4. This would have almost no effect, since it does not change the production of young.

5. a) sigmoid b) increased education and voluntary birth control

6. a) It will crash. b) The population will reach its carrying capacity, exceed it, and therefore be hit severely by some environmental factor, such as competition for food or nuclear war.

44

INTERACTIONS BETWEEN SPECIES

I. REVIEWING THE CHAPTER

A. CHAPTER HIGHLIGHTS

44.1 Introduction

Every organism is affected by the life of other organisms. Thus,
every change in the biotic environment has both direct and indirect
effects. Most interactions between species concern food, either through
predation or competition for another food source, and are of short
duration. Many other interactions are actually long-term associations
and are referred to as symbioses.

44.2 Predation

Predation is a major check on population size. Most heterotrophs
are predators. Because food plays a central part in the lives of
animals, many adaptations are geared to either increasing predator
efficiency or minimizing the risk of being another's prey. Those devices that minimize being preyed upon are: (1) camouflage—organism
blends in with its surroundings; (2) defense—structures and/or chemicals for repelling predators, often linked with protective coloration;
(3) mimicry, looking like another noxious species, can take two forms—
Batesian, in which a harmless species resembles a harmful one, and
Müllerian, in which an unpalatable species resembles another unpalatable one; (4) group behavior—cooperation in some form; (5) escape
responses—lead to a balance between predator and prey, neither growing
out of control nor going extinct.

44.3 Parasitism

Parasites are organisms that derive nourishment from the tissues
of another organism, doing their host some degree of harm. The
distinction between parasitism and predation is not always clear-cut,
but generally it is decided on the bases of the parasite/host size
relationship and the immediate degree of damage done. All organisms
on the earth are probably susceptible to parasitism at some time in
their existence. Damage to hosts can occur in two major ways: as a
result of tissue consumption, or by toxin production. In addition,
some parasites compete with their host for ingested food or essential
substances. In many cases, both host and parasite are mutually adapted
such that the parasite rarely kills its host. Many parasites have
lost nonuseful structures and functions and, in doing so, have become
more efficient, yet some have become incapable of living without their
hosts. This type of loss has been referred to as degeneration and may

be represented in the extreme by viruses since they lack most of the enzyme systems necessary for free life. Mutual evolutionary adaptations of host and parasite may lead to a situation in which less damage results to a more resistant host, as in the case of Australian rabbits infected with myxoma. In some instances evolutionary adaptation may produce a situation in which the host is not harmed at all under usual circumstances or even a situation in which both organisms benefit.

44.4 Commensalism

Commensalism is a symbiotic relationship in which one organism consumes the unused food of another. No harm results from such a relationship and only one party benefits. Examples of this relationship are the remora and shark and some of the intestinal bacteria and human associations.

44.5 Mutualism

A symbiotic relationship in which both partners benefit is called mutualism. Such relations between plants and fungi are very common. The fungus lives in secondary roots, an association referred to as a mycorrhiza. The plant's ability to absorb minerals from the soil is enhanced and the fungus derives nourishment from the sugar in the roots. Mutualistic relationships often involve adaptations of structure, function, and even behavior on the part of two species, as witnessed by the fungus-growing ants. One of the most important examples of mutualism, in terms of the biosphere's economy, is the relationship between the soil bacteria Rhizobium and legumes. This partnership accounts for a substantial amount of the nitrogen fixation that occurs on earth. Responding to a substance secreted by legumes, rhizobia invade the root cells through plant structures designed for that purpose and in response to bacterial presence. Ultimately, abnormal cell division on the part of the plant forms a nodule and the bacteria change shape and lose their motility, becoming bacteroids. Only at this point and in this association can the legume and rhizobia begin to fix nitrogen. The legume contributes energy and hemoglobin to the process while the bacteria possess the necessary metabolic machinery for fixation. Rhizobia are specific in their infection ability. Some strains will only infect certain species of legumes, a trait which is taken advantage of in agriculture by producing various strains of rhizobia and then infecting the appropriate crops. Genetic engineering has attempted to insert the genes responsible for nitrogen fixation into other types of bacteria (E. coli for research purposes). It would be extraordinarily beneficial to humans if such genes could be successfully introduced into their major food crops.

44.6 Interspecific Competition

The ecosystem in which an organism lives is its habitat; a more specific description of where it lives is its microhabitat. An organism's ecological niche is all its ecological requirements taken together. If two species both depend upon the same resource in an environment, then interspecific competition results. Usually such competition centers around food. Rarely is this competition of such a degree that the two species are competing for the same niche; therefore, the number of niches occupied by heterotrophs in any region is approximately equal to the number of species there. Where competition for a food supply exists, the K, or carrying capacity, for the various species is reduced, thereby keeping the populations in check. Often, when there is intense competition, character displacement occurs, reducing the niche overlap and reducing competition. This increased efficiency in feeding implies increased specialization and a progressive narrowing of a species niche. Competition for sunlight, water, and minerals is found among plant species and has led to the many and varied adaptations found within the plant kingdom.

44.7 How Many Species Can Coexist in One Area?

There appears to be a limit to the number of species that can coexist in a single area, a limit to the extent to which a habitat can be subdivided into niches. The climate of a region can restrict the number of species that can live there—regions of low productivity are less complex than regions of high productivity. Areas with a diverse topography can support more species than simple areas, and large regions have more species than small regions. Other things being equal, an area ten times larger than another will have twice as many species.

MacArthur and Wilson suggested that the number of species on an island (or other isolated habitat—pond, mountaintop) reflects the balance between the extinction of species on the island and the immigration of replacement species from the mainland. The more species on an island, the greater the extinction rate and the lower the immigration rate, because of interspecific competition.

A large island will support more species than a small one, because carrying capacities are large and populations are less susceptible to extinction. Islands near a mainland will have more species than distant island, because immigration is more often successful.

B. KEY TERMS

symbiosis	parasitism
commensalism	Batesian mimicry
mutualism	mycorrhiza
ecological niche	Mullerian mimicry
competitive exclusion	aggressive mimicry
cryptic coloration	group behavior
character displacement	escape response
camouflage	symbiotic nitrogen fixation
defense	habitat

II. MASTERING THE CHAPTER

A. LEARNING OBJECTIVES

When you have mastered the material in this chapter, you should be able to:

1. Define all key terms.

2. Describe how organisms are affected by other organisms both directly and indirectly.

3. Explain how predation has led to various adaptations. Give examples of the ways in which animals have adapted to minimize their role as prey.

4. Discuss parasitism, noting its universality and the major ways in which a host can be damaged. Explain why parasites are often referred to as degenerate. Discuss the importance of mutual evolutionary adaptations and the pathways which that process can take.

5. Define commensalism and give examples of it.

6. Discuss mutualism, noting the many mutualistic relations between plants and fungi, and give examples. Explain the importance and method by which nitrogen fixation occurs in legumes. Cite the interest of genetic engineering in some bacteria's ability to fix nitrogen.

7. Discuss interspecific competition in terms of niche overlap. Explain competitive exclusion and character displacement.

8. List three factors that can limit the number of species coexisting in an area.

9. Explain how island size and degree of isolation affect species diversity.

B. REVIEWING TERMS AND CONCEPTS

1. Most interactions between organisms concern (a) _____, either through (b) _____ or competition, and are of short duration. Other interactions are much longer lasting and are called (c) _____.

2. Cryptic coloration or (a) _____ is a means of avoiding (b) _____ by other species. The quills of a porcupine are a means of (c) _____. Some weapons, such as poison, may be accompanied by a conspicuous warning (d) _____ of the defender's body. The viceroy imitates the noxious monarch butterfly, thereby practicing a form of (e) _____ known as (f) _____ _____. Cooperation among members of the same species can also reduce the severity of (g) _____.

435

3. An organism that derives its nourishment slowly from the tissues of a (a) _____ organism at the expense of the latter is called a (b) _____. It can cause damage by actually consuming (c) _____ or by releasing (d) _____. Because some parasites have lost structures essential to free living, they are referred to as (e) _____; this state has arisen because of their extreme (f) _____. The ultimate form of degeneration may be represented by the (g) _____.

4. When one organism consumes the unused food of another, the relationship is referred to as (a) _____. Examples of this are the (b) _____ and the remora, as well as some of the bacteria in the human (c) _____ _____.

5. When both species benefit from a relationship, it is called (a) _____. Such relations between plant roots and (b) _____ _____ are very common. The association is called a (c) _____ _____. Soil bacteria of the genus (d) _____ invade roots of (e) _____ and cause swellings or (f) _____ to form. The bacteria then lose their motility in becoming (g) _____ and at this point (h) _____ _____ begins. The plant contributes (i) _____ and _____ to the relationship. Some biologists have attempted to implant the genes responsible for nitrogen fixation into other bacteria in an activity that has been labeled (j) _____ _____.

6. The ecosystem in which an organism lives is its (a) _____; a more specific description of where it lives is its (b) _____. An (c) _____ _____ is all its ecological requirements taken together. When two species compete for food it is called (d) _____ competition. If two species occupy exactly the same (e) _____, the less well-adapted species is eliminated by a process called (f) _____ _____. A reduction in the degree of niche overlap results when (g) _____ _____ _____ occurs.

7. A harsh climate leads to a low ecosystem productivity, a (a) _____ number of individual organisms, and a (b) _____ number of different species. A uniform topography supports a (c) _____ number of species, and the size of a region is correlated with a species diversity as well. Other things being equal, an area ten times larger than another will have (d) _____ times as many species.

8. The number of species on an island reflects a balance between the (a) _____ of resident species and the (b) _____ of replacement species. A large island will have (c) _____ species than a small one, because populations will tend to be (d) _____ and less susceptible to (e) _____. Islands near the mainland will also tend to have more species than distant ones, because (f) _____ is more likely to be successful.

C. TESTING TERMS AND CONCEPTS

1. Commensalism is an interaction
   a) in which both organisms benefit.
   b) that is likely to lead to character divergence.
   c) that is relatively brief and therefore unimportant to at least one participant.
   d) that will cause evolutionary change in one participant but not in both.

2. The distinction is not always clear between parasitism and
   a) predation.
   b) commensalism.
   c) mutualism.
   d) competition.

3. If character displacement has occurred, it implies that
   a) parasitism preceded the displacement.
   b) intraspecific competition has been occurring.
   c) one species must have given rise to the present two in response to competition.
   d) niche overlap was great.

4. Legumes do not provide rhizobia with
   a) hemoglobin.
   b) nutrients.
   c) nitrogenase.
   d) oxygen.

5. Which of the following does not occur after a legume's roots are infected by a rhizobia?
   a) A specialized tissue leads the bacteria into the root.
   b) The bacterial cell becomes polyploid.
   c) Nodules begin to form.
   d) The bacteria lose their motility.

6. When a less well-adapted species is replaced by interspecific competition
   a) intraspecific competition can be expected to intensify.
   b) competitive exclusion has occurred.
   c) predation has been extreme.
   d) mutualism results.

7. A harmless insect whose coloration resembles that of a harmful one is an example of
   a) Batesian mimicry.
   b) Müllerian mimicry.
   c) cryptic coloration.
   d) aposematic coloration.

8. A predator species that occupies a narrow niche
   a) can easily survive the extinction of a prey species.
   b) often causes the extinction of a prey species.
   c) closely follows the rise and fall (in numbers) of its prey species.
   d) often relies upon mimicry or aposematic coloration.

9. Which of the following is <u>not</u> a manner in which parasites can affect their hosts?
   a) consume host tissues
   b) consume host food
   c) produce vitamins for host
   d) release toxins

10. The association of a fungus and a root is called a
    a) mycorrhiza.
    b) rhizobia.
    c) bacteroid.
    d) nodule.

11. If the niches of two species are identical, then often
    a) one will prey on the other.
    b) competitive exclusion will occur.
    c) character displacement has occurred.
    d) they have coexisted for a long time.

12. Niche overlap tends to
    a) increase K for each species.
    b) increase $\bar{K}$ for the total environment.
    c) reduce K for each species.
    d) reduce $\bar{K}$ for the total environment.

13. Some parasites are said to be degenerate because of their
    a) loss of structures essential to living freely.
    b) negative effect on their hosts.
    c) method of obtaining nutrients.
    d) primative evolutionary position.

14. Parasites often
    a) have lost useful structure through evolutionary adaptation.
    b) lack some enzyme systems.
    c) place a low emphasis on reproduction.
    d) eventually kill their hosts.

15. If a few individuals of a new species were artificially introduced to an island, previously at species equilibrium, one would <u>not</u> expect to see
    a) greater competition for resources and a lower overall carrying capacity on the island.
    b) greater competition and consequent narrowing of niches of at least some species.
    c) extinction of new species <u>or</u> of one or more resident species.
    d) decreased incidence of natural immigration to the island.

D. UNDERSTANDING AND APPLYING TERMS AND CONCEPTS

1. Herbivores rarely eat so much of their food plants that they kill them. Is a cow an example of a predator or a parasite? (a)_____
   _____ Why? (b)_____
   _____

438

2. Label the following interactions using this key:

> A. predation
> B. parasitism
> C. commensalism
> D. mutualism
> E. interspecific competition
> F. other

____ a) A honeybee pollenates some flowers while collecting nectar.

____ b) a farmer milks a cow.

____ c) A pile of honeysuckle vine leans against an oak.

____ d) Tent caterpillars strip the leaves from a maple tree.

____ e) A pine seedling struggles to grow in the shade of taller oaks.

____ f) A mistletoe seed germinates on an oak branch and sinks its roots into the oak's sap flow.

____ g) An insect tries to mate with flowers that give off the insect's sex pheromone and so pollenates them.

____ h) A young robin is chased from an area by older robbins and is therefore unable to nest.

____ i) A person throws a rotten apply into the trash.

____ j) A person pets a dog.

____ k) A cockroach eats the crumbs under someone's toaster.

____ l) A person eats a piece of toast.

3. A wolf pack tends to take what sort of moose as food? (a)_____
Why? (b)_____

4. Cryptic coloration, aposematic coloration, Batesian mimicry, and Müllerian mimicry all share a common function that differs from that of aggressive mimicry. What is it? _____

5. Are parasites best adapted to their hosts when they eventually kill the host, only moderately damage the host, or when they have little noticeable effect on the host? (a)_____
Why? (b)_____

6. For each of the following pairs of terms, describe a <u>similarity</u> and a <u>difference</u> between them.

   commensalism -- mutualism

   a) smiliarity _____
   b) difference _____

   predation -- parasitism

   c) similarity _____
   d) difference _____

   aposematic coloration -- Müllerian mimicry

   e) similarity _____
   f) difference _____

   parasitism -- interspecific competition

   g) similarity _____
   h) difference _____

   cryptic coloration -- Batesian mimicry

   i) similarity _____
   j) difference _____

   habitat -- niche

   k) similarity _____
   l) difference _____

   competitive exclusion -- character displacement

   m) similarity _____
   n) difference _____

7. On the mainland, three seed-eating birds compete for the available seeds, one specializing on large, the second on medium, and the third on small seeds. Only one of these species has colonized a distant island. What can you hypothesize about that one species' niche? (a)_____

8. Each of the factors listed down the left has what effect on the immigration of new species to an island and the extinction of species on the island? Place a "+" for an increase, a "-" for a decrease, and a "o" for no effect.

	Immigration	Extinction
		b)
Large size	a)	

440

Great distance from mainland	c)	d)
Harsh climate	e)	f)
Diversity of topography	g)	h)

ANSWERS TO CHAPTER EXERCISES

1. a) food b) predation c) symbioses

2. a) camouflage b) predation c) defense d) coloration e) mimicry
    f) Batesian mimicry g) predation

3. a) host b) parasite c) tissues d) toxins e) degenerate
    f) specialization g) viruses

4. a) commensalism b) shark c) large intestine

5. a) mutualism b) fungi c) mycorrhiza d) Rhizobium e) legumes
    f) nodules g) bacteroids h) nitrogen fixation
    i) nutrients, hemoglobin j) genetic engineering

6. a) habitat b) microhabitat c) ecological niche d) interspecific
    e) niche f) competitive exclusion g) character displacement

7. a) low b) low c) lower d) two

8. a) extinction b) immigration c) more d) large e) extinction
    f) immigration

Testing Terms and Concepts

1.	d	6.	b	11.	b
2.	a	7.	a	12.	c
3.	d	8.	c	13.	a
4.	c	9.	c	14.	b
5.	b	10.	a	15.	a

Understanding and Applying Terms and Concepts

1. a) predator b) A parasite not only avoids killing its host, it
    participates in a relatively intimate and long lasting relation-
    ship with its host. This the cow does not do.

2. a) D b) D c) C d) A e) E f) B g) B h) F (intraspecific
    competition) i) E j) D k) C l) A

3. a) the young, old, or ill b) These tend to lag behind the rest
    of the herd; they are easier to catch.

4. They are all defenses against predation, while aggressive mimicry
    is one means of predation.

5. a) when they have little effect b) If a parasite can make its
    living without hurting the host it has the best chance of living a
    long life itself and producing many offspring. Destroying the
    host is like destroying one's home and livelihood.

6. a) In these interactions, at least one species benefits and neither
    species is hurt. b) In commensalism, only one species benefits;
    in mutualism, both species benefit. c) In both interactions, one
    species eats or otherwise harms the other. d) Predation is a brief
    interaction; parasitism is a symbiosis. e) Both hide harmless
    individuals. f) Cryptic coloration matches the organism to its
    surroundings; Batesian mimicry matches the organism to a well-de-
    fended species. g) Both advertise well-defended individuals. h)
    Aposematic coloration is simply obvious; Müllerian mimicry also
    matches the individual to other well-defended species. i) Both are
    long-term interactions in which at least one species is hurt j)
    Parasitism hurts only one species; interspecific competition hurts
    both. k) Both refer to the needs of an organism, where it lives.
    l) Habitat refers specifically to the area, the topography; niche
    refers to all the needs of the organism. m) Both are results of
    interspecific competition. n) Competitive exclusion involves
    local extinction of one species; character displacement implies
    that both species remain in the area.

7. On the island, the species' niche will be broader, maybe including
    seeds both larger and smaller than those the species takes on the
    mainland. b) The new species will shift the niche of the first;
    eventually, the two species might evenly divide the available
    seeds.

8. a) + b) - c) - d) o e) - f) + g) + h) -

HUMAN ECOLOGY I: PESTILENCE

I. REVIEWING THE CHAPTER

A. CHAPTER HIGHLIGHTS

45.1 Mortality: Famine, War, and Pestilence

Worldwide population is increasing at an unprecedented rate.
This is due not to birth rates, which have not increased, but to a
dramatic decline in death rates. The forces that normally operate to
keep death rates high in any population of organisms are: (1) the
limited capacity of the environment, (2) the efficiency of the predators, and (3) the inefficiency of parasites. In human terms these
are equivalent to famine, war, and pestilence. Agriculture and animal
husbandry have increased the food supply and, although wars kill many
individuals directly or indirectly, the effectiveness of our struggles
against parasites has led to a dramatic rise in population. Unfortunately, lowering death rates have not been matched by lowering birth
rates.

45.2 Parasitism: Invading the Host

Humans can act as hosts for a wide variety of parasites at almost
all body locations. The exteriors of vertebrate bodies are well protected against parasite invasion by virtue of the fact that, except
for the cornea, no living cells are directly exposed to the outside
environment. Metabolism requirements demand that a number of organs
have indirect access to the outside world. These living cells provide a means of entry for microorganisms. The organs do possess defenses, however. Lungs have cilia, phagocytes, and IgA, a class of
antibody. The gastrointestinal tract relies upon the low pH of gastric
juice, on phagocytes, and on IgA. The urinary system relies upon the
sterility of urine and the flusing action of excretion. Occasionally
the body's defenses are overcome. Once a parasite has gained entrance,
it may be destroyed, be tolerated for a long period of time, or kill
the host.

45.3 Host Resistance to Parasites

Host resistance to parasites depends upon the mechanisms employed
to combat the invasion. For instance, if bacteria enter through a
puncture wound, phagocytes soon attack them and other localized responses
occur, all of which constitute an acute inflammation. If these
measures fail to check the infection, bacteria may invade the lymphatic
and/or blood systems. In the case of the lymphatic system, the nodes
contain phagocytes to combat the infection. The bacteria reaching the

bloodstream are engulfed by phagocytes in the spleen and liver. With many infections, the production of antibodies is the deciding factor. These antibodies, which require time for production, may allow the phagocytes to engulf the bacteria more easily. If the immune system has encountered the infection before, a secondary response may occur so quickly that no symptoms are noted and the individual is considered immune to that parasite. Parasites that reside intracellularly are able to avoid antibodies. Some are able to change their antigenic determinants. Others incorporate host antigens into themselves, masquerading as body cells. Interferon is an antiviral agent that can be produced more rapidly than antibodies and is effective against all viruses. Research is attempting to find ways to enhance the body's production of interferon.

45.4 Interfering with the Transmission of Parasites

All parasites face the major problem of getting from one host to another. One method that actually improves some parasites' chances of accomplishing this is the use of an intermediate host prior to reentry of the primary host. The pig tapeworm is an example of this; another is the fish tapeworm with its two intermediate hosts. Transport is a greater problem for parasites that infect the blood and other tissues than it is for intestinal parasites. In this case the parasite uses a vector (animal) for transmission, usually by inoculation, into the tissues of the host. Even with elaborate life cycles, the chances of entering a host are small and, as a result, parasites are extremely fecund. The control of parasitic diseases is best accomplished by interrupting the parasite life cycle. This can be done by improved sanitation, careful preparation of food, and killing of vectors.

45.5 Density-dependence of Human Parasites

Many parasites are density dependent, i.e. they require a high density of susceptible individuals. Some examples of this type are measles, smallpox, and polio. The characteristics of density-dependent parasites are: (1) they cause an acute illness of short duration, (2) the period of the victim's contagiousness is brief, (3) they must reach susceptible individuals during this brief period of contagiousness, and (4) they survive best in cities in which new crops of susceptible individuals arrive periodically. These diseases are tailor-made for cities experiencing immigration and for military camps, hence their association with war. The seemingly poor adaptation of parasites that either kill or confer lifelong immunity upon their hosts is accounted for by the short time span in which humans have lived in cities. These parasites may not have existed in humans before that time. Humans have always been attacked by parasites but not always by density-dependent ones. Older human parasites produce generally chronic diseases such as malaria and tuberculosis. The parasite remains for a long period of time in its host and causes the host to be contagious for much or all of that time. Also, these diseases produce only weak immunity, and parasites can exist for long periods of time in isolated communities. Much more time has been available for the mutual evolutionary adaptation of parasites of this type and their human hosts. Because of their various adaptations, the control and/or elimination of these parasites poses very difficult problems.

45.6 Chemotherapy

The treatment of parasites within the host by chemotherapeutic
techniques is difficult, owing to the fact that host and parasite
share many of the same biochemical activities. A chemotherapeutic
agent must disrupt some essential metabolic process of the parasite
which is not shared by the host. The first such agent, Salvarsan, was
synthesized in 1910 by Paul Ehrlich to combat the spirochete responsible
for syphilis. In the 1930s sulfanilamide was developed, the first of
a large family of sulfa drugs, which interfere with the bacteria's
manufacture of the necessary vitamin, folic acid. Since humans cannot
manufacture folic acid, the drug has no effect on the host. The discovery of other agents effective against parasitic bacteria and protozoans has been a slow process. Only a few drugs have been found
that are effective against viruses.

45.7 Antibiotics

Antibiotics are identical with chemotherapeutic agents in terms
of their method and effects. The only distinction is that the antibiotics were first discovered as the secretions of a fungus or bacterium.
The first and best general antibiotic, penicillin, was discovered by
Alexander Fleming.

45.8 Passive Immunity

In some instances it is possible to inject infected persons with
antibodies produced by another animal rather than wait for the hosts
to develop their own antibodies. This antiserum therapy confers an
immediate immunity that is referred to as passive immunity. Such
sera can be produced from the antitoxins of animals or, as in the case
of the antiserum for hepatitis, from human gamma globulin.

45.9 Public Health Measures: The Outlook

The above successes against disease have produced some negative
side effects. Widespread use of antibiotics has led to the evolution
of resistant strains of once-susceptible microorganisms. Also, gains
in life expectancy have increased population to a point where the environmental capacity of the biosphere has become a limiting factor.

B. KEY TERMS

limited capacity of the
environment

predators

parasites

phagocytes

secondary response

immune

intermediate host

primary host

acute inflammation   vector

lymph nodes   sulfanilamide   .

spleen   antiserum

liver   antitoxin

antibodies

II.   MASTERING THE CHAPTER

A.   LEARNING OBJECTIVES

When you have mastered the material in this chapter, you should be
able to:

1.   Define all key terms.

2.   List the forces that keep death rates high in a population and re-
     late them to their human equivalents. Explain how the control of
     one of these factors has led to an accelerating human growth rate.

3.   Discuss the body's defenses against invasion by parasites and how
     the requirements of metabolism make certain pathways more suscepti-
     ble to infection. Cite the defenses these pathways possess. List
     the possible outcomes following invasion and the factors involved
     in determining them.

4.   Explain the ways in which a host can resist parasites, particularly
     bacterial infections. Describe how immunity results and discuss
     the body's defense against viruses.

5.   Discuss the life cycle and transmission of parasites and how these
     can be interrupted so as to reduce human infection.

6.   Discuss the density-dependence of human parasites. Note the
     requirements and characteristics of those parasites requiring high
     densities. Contrast the above group with those surviving in low-
     density conditions. Give examples of both groups.

7.   Explain the major difficulty involved in discovering effective
     chemotherapeutic agents. Give the history of such agents.

8.   Contrast chemotherapeutic agents and antibiotics, noting the dis-
     covery and properties of penicillin.

9.   Explain passive immunity, noting the sources of antisera.

10. Discuss the negative effects of public health and therapeutic
     measures on civilization as a whole.

B.   REVIEWING TERMS AND CONCEPTS

1. The death rate is kept high in any population by the limited capacity of the (a) _____, the efficiency of (b) _____, and the inefficiency of (c) _____. In human terms these translate into (d) _____, (e) _____, and (f) _____.

2. About the only place where cells are directly exposed to the external environment are the (a) _____ of the eyes. Here, the enzyme (b) _____ acts as a line of defense against bacteria. The lungs are protected from infection by (c) _____, (d) _____ cells, and the antibody (e) _____.

3. The localized responses to infection are referred to as (a) _____ _____. A second line of defense is the (b) _____ _____, which are lined with "fixed" (c) _____. Bacteria that reach the bloodstream are destroyed by the (d) _____ and _____. The production of (e) _____ by the host often quickly turns the tide of infection. If a reinfection occurs, the immune system can mount a secondary response so quickly that the individual is considered (f) _____ to that parasite. In the case of viruses, host cells produce (g) _____, which prevents contemporaneous infection by other viruses.

4. Parasites that are relatively new to humankind are generally (a) _____-dependent and require a sufficient supply of susceptible individuals, such as are found in (b) _____ and temporary aggregates of people. Fresh crops of individuals can be produced by (c) _____ and _____.

5. Using chemical agents to interfere with the metabolism of parasites is referred to as (a) _____. It was first accomplished by (b) _____ when he discovered (c) _____ was effective against the disease (d) _____. Later, the drug sulfanilamide gave rise to the large family of (e) _____ drugs. These drugs rely upon (f) _____ mimicry. Antibiotics are similar in action but have originally been derived from the secretions of (g) _____ or _____.

6. If sublethal quantities of tetanus toxin are injected into a horse, the animal develops (a) _____, which yields an (b) _____ that can confer passive immunity upon an individual infected by the (c) _____ bacillus.

C. TESTING TERMS AND CONCEPTS

1. Penicillin
   a) was discovered by Paul Ehrlich.
   b) is an effective antiviral agent.
   c) blocks synthesis of peptidoglycan.
   d) mimics PABA.

2. Sulfanilamide is capable of
   a) destroying folic acid.

b) blocking folic acid production.
c) destroying prokaryotic cell walls.
d) blocking prokaryotic wall production.

3. The development of antitoxin
    a) is an active response to infection.
    b) is an example of passive immunity.
    c) yielded the first chemotherapeutic agent.
    d) was first accomplished in a fungus.

4. Which of the following has the least effect upon the death rate?
    a) immunization
    b) elimination of war
    c) improved sanitation
    d) improved agricultural technique

5. In the case of Rocky Mountain spotted fever, the tick
    a) is considered the parasite.
    b) is considered the vector.
    c) causes acute inflammation.
    d) causes antibody production.

6. One of the defenses of the lungs is their
    a) keratinized cells.
    b) lysozyme.
    c) low pH.
    d) cilia.

7. Interferon
    a) binds to the bacterial cell wall and facilitates phagocytosis.
    b) interferes with bacterial cell division.
    c) blocks viral reproduction.
    d) is involved in the phenomenon of secondary immunity.

8. Passive immunity
    a) results from treatment with chemotherapeutic agents.
    b) results from treatment with antibiotics.
    c) is conferred by exposure of the host to weakened parasites.
    d) requires the use of an antiserum.

9. Which of the following is not a characteristic of diptheria?
    a) relatively recent parasite of humans
    b) relatively brief acute illness
    c) associated with large human populations
    d) low death rate

10. Which of the following is not a defense mechanism of the gastrointestinal tract?
    a) IgA
    b) phagocytes
    c) lysozyme
    d) low pH

11. When a parasite is inefficient,

a) the death rate of its host species is high.
b) it is referred to as degenerate.
c) the disease is only moderately contagious.
d) it reproduces only slowly.

12. Which of the following is <u>not</u> utilized by parasites to increase the chances of reentering a host?
a) intermediate host
b) fecundity
c) vector
d) virulence

13. Salvarsan was
a) the first person to discover a cure for syphilis.
b) the discoverer of sulfa drugs.
c) the first synthetic chemotherapeutic agent.
d) a drug similar to PABA in structure.

14. What is the best indication that a parasite has had a long evolutionary relationship with a host species?
a) Individual infections are of a relatively long duration.
b) It is highly specialized and associated with that host only.
c) It easily overcomes the host defenses.
d) It regulates host population size in a cyclic manner, closely following host increases and decreases.

15. Why is pestilence such a frequent companion to war?
a) Large groups of previously unassociated men especially invite contagious spread of already existent disease.
b) Large numbers of deaths and lack of sanitary practices serve as breeding grounds for disease.
c) Overstressed soldiers are especially susceptible to disease.
d) Civilians are often deprived of food and fuel and therefore unusually susceptible to disease.

D. UNDERSTANDING AND APPLYING TERMS AND CONCEPTS

1. Primitive man had a life expectancy of about 20 years. Why can you expect to live for 70 years or more?

a)_____

b)_____

c)_____

2. Many pathogens enter the body through the lungs and digestive tracts. Why aren't these areas of the body better protected against the entrance of foreign organisms? a)_____

What defenses <u>do</u> exist at these sites of entry into the body?
b)_____ c)_____ d)_____

449

3. What aspects of a local inflammation keep it local and prevent the spread of infection throughout the body?

    a)_____

    b)_____

    c)_____

4. Describe the similarity between a lymph node and the spleen. (a)
    _____
    What is the major difference between them? (b) _____
    _____

5. How would the progress of an infection differ if histamine were not produced? _____
    _____

6. Considering the life cycle of a virus, why is it so difficult to develop chemotherapeutic agents or antibiotics against them? ____
    _____
    _____
    _____

7. Their regular use in hospitals has resulted in a serious problem associated with chemotherapeutics, antibiotics, and disinfectant cleaners. What is that problem, and why has it arisen? _____
    _____
    _____

ANSWERS TO CHAPTER EXERCISES

Reviewing Terms and Concepts

1. a) environment b) predators c) parasites d) famine e) war
    f) pestilence

2. a) corneas b) lysozyme c) cilia d) phagocytic e) IgA

3. a) acute inflammation b) lymph nodes c) phagocytes
    d) liver, spleen e) antibodies f) immune g) interferon

4. a) density b) cities c) birth, immigration

5. a) chemotherapy b) Ehrlich c) Salvarsan d) syphilis e) sulfa
    f) molecular g) bacteria, fungi

6. a) antitoxin b) antiserum c) tetanus

Testing Terms and Concepts

1.	c	6.	d	11.	a
2.	b	7.	c	12.	d
3.	a	8.	d	13.	c
4.	b	9.	d	14.	a
5.	b	10.	c	15.	a

Understanding and Applying Terms and Concepts

1. a) Agriculture has increased the carrying capacity of the earth.
    b) Medicine and public sanitation have effectively combated disease.
    c) We have effectively constrained all predators on humans except
    ourselves.

2. a) Structured to admit nutrients, they can't help but admit other
    things as well. b) bacteriocides c) phagocytes d) antibodies

3. a) Stimuli from damaged tissue cause increased blood flow and
    phagocyte movement to that particular area. b) Blockage of
    lymphatic vessels reduces flow from the infected area. c) Phago-
    cytes wall off the infected area and also reduce flow from the
    infected area.

4. a) Both filter body fluids and house phagocytes that engulf
    bacteria. b) A lymph node filters lymph, and the spleen filters
    blood.

5. There would be reduced inflammation and phagocytic activity at
    local infections.

6. The viral strategy is to force host cells to manufacture more
    viruses. Most agents that would interfere with this activity,
    would interfere with other host activities as well.

7. Many of these compounds no longer kill parasites as they origin-
    ally did. The parasites have evolved a high degree of resistance
    in some cases, through artificial selection.

HUMAN ECOLOGY II: COMPETING FOR
FOOD

I. REVIEWING THE CHAPTER

A. CHAPTER HIGHLIGHTS

46.1 The Carrying Capacity of the Environment

The capacity of the environment of a given species depends ultimately upon the amount of energy that species can trap. A growing
human population has to seek out ways to harvest more energy for itself. This can be done by agritechnological means and by minimizing
the competition of other species. The major competitors are the insects, fungi, and weeds.

46.2 The Hazards of Monoculture

Efficient agriculture demands growing pure stands of a single
plant (monoculture). This procedure also increases the K of our
competitors for that crop. The high density and large amounts of a
single crop can lead to exponential population growth among the
competitors.

46.3 Early Pest Control Techniques

Early pest control amounted to removing insects from plants by
hand and washing plants with various innocuous materials. By the end
of the nineteenth century, chemical compounds, sometimes with high
toxicity, came into use. Organic pesticides such as pyrethrum and
rotenone were highly successful yet extremely expensive.

46.4 DDT

The scarcity of pyrethrum caused by the onset of World War II led
to the production of DDT in huge quantities and at very low cost.
After the war, DDT was used on a worldwide basis to control crop predators and disease vectors. Although such measures were successful,
serious drawbacks became evident, e.g. strains of insects were appearing that were resistant to DDT. To combat these resistant strains,
the search for other chlorinated hydrocarbons was accelerated and a
group of effective, yet highly toxic, chemicals was found. The use of
DDT is forbidden in the United States (except for public health uses)
because it does not decompose in the environment and becomes stored
in the fatty tissue of humans and animals. To date, there is no direct
evidence that such body burdens of DDT are harmful to humans. This is
not the case with other species such as fish, earthworms, and birds.

Part of the problem arises from the fact that DDT becomes more concentrated as it travels up the food chain. Another shortcoming of DDT is that it kills beneficial insects as well as harmful ones.

46.5 The Organophosphates and Carbamates

Alternatives to DDT and other chlorinated hydrocarbons were found among the organophosphates such as parathion and malathion. These engage in an irreversible reaction with the enzyme acetylcholinesterase, thereby attacking the nervous system. They are exceedingly toxic to humans yet are of value because they break down in the environment quite rapidly. The appearance of organophosphate-resistant pests led to the development of the carbamates. These work in a fashion similar to the organophosphates, yet their action on the nervous system is reversible and they are rapidly detoxified and excreted. Therefore, they are not especially hazardous to warm-blooded animals. There remain the problems that they kill beneficial insects and will probably give rise to carbamate-resistant strains. Diflubenzuron is a new insecticide that interferes with the synthesis of chitin.

46.6 "Third-generation" Pesticides

"Third-generation" pesticides are those that mimic the juvenile hormone of insects and may be especially useful since it is unlikely that insects will develop a resistance to something that resembles a normal body constituent. One such compound has been marketed, with more likely to follow. A possibility for "fourth-generation" pesticides lies in the discovery of anti-JH activity substances called precocenes.

46.7 Biological Controls

Many pests posing the greatest threat to agriculture are those that have been imported into a region. This is because the biological controls (other organisms) found in the pest's native habitat have not been imported along with it. Such controls, in the form of insects, bacteria, and viruses, are now being used to control destructive species of insects as well as some weeds.

46.8 Breeding-Resistant Species

One approach to increasing crop yields is to breed resistant strains of the plant. Many resistant strains of vegetables and cereal grains exist today. Such measures are temporary because new strains of parasites that can overcome the plant's resistance develop as evidenced by the southern corn leaf blight in the United States. This epidemic also emphasized the fact that the more widespread the cultivation of a genetically uniform crop, the greater the potential for epidemics.

46.9 Other Approaches to Pest Control/46.10 The Sterile Male Technique

Other methods of pest control involve crop rotation, insect repellants and attractants, and manipulating diapause by artificial illumination. One of the cleverest and most successful techniques involves the utilization of sterile males. These males mate with fertile females who then lay sterile eggs, thereby producing no offspring. This process has been used successfully in controlling pink bollworm, screwworm, and harmful fruit flies. Sterility is produced either by irradiation of the males or by the development of males carrying abnormal chromosomes that give rise to nonviable offspring.

46.11   What Does the Future Hold?

The human population depends upon the deliberate upsetting of the natural balance of other species in our favor. With our increasing population, the margin for error in such activities grows smaller. In order to avoid catastrophes, our population and life styles must be brought into balance with the long-term capacity of our environment. We must learn to use our environment, but not to use it up.

B. KEY TERMS

juvenile hormone   JH

diapause   chlorinated hydrocarbon

II. MASTERING THE CHAPTER

A. LEARNING OBJECTIVES

When you have mastered the material in this chapter, you should be able to:

1. Define all key terms.

2. Note the ways in which we have endeavored to increase the carrying capacity of the environment for our own species.

3. Explain the danger involved in monoculture.

4. Discuss early forms of pest control and the advent of chemical controls.

5. Trace the history of the development and use of DDT. Discuss the reasons why DDT use is banned in the United States.

6. Discuss the development and functioning of the organophosphate and carbamate insecticides.

7. Explain the functioning and possibilities inherent in "third-" and "fourth-generation" pesticides.

8. Discuss the rationale for biological controls and the various forms they can take. Give examples.

9. Discuss other methods of pest control, particularly the sterile male technique and its successes.

10. Discuss the direction humans must take in the future with regard to energy and other resources.

B. REVIEWING TERMS AND CONCEPTS

1. The capacity of the environment of a given species depends ultimately upon the amount of (a) _____ it can trap. Efficient crop raising demands (b) _____, a situation that greatly increases the capacity of the environment to support our competitors for that crop.

2. By the end of the nineteenth century, the use of (a) _____ to control insect pests was widespread. These were (b) _____ substances like lead arsenate and (c) _____ substances like pyrethrum and rotenone.

3. With the onset of World War II, pyrethrum became scarce and the substitute (a) _____ was introduced. It was used against crop predators and (b) _____, but soon produced (c) _____ strains of parasites. This led to the search for other insecticides belonging to the chemical family of (d) _____ _____.

4. Parathion and malathion are (a) _____ that irreversibly react with the (b) _____ acetylcholinesterase, thereby affecting the (c) _____ system. Another group similar in action is the (d) _____. Their value lies in the fact that their inhibitory action is (e) _____ in warm-blooded animals.

5. Substances that are analogues of juvenile hormone are referred to as (a) _____ pesticides. Substances with anti-JH activity are called (b) _____ and may give rise to "fourth-generation" pesticides.

6. Using lady bird beetles to control other insects is a form of (a) _____ control. In addition to insects, (b) _____ and _____ have been used to play a similar role. Artificial lighting can be used to prevent (c) _____ in larvae.

7. The sterile male technique relies upon the (a) _____ of males, which are then released into the pest population with the result that the female lays (b) _____ eggs. Another related technique uses (c) _____ abnormalities which result in (d) _____ offspring.

8. In order to avoid catastrophe, we must bring the long-term capacity of our environment into balance with our (a) _____. We must therefore shift our dependence to resources that are (b) _____ and conserve those that are not.

C. TESTING TERMS AND CONCEPTS

1. The diapause of insects can be prevented by
a) artificial illumination.
b) organic pesticides.

c) the sterile male technique.
d) the use of precocenes.

2. The more widespread the cultivation of a genetically uniform crop, the
    a) more effective the use of pesticides.
    b) the more effective the use of biological controls.
    c) greater the resistance to fungi infection.
    d) greater the risk of parasite epidemics.

3. A second group of chlorinated hydrocarbons were discovered that
    a) decompose rapidly in the soil.
    b) do not spread as far up food chains.
    c) were useful against some DDT-resistant organisms.
    d) were less toxic than DDT.

4. If an insect becomes resistant to a pesticide,
    a) it will be able to develop resistance to other pesticides more easily.
    b) it will be more susceptible to other pesticides than it would have been.
    c) switch to a pesticide that attacks a different aspect of the insect's biology.
    d) increase the dosage of the same pesticide.

5. "Third-generation" pesticides
    a) are analogues of juvenile hormones.
    b) are artificial sex attractants.
    c) have yet to be produced.
    d) are called precocenes.

6. Which of the following is not true of inorganic pesticides?
    a) came into use in the late 1800s
    b) high toxicity to organisms other than target organism
    c) break down rapidly in the soil
    d) can cause the evolution of resistance

7. Resistant species of crops
    a) are mostly resistant to various types of fungi.
    b) remain effective much longer than pesticides can.
    c) generally lead to the development of resistant parasites.
    d) have the advantage of being relatively genetically diverse.

8. Which of the following is not a form of biological control?
    a) use of certain parasites
    b) introduction of juvenile hormone into a field
    c) infection with selective viruses
    d) use of the lady beetle or praying mantis

9. The sterile male technique
    a) results in sterile eggs or nonviable offspring.
    b) leaves poisonous residues in the soil.
    c) leaves radiation by-products in the soil.
    d) was used to eliminate the Anopheles in some parts of the world.

10. Organophosphates
    a) are less toxic to humans than DDT.
    b) produce a reversible reaction.
    c) are substances such as lead arsenate.
    d) are susceptible to rapid breakdown.

11. Monoculture is effective in
    a) reducing pest buildup.
    b) increasing agricultural efficiency.
    c) decreasing soil stress.
    d) primitive farming areas.

12. Which of the following is not true of DDT?
    a) It is selective in its ability to kill insects.
    b) It has produced DDT-resistant strains.
    c) It is concentrated at the end of food chains.
    d) It is a chlorinated hydrocarbon.

13. Using lady bird beetles to control insect populations
    a) has met with little success.
    b) can give 100% control of certain pests.
    c) is a form of biological control.
    d) gave rise to the term "fourth-generation" pesticide.

14. Pyrethrum and rotenone
    a) are synthetic compounds.
    b) do not decompose rapidly.
    c) are organic pesticides.
    d) do not threaten other members of the ecosystem.

15. Precocenes
    a) have the problem of magnification up the food chain.
    b) are "third-generation" pesticides.
    c) operate in a fashion similar to lead arsenate.
    d) display anti-JH activity.

D. UNDERSTANDING AND APPLYING TERMS AND CONCEPTS

1. Pest controls must somehow interfere with the normal functioning
   of the pest. Match the various means of interference on the right
   with the various agents or techniques listed at left.

   (a)_____ whale oil soap   A. make it difficult for pest
                                        to find its food
   (b)_____ organophosphates
                                     B. make crop unpalatable to
   (c)_____ diflubenzuron   pest

   (d)_____ sterile male technique   C. make it difficult for pest
                                        to find mates
   (e)_____ precocenes
                                     D. produce infertile eggs
   (f)_____ crop rotation
                                     E. prevent diapause

457

(g)_____ juvenile hormone

(h)_____ insect repellents

(i)_____ sex attractants

(j)_____ lady beetle

(k)_____ Bacillus thuringiensis

F. parasitize the insect

G. prey upon the insect

H. causes premature meta-morphosis

I. prevents metamorphosis

J. suffocate insect

K. stimulate synaptic activity to lethal degree

L. interfere with synthesis of chitin

2. A pesticide that is broadcast over some farmland might be quite ineffective because of (a) _____ _____ and quite dangerous to higher trophic levels because of (b) _____ _____. If the pesticide were not broadcast but placed at discrete stations with a natural attractant to draw the pest to the poison, it would be more effective, because (c)_____
_____,
and it would endanger higher trophic levels less, because (d)_____
_____.
The pest might even be unable to evolve a resistance to the poison, if (e)_____.

3. In what ways were second-generation pesticides better than first-generation pesticides?
(a)_____
(b)_____
Third-generation pesticides were still better in more important ways. What were these improvements? (c)_____
(d)_____

4. In what negative way are antibiotics, chemical disinfectants, and second-generation pesticides all the same? _____
_____

5. Artificial pesticides lose power as they are used year after year. Does the sterile male technique similarly lose effectiveness, does it remain equally powerful year after year, or does it become more powerful as it is repeated again and again? (a)_____
Why? (b)_____
_____

6. What category of pest control does each of the following best fit?
(a) You hand-pick tomato hornworms from your plants. _____
_____
(b) You put mineral oil on your corn silks to prevent earworm damage. _____

(c) You put out arsenic-laced peanuts to kill moles. _____

(d) You purchase a bacterium to spread over your lawn against Japanese beetles. _____
(e) You pasture a goat on some property to get rid of poison ivy. _____

ANSWERS TO CHAPTER EXERCISES

Reviewing Terms and Concepts

1. a) energy b) monoculture

2. a) pesticides b) inorganic c) organic

3. a) DDT b) vectors c) resistant d) chlorinated hydrocarbons

4. a) organophosphates b) enzyme c) nervous d) carbamates
    e) reversible

5. a) third-generation b) precocenes

6. a) biological b) bacteria, viruses c) diapause

7. a) irradiation b) sterile c) genetic d) nonviable

8. a) population b) renewable

Testing Terms and Concepts

1. a	6. c	11. b
2. d	7. c	12. a
3. c	8. b	13. c
4. c	9. a	14. c
5. a	10. d	15. d

Understanding and Applying Terms and Concepts

1. a) J b) K c) L d) D e) H f) A g) I h) B i) C j) G k) F
    l) E

2. a) evolved resistance b) food-chain concentration c) the pests
    would receive a dosage much higher than could be achieved by
    broadcasting d) the poison would not find its way into plants and
    from there be magnified in herbivores and carnivores e) 100% of
    those attracted to the stations are killed

3. a) cheaper b) less toxic to humans c) biodegradable d) normal
    body constituent, so evolution of resistance unlikely

4. All encourage the evolution of resistance in their targets.

5. a) It becomes more powerful. b) As the wild population dwindles
    and the ratio of wild male to sterile male decreases, the chances

459

of normal reproduction grow less and less.

6. a) biological control   b) repellents   c) first-generation poison
    d) biological control   e) biological control